METHODS IN MOLECULAR BIOLOGY™

Series Editor
John M. Walker
School of Life Sciences
University of Hertfordshire
Hatfield, Hertfordshire, AL10 9AB, UK

For further volumes:
http://www.springer.com/series/7651

Leucocytes

Methods and Protocols

Edited by

Robert B. Ashman

School of Dentistry, University of Queensland, Brisbane, QLD, Australia

Editor
Robert B. Ashman, Ph.D.
School of Dentistry
University of Queensland
Brisbane, QLD, Australia
r.ashman@uq.edu.au

ISSN 1064-3745 e-ISSN 1940-6029
ISBN 978-1-61779-526-8 e-ISBN 978-1-61779-527-5
DOI 10.1007/978-1-61779-527-5
Springer New York Dordrecht Heidelberg London

Library of Congress Control Number: 2011945167

Printed on acid-free paper

Humana Press is part of Springer Science+Business Media (www.springer.com)

Preface

The innate immune response is a crucial component of early resistance to infection, and it is now revealing increasing levels of complexity, both in itself, and in its interactions with the adaptive immune system. As technology has advanced, many important questions, long thought to have been resolved, have been revisited, often yielding unexpected or novel insights.

The ability to modify the genome in vivo, as in the creation of transgenic mice, has facilitated understanding of complex interactions between leucocytes and other components of the immune system, and phenotype-driven strategies using chemical mutagenesis have placed another powerful weapon in the armamentarium. More conventional approaches, such as flow cytometry, have also become increasingly sophisticated.

This edition of "Methods in Molecular Biology" provides detailed protocols and practical advice on a variety of modern approaches to the study of leucocytes and their products. It should be of use not only to the working scientist, but also to those entering the field, who appreciate the type of advice most often provided by their peers.

I would like to thank the Chief Editor, Professor John Walker, for his advice, and the contributors for their cooperation and, in many cases, their forbearance, during the preparation of this book.

Brisbane, QLD, Australia *Robert B. Ashman, Ph.D.*

Contents

Contributors

SYLVIE ATTUCCI • *INSERM U-618 "Protéases et Vectorisation Pulmonaires", Tours, andUniversité François Rabelais, Tours 37032, France*

VALÉRIE BARBIER • *Stem Cell and Cancer Group, Mater Medical Research Institute, South Brisbane, QLD, Australia*

BIBIANA BECKMANN • *Institute of Clinical Pharmacology, Hannover Medical School, Hannover 30623, Germany*

CATHERINA H. BIRD • *Department of Biochemistry and Molecular Biology, School of Biomedical Sciences, Monash University, Clayton, VIC, Australia*

PHILLIP IAN BIRD • *Department of Biochemistry and Molecular Biology, School of Biomedical Sciences, Monash University, Clayton, VIC, Australia*

DAWN M.E. BOWDISH • *Department of Pathology and Molecular Medicine, McMaster University, Hamilton, ON, Canada*

JOHAN BYLUND • *Phagocyte Research Lab, Department of Rheumatology and Inflammation Research, Sahlgrenska Academy at University of Gothenburg, Gothenburg, Sweden*

ARTURO CASADEVALL • *Departments of Microbiology and Immunology and Medicine, Albert Einstein College of Medicine of the Yeshiva University, 1300 Morris Park Avenue, Bronx, NY, USA*

ANTONIO CELADA • *Macrophage Biology Group, Institute for Research in Biomedicine (IRB Barcelona), and Departament de Fisiologia i Immunologia, Universitat de Barcelona, Barcelona 08028, Spain*

YU CHEN • *Department of Surgery, Beth Israel Deaconess Medical Center, Harvard Medical School, Boston, MA, USA*

EVAN L. CHISWICK • *Department of Pathology and Laboratory Medicine, Boston University School of Medicine and Boston Medical Center, Boston, MA, USA*

KARIN CHRISTENSON • *Phagocyte Research Lab, Department of Rheumatology and Inflammation Research, Sahlgrenska Academy at University of Gothenburg, Gothenburg, Sweden*

MÒNICA COMALADA • *Macrophage Biology Group, Institute for Research in Biomedicine (IRB Barcelona), and Departament de Fisiologia i Immunologia, Universitat de Barcelona, Barcelona 08028, Spain*

BEN A. CROKER • *The Walter and Eliza Hall Institute, 1G Royal Parade, Parkville, and Department of Medical Biology, University of Melbourne, Parkville, VIC, Australia*

ELIZABETH DUFFY • *Department of Pathology and Laboratory Medicine, Boston University School of Medicine and Boston Medical Center, Boston, MA, USA*

CHRISTOPHE EPINETTE • *INSERM U-618 "Protéases et Vectorisation Pulmonaires", Tours, and Université François Rabelais, Tours 37032, France*

RAJARAMAN ERI • *Mater Medical Research Institute, South Brisbane, QLD, and the University of Tasmania, Launceston, TAS, Australia*

Francis Gauthier • *INSERM U-618 "Protéases et Vectorisation Pulmonaires", Tours, and Université François Rabelais, Tours 37032, France*
Olaf Groß • *Department of Biochemistry, University of Lausanne, Chemin des Boveresses 155, Epalinges, CH 1066, Switzerland*
Frank-Mathias Gutzki • *Institute of Clinical Pharmacology, Hannover Medical School, Hannover 30623, Germany*
Ian Harper • *Monash Micro Imaging, Monash University, Clayton, VIC, Australia*
Corrine Hitchen • *Department of Biochemistry and Molecular Biology, School of Biomedical Sciences, Monash University, Clayton, VIC, Australia*
Yan Hu • *Guanghua School of Stomatology, Sun Yat-Sen University, Guangzhou, Guangdong, China*
Brian Japp • *Department of Pathology and Laboratory Medicine, Boston University School of Medicine and Boston Medical Center, Boston, MA, USA*
Marie-Lise Jourdan • *INSERM U-921 "Nutrition, Croissance et Cancer", Tours 37032, France*
Luiz Juliano • *Departamento de Biofísica, Escola Paulista Medicina, Universidade Federal, São Paulo 04044-20, Brazil*
Wolfgang G. Junger • *Department of Surgery, Beth Israel Deaconess Medical Center, Harvard Medical School, Boston, MA, USA, and Ludwig Boltzmann Institute for Experimental Traumatology, Vienna 1200, Austria*
Dion Kaiserman • *Department of Biochemistry and Molecular Biology, School of Biomedical Sciences, Monash University, Clayton, VIC, Australia*
Philip L. Kong • *Kennedy Institute for Rheumatology, Imperial College, London, UK*
Brice Korkmaz • *INSERM U-618 "Protéases et Vectorisation Pulmonaires", Tours, and Université François Rabelais, Tours 37032, France*
Jean-Pierre Lévesque • *Stem Cell Biology Group, Mater Medical Research Institute, South Brisbane, and School of Medicine, University of Queensland, Brisbane, QLD, Australia*
Jorge Lloberas • *Macrophage Biology Group, Institute for Research in Biomedicine (IRB Barcelona), and Departament de Fisiologia i Immunologia, Universitat de Barcelona, Barcelona 08028, Spain*
Matthew Stephen James Mangan • *Department of Biochemistry and Molecular Biology, School of Biomedical Sciences, Monash University, Clayton, VIC 3800, Australia*
Silvia Manzanero • *School of Biomedical Sciences, University of Queensland, Brisbane, QLD, Australia*
Michael A. McGuckin • *Mater Medical Research Institute, South Brisbane, and the University of Queensland, Brisbane, QLD, Australia*
Anja Mitschke • *Institute of Clinical Pharmacology, Hannover Medical School, Hannover 30623, Germany*
Manuel Modolell • *Department of Cellular Immunology, Max Planck Institute for Immunobiology, Freiburg 79108, Germany*
Andre Moraes Nicola • *Departments of Microbiology and Immunology and Medicine, Albert Einstein College of Medicine of the Yeshiva University, 1300 Morris Park Avenue, Bronx, NY, USA*
Nicos A. Nicola • *The Walter and Eliza Hall Institute, 1G Royal Parade, Parkville, and Department of Medical Biology, University of Melbourne, Parkville, VIC, Australia*

BIANCA NOWLAN • *Stem Cell Biology Group, Mater Medical Research Institute, South Brisbane, QLD, Australia*
ELODIE PITOIS • *INSERM U-618 "Protéases et Vectorisation Pulmonaires", Tours, and Université François Rabelais, Tours 37032, France*
PAUL K. POTTER • *Mammalian Genetics Unit, Medical Research Council Harwell, Oxfordshire, UK*
MARK PRESCOTT • *Department of Biochemistry and Molecular Biology, School of Biomedical Sciences, Monash University, Clayton, VIC, Australia*
ALICJA PUCHTA • *Department of Pathology and Molecular Medicine, McMaster University, Hamilton, ON, Canada*
DANIEL REMICK • *Department of Pathology and Laboratory Medicine, Boston University School of Medicine and Boston Medical Center, Boston, MA, USA*
VERA M. RIPOLL • *Mammalian Genetics Unit, Medical Research Council Harwell, Oxfordshire, UK*
ANDREW W. ROBERTS • *The Walter and Eliza Hall Institute, 1G Royal Parade, Parkville, and Department of Medical Biology, and Faculty of Medicine, University of Melbourne, Parkville, and Royal Melbourne Hospital, Parkville, VIC, Australia*
R. TEDJO SASMONO • *Eijkman Institute for Molecular Biology, Jl. Diponegoro 69, Jakarta 10430, Indonesia*
SARAH ELIZABETH STEWART • *Department of Biochemistry and Molecular Biology, School of Biomedical Sciences, Monash University, Clayton, VIC, Australia*
MARIA-THERESIA SUCHY • *Institute of Clinical Pharmacology, Hannover Medical School, Hannover 30623, Germany*
FREDRIK B. THORÉN • *Section of Hematology, Department of Internal Medicine, Sahlgrenska Academy at University of Gothenburg, Gothenburg, Sweden*
DIMITRIOS TSIKAS • *Institute of Clinical Pharmacology, Hannover Medical School, Hannover 30623, Germany*
CHRIS P. VERSCHOOR • *Department of Pathology and Molecular Medicine, McMaster University, Hamilton, ON, Canada*
DIPTI VIJAYAN • *Australian Institute for Bioengineering and Nanotechnology, The University of Queensland, St Lucia, QLD, Australia*
ROBERT WADLEY • *Mater Medical Research Institute, South Brisbane, QLD, Australia*
ELIZABETH WILLIAMS • *Transgenic Animal Service of Queensland, University of Queensland, St Lucia, QLD, Australia*
INGRID G. WINKLER • *Stem Cell and Cancer Group, Mater Medical Research Institute, South Brisbane, QLD, Australia*
ANDREE YERAMIAN • *Macrophage Biology Group, Institute for Research in Biomedicine (IRB Barcelona), and Departament de Fisiologia i Immunologia, Universitat de Barcelona, Barcelona 08028, Spain*

Chapter 1

ENU-Based Phenotype-Driven Screening

Vera M. Ripoll, Philip L. Kong, and Paul K. Potter

Abstract

Deciphering the contribution of individual genes and in turn pathways to cellular processes can be complicated and is often based on prior knowledge or assumptions of gene function. Phenotype-driven mutagenesis screens based around *n*-ethyl-*n*-nitrosurea (ENU) have been successful in a wide range of physiological systems in identifying novel genes that contribute to a given phenotype. Here, we describe methodologies we have employed in analysing cellular phenotypes in pipelines of mutagenised mice. Examples of primary screens to identify outliers, and secondary screens to provide a more detailed characterisation are outlined.

Key words: ENU, Phenotype, Mutagenesis, Leukocyte screening

1. Introduction

A phenotype-driven screen is simple in concept. In essence, one generates random mutations in the genome of mice and screen these pipelines of mice for phenotypes or characteristics of interest; nonetheless, this can be complex in practice. Commonly, these screens are designed to generate novel mouse models of disease and thereby identify new genes associated with disease (reviewed in ref. 1, 2). The strategy of identifying a distinct phenotype and then working back to the underlying mutation is a powerful way of identifying novel genes and pathways contributing to disease as no assumptions are made about the underlying genetic contribution; mice are identified purely by phenotype. ENU creates point mutations which can result in hyper-, neo-, or hypomorphs and also affect the function of individual domains, thus having advantages over complete null alleles.

Male mice are injected with ENU causing DNA adducts which in turn result in point mutations (3). Each sperm, and hence every offspring, of these male mice contain a unique

Robert B. Ashman (ed.), *Leucocytes: Methods and Protocols*, Methods in Molecular Biology, vol. 844,
DOI 10.1007/978-1-61779-527-5_1, © Springer Science+Business Media, LLC 2012

range of random point mutations throughout the genome. The offspring or subsequent generations of mice are then available for phenotypic analysis. To provide enough coverage of the genome these phenotype-driven screen typically employ large numbers of mice that are screened using several methodologies. This usually necessitates employing a high-throughput primary screen followed by a more detailed characterisation of the phenotype and mapping of the causative mutation by further outcrossing.

Several groups have applied ENU-mutagenesis to identify genetic modulators of the immune response pathways, notably the groups of Chris Goodnow and Bruce Beutler. (reviewed in refs. 4 and 5).

Peripheral blood is the primary source of mono- and poly-morphonuclear cells, such as lymphocytes, monocytes, and neutrophils, also known as white cells or leukocytes. These blood cells are a critical component of the immune system to fight infection. This section describes two methods for the investigation of blood leukocytes phenotypes in ENU-mutagenised mice. The first method uses a haematology analyser to obtain full blood counts and differentials. The second approach utilises direct immunofluorescence-label antibodies and flow cytometry analysis for the detection of cell surface markers of the different populations of peripheral blood cells. We also describe the analysis of macrophages that have been differentiated in vitro from bone marrow samples obtained from mutant mice.

2. Materials

2.1. Mutagenesis Protocols

A detailed description of the protocols used to generate pipelines of mutagenised mice for phenotypic interrogation lies outside the remit of this article. Several protocols have been developed but at MRC Harwell; we have been using C57BL/6J males injected once a week for 3 weeks with a dose of 100 mg/kg of ENU. Dosages can vary according to the strain employed (see Note 1). Once these mice have regained fertility they are crossed C3H/HeH females and bred as outlined in Fig. 1 to screen for dominant and/or recessive mutations (see also ref. 1). In the case of recessive mutations, we generally generate pedigrees of about 20 G3 offspring which are all related and carry various combinations of the same mutations and from which you would expect one in eight mice to be a homozygous mutant. Using recessive pedigrees, we have been able to screen the mice using post-mortem samples as the founders are either still able to breed or have been archived and could be re-derived to regenerate the pedigree. Below we describe methodologies that we have employed to analyse leukocytes from mutagenised mice, with the aim of identifying models

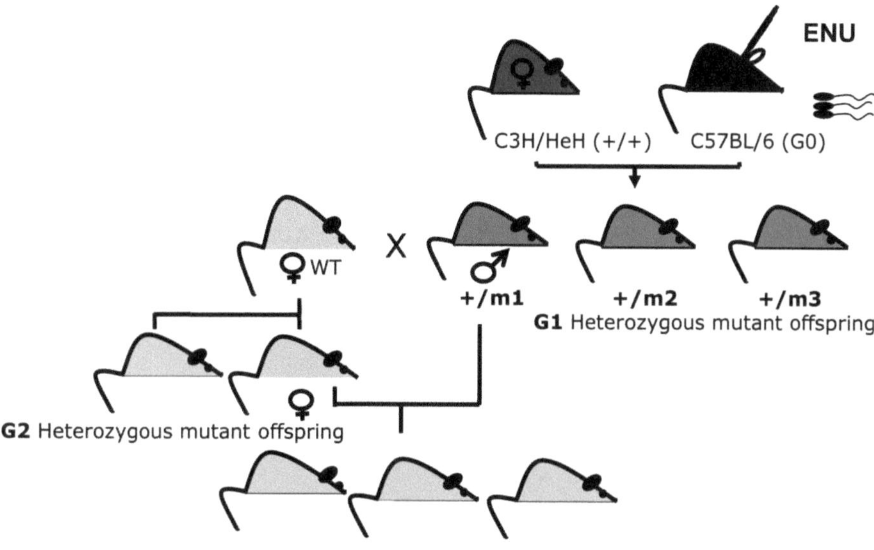

Fig. 1. The breeding scheme commonly used for the MRC Harwell phenotype-driven screens. Male C57BL/6 mice are treated with *N*-ethyl-*N*-nitrosourea (ENU) using a fractionated dose of 100 mg/kg once a week for 3 weeks. Mice are then outcrossed to C3H/HeH females to produce G1 offspring. Each G1 male is then outcrossed to C3H/HeH females and then crossed to some of his daughters to find a pedigree of siblings which will be heterozygous, homozygous, or wild-type for a particular ENU induced mutation. Mapping is carried out by identifying a region of C57BL/6 genome that is always inherited with the phenotype of interest.

of inflammatory dysregulation. Almost any relatively high-throughput methodology can be employed to interrogate pipelines of mutagenised mice; the limiting factor is usually manpower. Both of these screens were applied as terminal investigations to add value to mice that had already been through a comprehensive screening programme.

2.2. Haematology Analyser

1. EDTA collection tubes.
2. Rotary agitator.
3. ADVIA haematology analyser 2120 system (Bayer Health Care Division). Other automated cell analysers are available and suitable for such screens.

2.3. FACS Analysis

1. EDTA mouse peripheral blood samples.
2. Fixative-free Lysing solution.
3. FACS buffer (0.5% BSA 0.1% NaN_3 in PBS, pH 7.45).
4. Monoclonal antibodies against mouse leukocytes (see Table 2).
5. Plate shaker.
6. Rat anti-mouse CD16/CD32 (Fc block).
7. 96-Well microplates, round-bottom.
8. Tabletop centrifuge and rotor, 4°C.

2.4. Bone Marrow Cell Extraction and Macrophage Differentiation

1. Serum-free D-MEM (SF D-MEM).
2. Mouse complete culture medium: D-MEM, 10% foetal calf serum (FCS) that has been batch-tested, 1% L-glutamine, 1% antimyotic–antibiotic, 0.1% beta-mercaptoethanol.
3. Macrophage-colony stimulating factor (M-CSF) that has been batch-tested dissolved in complete culture medium at 100 μg/ml, aliquoted, and stored at −80°C (stable for at least 6 months).
4. 2× Macrophage freezing medium: 50% SF D-MEM, 20% DMSO, 30% FCS.
5. 2× Bone marrow (BM) freezing media: Same as macrophage freezing medium, with 160 μg/ml ascorbic acid.
6. Red blood cell lysis buffer.
7. Disposable syringes and 26 G needles.
8. Petri dishes (10 cm diameter).
9. Cryovials.
10. Cell lifters.

In vitro stimulation of macrophages and cytokine analysis by Luminex

1. 96-Well flat-bottom tissue culture plates.
2. Reagents for stimulating macrophages.
3. Luminex 100/200 System (Luminex, Austin, TX, USA).
4. 96- Well multiscreen HTS plates with filter (Millipore, Bellirica, MA, USA).
5. Bio-plex microbeads and amine coupling kit (Bio-Rad, Hercules, CA, USA).
6. Primary and secondary antibody pairs for cytokine detection (typically, antibodies that work for ELISA are suitable, but will need to be tested for multiplexing).
7. Streptavidin-PE.

3. Methods

3.1. Quantification of Mouse Peripheral Blood Leukocytes Using a Haematology Analyser

The haematology analyser is a simple and comprehensive system that uses a combination of light scatter, cytochemical staining (peroxidase method), and nuclear density to measure the total and differential white blood cells count. Blood leukcoytes are first clustered and counted according to their size and light absorption properties. Consequently, a two-stage peroxidase chemistry reaction is employed to enable the differentiation of peroxidase positive and negative populations. Lastly, cells are grouped according to

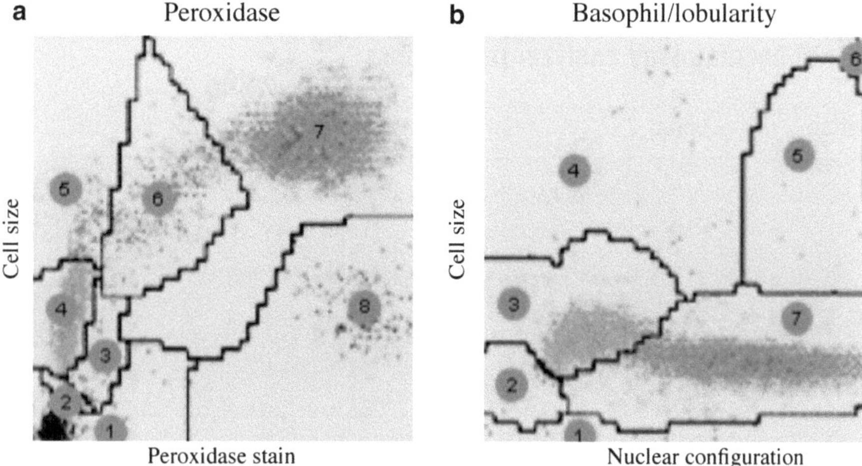

Fig. 2. Profiles of white blood cells on the peroxidase (**a**) and basophil/lobularity (**b**) cytograms. (**a**) Light scatter and absorption is used to determine each cell's size and level of peroxidase staining. 1.Noise, 2. Nucleated red blood cells, 3. Platelets, 4. Lymphocyte and Basophils, 5. Large unstained cells, 6. Monocytes, 7. Neutrophils, 8. Eosinophils. (**b**) After the lysis of the cytoplasm of all white cells except basophils, cluster analysis is used to identify and count cells and nuclei. 1. Noise, 2. Blast cell nuclei, 3. Monocytes and Lymphocyte nuclei, 4.Basophils, 5. Baso suspect, 6. Saturation, 7. Neutrophils, eosinophils. Adapted from http://www.medical.siemens.com.

their lobularity and number of nuclei. Figure 2 illustrates an example of the methods used by the haematology analyser (see Note 2).

3.1.1. Retro-Orbital Blood Collection

The peri-orbital sinus of the mouse can be used as a supply of venous blood (see Note 3). Blood can be collected from the medial or lateral canthus by the use of a capillary. This procedure is normally carried out under general anaesthesia and should only be conducted by experienced personnel. Local regulations may vary, but in the UK this can only be carried out as a terminal procedure.

1. Weigh and anaesthetise each mouse according to the method of choice.
2. Insert a capillary through the conjunctiva and into the orbital sinus by quickly rotating the tube. Allow the blood to flow by capillary action into an EDTA collection tubes. Flow may increase by changing the angle of capillary.
3. Invert the tubes 4–5 times. Keep the blood at room temperature until the time of analysis.

3.1.2. Blood Cell Count

1. Immediately after collection, blood samples are mixed on a rotary mixer for at least 30 min at room temperature.
2. Blood samples must be analysed within 2 h of collection in accordance with the manufacturer's instructions.
3. Data is exported into an excel file with the help of a macro. The set of parameters that can be collected from the analyser is shown in Table 1.

Table 1
ADVIA 2120 haematology analyzer parameters

Abbreviation	Units	Meaning
WBCP	×10^3 cells/μl	White blood cells count—Peroxidase method
WBCB	×10^3 cells/μl	White blood cells count—Basophil method
RBC	×10^6 cells/μl	Red blood cells
meanHGB	g/dl	Haemoglobin
HCT	L/L	Hematocrit
MCV	fl	Mean corpuscular volume
MCH	pg	Mean corpuscular haemoglobin
MCHC	g/dl	Mean corpuscular haemoglobin concentration
CHCM	g/dl	Cellular haemoglobin concentration mean
RDW	%	Red blood cell distribution width
HDW	g/dl	Haemoglobin distribution width
PLT	×10^3 cells/μl	Platelet count
MPV	fl	Mean platelet volume
PDW	%	Platelet distribution width
PCT	%	Platelet concentration
%NEUT	%	% of neutrophils
%LYM	%	% of lymphocytes
%MONO	%	% of monocytes
%EOS	%	% of eosinophils
%LUC	%	% of large unstained cells
%BASO	%	% of basophils
abs_neuts	×10^3 cells/μl	Absolute count of neutrophils
abs_lymphs	×10^3 cells/μl	Absolute count of lymphocytes
abs-mono	×10^3 cells/μl	Absolute count of monocytes
abs_eos	×10^3 cells/μl	Absolute count of eosinophils
abs_lucs	×10^3 cells/μl	Absolute count of large unstained cells
abs_basos	×10^3 cells/μl	Absolute count of basophils
#Retic	×10^9 cells/l	Absolute count of reticulocytes
%Retic	%	% of reticulocytes
Large_PLT	×10^3 cells/μl	Large platelets
%NRBC	%(#NRBC/100 WBC)	% nucleated red blood cells
abs NRBC	×10^9 cells/l	Absolute count of nucleated red blood cells

3.2. Flow Cytometry of Mouse Whole Blood Samples

Flow cytometry analysis is a widely used technique to obtain information about peripheral blood leukocyte populations. White blood cells are purified and stained with a panel of fluorescent-conjugated monoclonal antibodies specific for receptors or surface proteins solely expressed in each population. Cells are then processed in a flow cytometry analyser. The key advantage of using FACS analysis is that samples can be obtained from live mice and hence outliers identified and further breeding carried out from these mice. We have employed the Advia analyser as a primary screen to identify pedigrees that warranted further investigation. More mice were bred from the founders and then their peripheral blood cells analysed by FACS.

3.2.1. Tail Bleed Collection

The tail arteries and veins can be used to collect small samples of blood. This technique may require the animals to be warmed in order to dilate the blood vessel prior to taking the sample.

1. Apply a local anaesthetic cream to the base of the tail in the area where the incision will be made. Leave for 15 min to allow the local anaesthetic to work.
2. Restrain mouse in a rodent restrainer and using a razor blade, nick the tail vein.
3. Collect the blood into EDTA coated tubes.

3.2.2. Red Blood Cell Lysis

Analysis of the results is usually much simpler and more accurate when the red blood cells are not included in the sample.

1. 50 μl of whole blood sample is transferred to a 96-well round-bottomed plate. Red cells in whole blood samples are lysed using 200 μl of fixative-free lysing solution.
2. Samples are incubated on a shaker at 300 rpm for 10 min.
3. Centrifuge the plate at 500×*g* for 5 min at 4°C. Discard the supernatant.
4. Resuspend the pellet in 200 μl of fixative-free lysing solution. Repeat the incubation and centrifugation Subheadings 2.2 and 2.3.
5. If pellet is still red, resuspend it in another 200 μl of lysis buffer and repeat both the incubation and centrifugation steps.
6. Wash the pellet at least two times in 200 μl of FACS buffer, spinning down the plate and discarding the supernatant.

3.2.3. Blocking and Staining

1. The purpose of this is to block Fc receptors and thereby reduce non-specific binding of the monoclonal antibodies used in the subsequent steps. Incubate the leukocyte pellet with 50 μl of (1:250) rat anti-mouse CD16/CD32 antibody for 20 min at 4°C.
2. Wash off the unbound antibody by adding 150 μl of FACS buffer to each well, mixing cells by pipetting up and down.

Table 2
Optimised monoclonal antibodies against mouse leukocyte populations

Antibody	Optimised concentration	Supplier
Anti-mouse CD45 PE-Cy7	1:2,000	BD Bioscience
Anti-mouse CD3 Pacific Blue	1:50	Serotec
Anti-mouse CD11b PerCP-Cy5.5	1:200	eBioscience
Anti-mouse Ly6G APC	1:1,000	BD Bioscience
Anti-mouse F4/80 FITC	1:200	Serotec
Anti-mouse CD19 APC-Cy7	1:400	BD Bioscience

3. Spin down the plate at $500 \times g$ for 5 min at 4°C. Discard the supernatant.
4. Add optimised amount of the monoclonal antibody master mix (Table 2) to each well. Mix gently and incubate for 30 min on ice in the dark.
5. After incubating for 25 min add 5 μl of propidium iodide to all the wells to allow the discrimination of necrotic cells, shake briefly and incubate for 5 min in the dark at 4°C.
6. Wash at least three times with FACS buffer, spinning down the plate and discarding the supernatant.
7. If flow cytometry is to be carried out the same day, resuspend cells in 200 μl of FACS buffer; if analysis is to be delayed, resuspend cells 200 μl of FACS buffer containing 2% PFA.
8. Analyse cells by flow cytometry. Set the FSC and SSC voltage to recover the maximum population in the dot plot, then gate between 20,000 and 50,000 CD45 positive events.
9. Select different populations of interest by gating on physical parameters and positive fluorescence of each of the antibody used.

3.3. Bone Marrow Cell Extraction

When the animals are euthanised the tissues can be used in terminal screens. We are interested in the role of macrophages in autoimmune diseases; in particular, the regulation of macrophages in inflammation. To this end, we have carried out a screen using bone marrow cells as the source material.

1. Euthanise mouse and carefully remove both legs in their entirety. Remove skin and muscles gently and isolate femurs and tibias.

2. Soak bones in 70% ethanol for 2 min then place in SF D-MEM (we use 12-well plates, one well per mouse) (see Note 4).
3. Cut off both ends of each bone using a scalpel blade or scissors sterilised with 70% ethanol.
4. Flush 1–2 ml SF D-MEM through the lumen of each bone into 15 ml falcon tubes with 26 G needle and syringe.
5. Spin down at $500 \times g$ for 5 min. Pour off supernatant carefully.
6. Resuspend pellet in 1 ml red blood cell lysis buffer. Leave to stand for no more than 5 min at room temperature.
7. Add 10 ml SF D-MEM and mix gently. Spin down at $500 \times g$ for 5 min.
8. Resuspend in 1.5 ml mouse complete culture medium.
9. Aliquot 0.5 ml of the cells into two cryovials, each containing 0.5 ml of 2× BM freezing media. Transfer cryotubes to –80°C for 24 h for slow freezing and transfer to liquid nitrogen afterwards for long-term archiving (see Note 5).
10. Culture the remaining bone marrow cells in 10 ml mouse complete medium, supplemented with M-CSF at 100 ng/ml, in a 10 cm petri dish (see Note 6).
11. Leave for 7 days at 37°C.

3.4. Harvesting and Stimulating Macrophages

1. After 7 days incubation at 37°C, observe cells and carefully remove media.
2. Wash dish 1× gently with 10 ml serum-free D-MEM.
3. Add 5 ml SF D-MEM. Scrape entire plate gently with cell lifter.
4. Pipette cells into falcon tube. Spin down at $500 \times g$ for 5 min.
5. Resuspend cells in 5 ml mouse complete culture medium. Count cells carefully (see Note 7).
6. Adjust cell concentration to 5×10^5 cells per ml with mouse complete culture medium.
7. Plate cells out in 96-well flat bottom plates, 100 μl per well for 5×10^4 cells per well in triplicate.
8. (Optional: freeze remaining cells in macrophage freezing medium.)
9. Leave to rest overnight then stimulate with ligands of interest (see Note 8) (Table 3).

3.5. Multiplex Analysis of Cytokine by Luminex

Macrophages are potent producers of pro-inflammatory cytokines, a key factor in their important role in autoimmune diseases.

From both scientific and logistics viewpoints, it is beneficial to analyse multiple cytokines if possible. Any given mutation may affect the multitude of pathways utilised by immune cells for

Table 3
Stimuli used in macrophage in vitro assay

Ligand	Concentration	Supplier
LPS	10 ng/ml	Axxora, San Diego, CA, USA
R848	250 ng/ml	Axxora, San Diego, CA, USA
IL-10	1 ng/ml	Peprotech, Rocky Hill, NJ, USA
Dexamethasone	200 nM	Sigma-Aldrich, St. Louis, MO, USA

controlling activation differently, and the plethora of multiplexing systems for cytokine analysis makes detecting additional cytokines relatively straightforward. We use Luminex to analyse the levels of five cytokines simultaneously. We measure tumour necrosis factor alpha (TNFα), C-X-C motif chemokine ligand 1 (CXCL1/KC), interleukin-6 (IL-6), interleukin-10 (IL-10), and C-X-C motif chemokine ligand 10 (CXCL10/IP-10), for their biological importance as well as compatibility with the Luminex platform.

Luminex microspheres can be either acquired, labelled, and ready-to-use, or can be labelled in-house. Latter is time-consuming to set up and optimise, but is hugely more economical in the long run. The route any given laboratory chooses depends on its individual situation.

1. Centrifuge labelled microspheres for 1 min at 10,000×*g*. Sonicate the pellet for 15–30 s and vortex to disperse aggregates. Dilute microspheres to 2.5×10^4 beads/ml PBS+1% BSA. If measuring more than one cytokine, then dilute all the microspheres in the same vial.
2. Prepare the standards in PBS+1% BSA. If measuring more than one cytokine, make up all standards in same tubes.
3. Dilute samples if necessary in PBS+1%BSA.
4. Pre-wet plates with 100 μl of PBS. Apply vacuum to remove buffer.
5. Add 50 μl standards/samples and then 50 μl of diluted beads to each well.
6. Agitate for or overnight at 4°C in the dark.
7. Remove supernatants by low vacuum filtration.
8. Wash plates once using 200 μl PBS+0.01% Tween.
9. Dilute biotinylated secondary antibodies in PBS+1% BSA. Add 50 μl/well and agitate in the dark for 2 h at room temperature.
10. Remove secondary antibodies by vacuum and wash plates as before.

11. Add 50 μl/well of streptavidin-PE diluted to 2 μg/ml in PBS + 1% BSA. Agitate for 1 h in the dark at room temperature.
12. Remove streptavidin-PE and wash as before.
13. Resuspend microspheres in 140 μl/well of Luminex buffer.
14. Agitate for a few minutes to resuspend the beads and read in the Luminex machine.

3.6. Thawing Archived Macrophages for Rescreening

1. Remove cryovials from liquid nitrogen carefully and thaw rapidly in a 37°C water bath.
2. With minimal delay, transfer the content into a tube containing 10 ml pre-warmed SF DMEM.
3. Spin down at 500 × *g* for 5 min.
4. Plate cells out in a petri dish with mouse complete culture medium supplemented with 100 ng/ml M-CSF.
5. Allow cells to recover for 2–3 days. Then, harvest and stimulate as usual.

3.7. Thawing Archived Bone Marrow Cells for Further Phenotypic Analysis

1. Remove cryovials from liquid nitrogen carefully and thaw rapidly in a 37°C water bath.
2. With minimal delay, transfer the content into a tube containing 10 ml pre-warmed SF DMEM.
3. Spin down at 500 × *g* for 5 min.
4. Plate cells out in a petri dish with mouse complete culture medium supplemented with 20% FCS and 100 ng/ml M-CSF.
5. Differentiate cells for 7 days. Observe under microscope to see if a sufficient number of cells are generated. If so then harvest and stimulate as usual. If not, remove medium and feed cells with fresh mouse complete culture medium supplemented with 20% FCS and 100 ng/ml M-CSF. Allow cells to grow for up to a further 7 days before harvesting (see Note 9).

4. Notes

1. The choice of strains for the mutagenesis protocols is dictated by several factors, including sensitivity to ENU (6) and phenotypes, inherent in the inbred strains. Screens can be biased towards certain phenotypes by employing inbred or genetically altered strains which are predisposed to a certain phenotype. Several modifier screens are also underway at various institutes using strains that develop a defined pattern of disease in an attempt to identify mutations that affect the development of disease in either a beneficial manner (suppressors) or detrimentally (enhancers).

2. The high-throughput nature of this technology has the advantage of being a simple but informative screen. However, the volume of blood required generally necessitates that terminal blood samples are used which has the disadvantage of requiring further breeding. We employed the Advia as a terminal screen to extract further phenotypic data from mice that had come to the end of their phenotyping schedule, thus avoiding complication of taking a large volume of blood in the middle of a phenotyping pipeline which may affect other parameters and would require a recovery period thus adding expense to the screen.
3. Any method of blood collection is suitable. We have, however, found the retro-orbital method to produce large volumes of high-quality blood very simply and quickly. It is a technique that is easy to learn. A minimum of 200 μl of whole blood samples is required for the Advia analyser.
4. It is important to avoid breaking the bones before this stage as ethanol entering the lumen will reduce the viability of the cells.
5. Where possible freeze cells slowly and thaw cells quickly. A slower rate of freezing can be achieved by using commercially available cell freezing containers, or large polystyrene insulating containers. Progenitor cells are fragile and do not freeze well; ascorbic acid helps promote their viability. It is also important to transfer the cells to liquid nitrogen from –80°C as soon as possible.
6. We use sterile petri dishes for growing macrophages, as these cells attach very firmly to normal tissue culture plates. Nonetheless, cell lifters are used to gently scrape cells off the plates after the differentiation period.
7. It is important to determine cell numbers accurately, as the level of cytokine in the well is sensitive to cell numbers. We recommend investing in an automatic cell counter.
8. Obviously, the method of stimulation depends on the readout desired. Extensive optimisation of the dosage, timing, and other technical details should be carried out before the screen. Where possible, purchase and aliquot large batches of reagents, to be used within their shelf life, will help reduce variability.
9. The macrophage yield from frozen bone marrow cells is considerably less than that from fresh cells. Using more FCS and allowing the cells to grow for longer will improve yield. As a result of the altered growth conditions, macrophages may be subtly different. In our hands, for example, these cells produce more TNF-α after stimulation.

References

1. Acevedo-Arozena, A., Wells, S., Potter, P. et al. (2008) ENU mutagenesis, a way forward to understand gene function, *Annu Rev Genomics Hum Genet* **9**, 49–69.
2. Hoyne, G.F., Goodnow, C.C. (2006) The use of genomewide ENU mutagenesis screens to unravel complex mammalian traits: identifying genes that regulate organ-specific and systemic autoimmunity, *Immunol Rev* **210**, 27–39.
3. Shibuya, T., Morimoto, K. (1993) A review of the genotoxicity of 1-ethyl-1-nitrosourea, *Mutat Res* **297**, 3–38.
4. Hoebe, K., Beutler, B. (2005) Unraveling innate immunity using large scale N-ethyl-N-nitrosourea mutagenesis, *Tissue Antigens* **65**, 395–401.
5. Papathanasiou, P., Goodnow, C.C. (2005) Connecting mammalian genome with phenome by ENU mouse mutagenesis: gene combinations specifying the immune system, *Annu Rev Genet* **39**, 241–62.
6. Justice, M.J., Carpenter, D.A., Favor, J. et al. (2000) Effects of ENU dosage on mouse strains, *Mamm Genome* **11**, 484–8.

Chapter 2

Detection and Quantification of Cytokines and Other Biomarkers

Evan L. Chiswick, Elizabeth Duffy, Brian Japp, and Daniel Remick

Abstract

Accurate measurement of cytokine concentrations is a powerful and essential approach to the study of inflammation. The enzyme-linked immunosorbent assay (ELISA) is a simple, low-cost analytical tool that provides both the specificity and sensitivity required for the study of cytokines in vitro or in vivo. This communication describes a systematic approach to develop an indirect sandwich ELISA to detect and quantify cytokines, or other biomarkers, with accuracy and precision. Also detailed is the use of sequential ELISA assays to analyze multiple cytokines from samples with limited volumes. Finally, the concept of a multiplex ELISA is discussed with considerations given to cost and additional time required for development.

Key words: Cytokines, Sandwich ELISA, Sequential ELISA, Multiplex ELISA, Antibodies

1. Introduction

Cytokines are a cornerstone of any study that deals with inflammation, whether it is an in vitro cell culture system or an in vivo animal model (1). The cytokine profile as a whole and the relative abundance of one cytokine, and the endogenous inhibitors, define an inflammatory process that is in motion (2). Cytokines may be used to describe the nature of the insult, infection, or injury (3), and may even be used to stage the disease process (4). These studies revolve around the ability to detect, quantify, and discriminate a single cytokine from a multitude of biomolecules present in any given sample. One such method that is routinely used is the indirect sandwich enzyme-linked immunosorbent assay (ELISA).

The ELISA exploits the specificity of antibodies (Abs) and uses them to capture and quantify an analyte of interest from a given

Robert B. Ashman (ed.), *Leucocytes: Methods and Protocols*, Methods in Molecular Biology, vol. 844,
DOI 10.1007/978-1-61779-527-5_2, © Springer Science+Business Media, LLC 2012

volume of sample, and it does this with remarkable sensitivity (pg/mL or ~0.5 pM for a 15 kDa protein) (5).

There are four basic steps involved in an indirect sandwich ELISA. (1) Capturing analyte from sample with capture antibody. (2) Detecting captured analyte with detection antibody (also specific for captured analyte) that is labeled with biotin. (3) Detection amplification with streptavidin that has been conjugated with an enzyme, in most instances horseradish peroxidase (HRP). Each streptavidin molecule has multiple HRP molecules attached, and each detection antibody has multiple biotins attached. Alternatively, detection antibodies may be directly conjugated to HRP, thereby eliminating an incubation step, but at the cost of sensitivity. (4) Substrate addition and signal measurement via optical density (OD) with a microplate reader.

After each incubation step throughout the assay, unbound reactants are washed away. Also, due to the binding properties of the microplates used, the plates must be blocked with an inert protein, after coating with capture antibody, in order to prevent nonspecific binding (NSB) of the sample.

This simple method can be expanded to measure several analytes from a single aliquot of sample via the sequential ELISA (6). In this instance, a single sample is removed from one ELISA plate and then incubated in a separate plate, since the first plate should have only captured the cytokine detected by the antibody used for capturing the cytokine. An alternative to the sequential ELISA is the multiplex ELISA, which offers a more rapid and cost-effective alternative to sequential analysis. In this approach, multiple capture antibodies, with differing specificities, are printed into a single well in a microplate. Each printed "spot" of antibody is exclusive from the others (see Subheading 3.5, Fig. 4). In this manner, it is possible to measure 17, or more, analytes from a single sample all at once.

This chapter details the steps involved in developing an ELISA by way of checkerboard titrations of antibodies, as well as selecting the optimal blocking buffer and diluent via dose–response curve analysis. Also discussed is the use of a spike-recovery method to determine if cross-reactivity or loss of analyte exists between the separate ELISAs used in the sequential ELISA format. Finally, the multiplex ELISA is given consideration in terms of the principle, determination of cross-reactivity, and cost efficiency compared to the standard ELISA.

2. Materials and Equipment

2.1. Standard and Sequential ELISA

1. Matched antibody pairs: Monoclonal for capture and either mono- or polyclonal biotinylated antibody for detection.
2. Recombinant proteins for cytokine standards.

3. Blocking buffers: There are several different blocking buffers. Trial and error must be used to determine the optimal buffer.
   (a) 2% (w/v) Bovine Serum Albumin (BSA) (Sigma Chemical Company, St Louis, MO) in 1× phosphate-buffered saline (PBS; 120 mM sodium chloride, 1.2 mM sodium phosphate monobasic, 2.8 mM potassium chloride, 8.8 mM sodium phosphate dibasic, pH 7.4).
   (b) Blocker™ Blotto in TBS (Pierce, Rockford, IL).
   (c) Blocker™ Casein in PBS (Pierce).
   (d) Superblock® Blocking Buffer in PBS (Pierce).
4. Dilution buffer.
   (a) 10% blocking buffer, 0.1% BSA, and 0.005% Tween-20 in 1× PBS.
   (b) Standard dilution buffer: 1× PBS, 0.005% Tween-20 (Pierce), and 2% (v/v) fetal calf serum (FCS).
5. Wash buffer: 1× PBS with 0.05% Tween-20, pH7.4.
6. Streptavidin-conjugated HRP (SA–HRP) diluted 1:20,000 in 1× PBS, 0.1% BSA, and 0.005% Tween-20.
7. Substrate: 1% 3, 3′,5,5′-tetramethylbenzidine (TMB) dissolved in DiMethyl Sulfoxide (DMSO), diluted 1:100 in 0.1 M sodium acetate (pH 6.0) and 0.005% hydrogen peroxide (H_2O_2). Do not add in the H_2O_2 until just prior to the use of the substrate. It is suggested to freeze aliquots of TMB. WARNING: TMB is a known carcinogen.
8. 1.5 N sulfuric acid.
9. 96-well Microtiter plates: High binding capacity (Nunc Immunoplate, Neptune, NJ) (see Note 1).
10. 96-well deep-well polypropylene for storage of diluted samples (source plate).
11. 96-well microtiter plates, nonbinding, for making of standards.
12. Multichannel pipette and single-channel pipettes.
13. Automatic plate washer.
14. MicroPlate optical density scanner that can read between 450 and 600 nm.

2.2. Multiplex Materials and Equipment

1. All materials listed in 2.1 are required, except for the following: 1.5 N sulfuric acid, MicroPlate reader, substrate (TMB), SA–HRP, all dilution and blocking buffers listed.
2. Blocking buffer: May be specific for the imaging system used for analysis. This analysis uses the Odyssey blocking buffer from Li-Cor.
3. Dilution buffer: 1:2 dilution of the Odyssey blocking buffer with 1× PBS, pH 7.4.

4. Streptavidin-conjugated infrared (IR) dye: This is applicable for the Li-Cor scanning system, see below, but may be applicable to other systems as well.
5. Microarray printer: This is used to "print" capture antibody onto the bottom of the well of the microplate. This communication is based on our experience with a noncontact spotter, the Perkin Elmer Piezorray.
6. Detection/imaging system: A system with high resolution is needed to resolve and detect the signal from the individual prints (150 nm diameter) in each well. We use a Li-Cor scanner which can also be used for other assays. This detects the infrared fluorescence of each spotted cytokine which can then be quantified using various software packages.
7. Automatic plate washer: An automatic plate washer is especially important for this multiplex assay. If manually washing, any contact of pipette tips to the spots in the well could result in a failed assay or even false-positive or -negative results.

3. Methods

3.1. Optimizing Antibody Concentrations (the Checkerboard)

1. *Day #1*: Dilute the capture antibodies in a suitable volume of PBS to make four dilutions as shown in Fig. 1 (use the manufacturer's suggested dilution as a starting point).
2. Coat a 96-well high-binding microplate with 50 μL/well of the appropriate dilution of capture Ab. Figure 1 shows the layout with the proper dilutions. Tap plate gently to ensure complete coverage of the well bottom. Incubate plate overnight at 4°C.
3. *Day #2*: Wash the coated plate with an automatic plate washer (five washes, 250 μL/wash, 15-s soaks between washes, *repeat for all wash steps*, see Note 2). After the last wash is complete, invert and tap the plate on a paper towel to remove excess liquid. (Once wash buffer is emptied from the plate, it is important to work quickly so that the plate does not dry.)
4. Add blocking buffer, 150 μL/well, and incubate for 1 h at ambient temperature on an orbital shaker (all subsequent incubations are at ambient temperature on an orbital shaker unless noted otherwise).
5. Using dilution buffer, prepare a suitable volume of high, medium, low, and zero standards, as shown in Fig. 1. (Standard values in Fig. 1 are only examples; however, a three-log range between high and low is usually sufficient.)
6. After the plate has been blocked, wash and tap the plate.

	1	2	3	4	5	6	7	8	9	10	11	12	Capture Ab	Strep:HRP
A	10	1	0.1	0	10	1	0.1	0	10	1	0.1	0		1:20,000
B	Standard conc. ng/mL				Standard conc. ng/mL				Standard conc. ng/mL				4ug/mL	*1:20,000
C	10	1	0.1	0	10	1	0.1	0	10	1	0.1	0		1:20,000
D	Standard conc. ng/mL				Standard conc. ng/mL				Standard conc. ng/mL				2ug/mL	*1:20,000
E	10	1	0.1	0	10	1	0.1	0	10	1	0.1	0		1:20,000
F	Standard conc. ng/mL				Standard conc. ng/mL				Standard conc. ng/mL				1ug/mL	*1:20,000
G	10	1	0.1	0	10	1	0.1	0	10	1	0.1	0		1:20,000
H	Standard conc. ng/mL				Standard conc. ng/mL				Standard conc. ng/mL				0.5ug/mL	*1:20,000
Detection Ab	0.2ug/mL				0.1ug/mL				0.05ug/mL				Ab Conc.	

Fig. 1. The checkerboard titration plate map: Capture and detection antibodies are titrated against each other over a three-log range of standard concentrations. All parameters are assayed in duplicate wells.* An additional titration for strep:HRP may be examined by alternating strep:HRP dilutions across rows.

7. Add 50 μL/well of each standard to the plate, as depicted in Fig. 1. Incubate for 2 h (see Note 3).
8. Prepare a suitable volume of three dilutions of biotinylated detection Ab as shown in Fig. 1.
9. Wash plate, and then add 50 μL/well of biotinylated detection Ab as shown in Fig. 1. Incubate for 2 h.
10. Prepare a suitable volume of dilute SA–HRP as normal. (If necessary, two dilutions of SA–HRP can be analyzed here as well by adding two different dilutions to alternating rows of the plate; see Note 4.)
11. Wash plate. Add SA–HRP, 50 μL/well, as shown in Fig. 1. Incubate for 30 min.
12. Wash plate: Add TMB substrate, 100 μL/well, and incubate in the dark without shaking for 20–30 min. Periodic measurements at 590 nm should be performed to determine when to stop the reaction. The high standards, for each antibody titration, should have an OD_{590} of at least 0.4–0.5 prior to stopping color development (see Note 5). It is important to protect TMB substrate from light in order to limit nonenzyme-mediated substrate catalysis (see Note 6 for troubleshooting color development).
13. Stop the reaction with 1.5 N sulfuric acid, 100 μL/well, and tap plate with hand to gently mix the solutions. The solution should turn yellow.

14. Scan plate with plate reader at 465 and 590 nm and use the delta OD (OD_{465}–OD_{590}) as the OD data point.
15. To determine the optimal antibody concentrations, compare the signal-to-noise ratio for each antibody pairing by dividing the OD of each standard by the OD of its corresponding zero standard, i.e.: for Fig. 1, divide the average OD of column 2, rows A and B, by the average OD of column 4, rows A and B. The antibody dilution pairs that yield the highest overall ratios for each standard represent the optimal antibody concentrations (see Note 7). It is highly recommended to use a spreadsheet to simplify the calculation process.

3.2. Determining the Optimal Blocking and Dilution Buffer for the Sample Matrix (the Dose–Response)

Certain sample matrices, such as serum, plasma, or tissue homogenates, may produce high nonspecific background (NSB) levels of interference which may affect the range, precision, and accuracy of the assay. Often, the blocking and dilution buffers used can exacerbate or diminish the level of NSB that a sample or the antibodies used impart on the assay. These effects are important to account for when selecting the blocking/dilution buffer, and the extent to which the sample should be diluted. The actual matrix dilutions to test are also dependent upon the sensitivity of the assay and the prevalence of the analyte in the sample. A dose–response test example is provided in Fig. 2.

The optimal buffer choice is that which yields the least amount of difference in OD between the 0, 10, and 50% sample

	1	2	3	4	5	6	7	8	9	10	11	12
A	10ng/mL Std			10ng/mL Std			10ng/mL Std			10ng/mL Std		
B	3ng/mL Std			3ng/mL Std			3ng/mL Std			3ng/mL Std		
C	1ng/mL Std			1ng/mL Std			1ng/mL Std			1ng/mL Std		
D	0.3ng/mL Std			0.3ng/mL Std			0.3ng/mL Std			0.3ng/mL Std		
E	0.1ng/mL Std			0.1ng/mL Std			0.1ng/mL Std			0.1ng/mL Std		
F	0.03ng/mL Std			0.03ng/mL Std			0.03ng/mL Std			0.03ng/mL Std		
G	0.01ng/mL Std			0.01ng/mL Std			0.01ng/mL Std			0.01ng/mL Std		
H	0			0			0			0		
% Sample Matrix	0	10%	50%	0	10%	50%	0	10%	50%	0	10%	50%
	Blocker #1			Blocker #2			Blocker #3			Blocker #4		

Fig. 2. Dose–response plate map: Each standard conc., in its respective buffer with or without sample matrix, is loaded into a single well.

matrix + standards for the entire range tested. For visual comparison, it may help to plot standard concentrations (x-axis) against ODs (y-axis). A brief procedure is listed below.

1. Coat a 96-well plate with capture Ab using the optimal concentration determined from the checkerboard (Subheading 3.1) and incubate overnight at 4°C.
2. Wash plate and divide the plate into 4 three-column sections. Add different blocking buffers (i.e.: Blotto, Casein, Superblock, or Lab Blocking Buffer) to separate sections as shown in Fig. 2. Incubate for 1 h.
3. Prepare stocks of each dilution buffer at 0, 10, and 50% sample matrix concentrations (v/v). Dilute the recombinant standard with each stock solution prepared and add them to the plate as shown in Fig. 2. Incubate for 2 h.
4. Prepare the detection Ab in each dilution buffer (without sample matrix!) using the optimal dilution determined in Subheading 3.1. Add to plate as depicted in Fig. 2.
5. Add SA–HRP, TMB, and sulfuric acid in the appropriate sequence and read as normal.
6. Again, the optimal buffer system contains the smallest differences between the dilution buffer + standards and the dilution buffer + standards + sample matrix.

3.3. Determining an Analyte's Concentration in a Given Sample (the Standard ELISA)

All the basic ELISA steps covered in the previous section apply here for analyte quantification. Expounded upon herein are the construction, modeling, and evaluation of a standard curve.

ELISAs have a dose–response curve shape that is sigmoidal; therefore, linear curve fitting models are inappropriate. The four-parameter logistic (4PL) fit is generally acknowledged to be the reference model of choice for ELISAs (7). As the name suggests, the 4PL model calculates on the basis of four parameters:

(a) The lower, minimum asymptote, where the OD from decreasing concentrations of standard approaches that of the zero standard
(b) The slope factor
(c) The inflection point, or IC50, where the concavity of the sigmoidal curve changes
(d) The upper maximum asymptote, where increasing standard concentration results in minimal to no net increase in OD

The 4PL model calculates $f(x) = D + ((A - D)/(1 + ((x/C) \wedge B)))$.

There are numerous commercial software packages that perform the calculations. Also, analysis software is usually available from the manufacturer of the plate reader.

1. Standard curve: An eight-point standard curve, including the blank, is usually sufficient for sample analysis. The high standard concentration, used in the checkerboard stage of development, is a good point from which to start the standard curve. Typically, threefold serial dilutions are made, beginning from the highest standard. All standards, including the blank, are assayed in duplicate.
2. Curve evaluation: A good curve is symmetrical and sigmoidal in shape with both upper and lower asymptotes (see Note 8), and a top OD within the linear range of the plate reader (see Note 8).

Sample ODs should fall within the steeper area of the curve for the most accurate quantitation. If the sample OD is on or near the top plateau of the curve, the samples should be assayed again, but at a higher dilution in order to place them in the steep portion. If the sample OD is at the lower end of the curve, determine if the sample can be run more concentrated based off the results of the dose–response data generated in Subheading 3.2.

Ultimately, the users should define the range of acceptable accuracy for their curve by preparing standards of known concentrations that differ from those used in the curve, and treating these samples as unknowns. The calculated value should be within 80–120% of the expected value. This process is frequently referred to as "spike and recovery."

3.4. The Sequential ELISA

This method uses the same protocol as the standard ELISA; however, the same samples can be run, sequentially, on multiple cytokine ELISA plates. This is particularly useful when the amount of sample available is limited.

A word of caution: Prior to assaying samples, it should be determined if there is cross-reactivity between the different ELISAs. To determine this, perform a spike and recovery, in which a known amount of each protein to be tested is spiked into a normal control sample and tested in the sequential ELISA format. An acceptable recovery range is 80–120%. Furthermore, samples should *only* be sequentially assayed in the same order as for which the spike–recovery performance was determined.

1. Coat the various cytokine ELISA plates as normal.
2. Wash only one plate, block, and add standard and samples as normal.
3. During the second hour of sample/standard incubation for the first plate, wash and block the second plate.
4. Before washing the first plate, transfer the samples from the first plate back to the source plate, as shown in Fig. 3, and then wash the plate as normal. Add detection Ab to the first plate

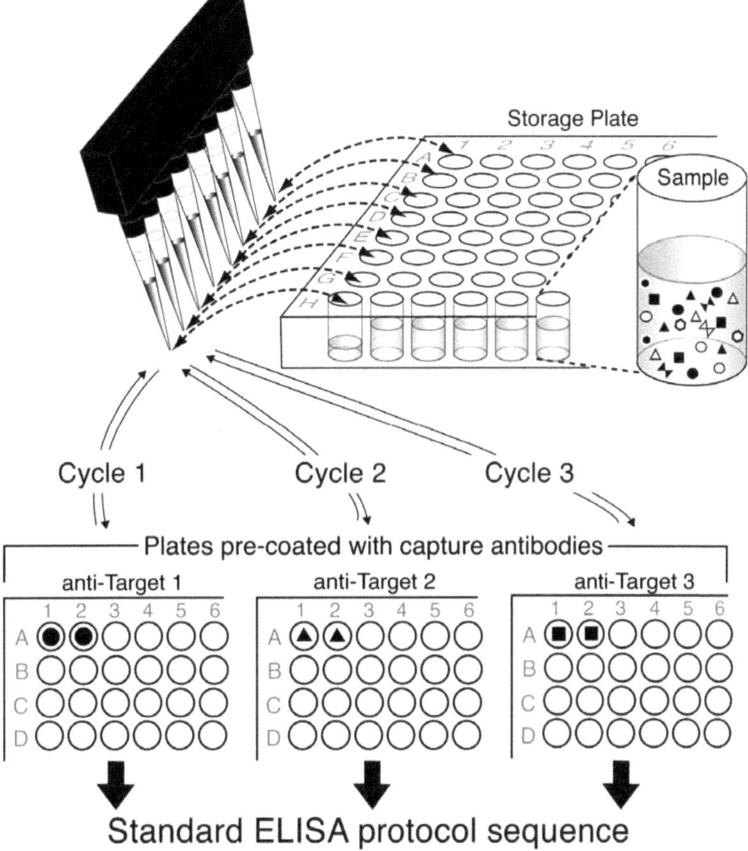

Fig. 3. A conceptual Illustration of the sequential ELISA. After the samples are subjected to one ELISA for cytokine 1, they are transferred back to a source plate for storage until they are used in the ELISA for cytokine 2, so on and so forth. The steps after the sample incubation are as normal for the particular ELISA adapted from ref. 6.

and the standard and samples to the second plate, as shown in Table 1.

5. Wash the first plate and add the SA–HRP. Transfer the samples from the second plate back to the source plate and wash the second plate. Then, add detection Ab to the second plate. We have measured up to three separate cytokines using the sequential ELISA (see Note 9).

3.5. The Multiplex ELISA

Two possible ways to measure multiple analytes from one sample include the cytometric bead array (CBA) and the multiplex ELISA. The CBA is a flow cytometry-based method in which specialized beads are coated with specific antibodies to the analytes of interest. Commercial kits are available that can provide measurement of up to 30 proteins from 25 to 50 μL of sample. If one desires, beads may be purchased without specific antibodies attached. This allows the user to customize the assay according to need. The general

Table 1
Schematic depicting the time arrangement of a simplified sequential ELISA protocol of exemplary targets analyzed in three subsequent cycles

	Previous day capture incubation overnight		
Time (h)	**Regular ELISA cycle # 1**	**Sequential ELISA cycle # 2**	**Sequential ELISA cycle # 3**
0.0	Blocking		
1.0	Samples		
2.0	Incubation	Blocking	
3.0	Detection ⇒TRANSFER SAMPLES[a]	TO CYCLE # 2 ⇒Samples	
4.0	Incubation	Incubation	Blocking
5.0	Streptavidin–HRP	Detection ⇒TRANSFER SAMPLES[a]	TO CYCLE # 3 ⇒Samples
5.5	TMB	Incubation	Incubation
6.0	Reading	Incubation	Incubation
7.0		Streptavidin–HRP	Detection ⇒REMOVE AND STORE SAMPLES[b]
7.5		TMB	Incubation
8.0		Reading	Incubation
9.0			Streptavidin–HRP
9.5			TMB
10.0			Reading

[a] Samples are removed and transferred to the storage plate (master mix) before washing and detection step
[b] Samples are removed before washing, transferred to the master mix, and stored overnight at 4°C for consecutive cycles

method is as such: Beads are conjugated in various intensities with a dye which fluoresces strongly within a specific channel filter (i.e., FL3). This creates distinct populations for data acquisition. Next, each specific population of these beads is coated with a specific capture antibody for the target of interest. Samples are then incubated with the "capture beads," followed by incubation with a detection antibody conjugated to a fluorophore that emits light into a different filter channel than that of the capture bead (i.e., FL2). The FL2 fluorescence is in direct proportion to the amount of analyte present while the FL3 fluorescence indicates the specific analyte present (8). This method uses less sample volume than a single ELISA; however, the cost may be prohibitive because the user is required to buy kits. The multiplex ELISA is another way to measure multiple analytes from one sample and is further discussed below.

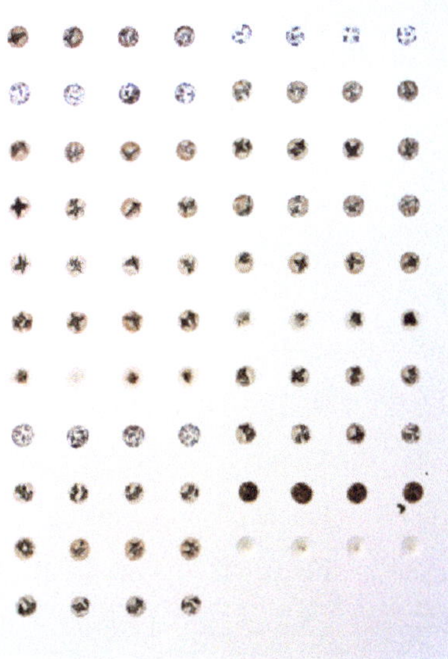

Fig. 4. Example array of "Printed Capture Antibodies for the Multiplex ELISA".

Due to the nature of the multiplex ELISA, specialized equipment is required. This necessitates a sizable initial investment; however, because of the high throughput and reduced labor, the cost per assay is substantially lower. There are significant differences in the printing apparatuses between manufacturers of the specialized equipment; as such, this section covers general considerations in developing a multiplex ELISA.

Due to the similarities between the standard single ELISA and the multiplex, similar optimization techniques can be used to develop the assay. The checkerboard described in Subheading 3.1 can be used to determine the optimal concentrations of antibodies and standards for the multiplex assay. In this instance, however, the standard and detection Ab must be added to each well as a cocktail because in each well there is essentially multiple, simultaneous ELISAs occurring (see Fig. 4 for illustration and Table 2 for list of analytes in the multiplex).

3.5.1. General Method

1. The ELISA plates must first be printed with the capture antibody specific for the cytokines of interest. Again, the printing process is specific to the printer used, and the manufacturer's recommendations should be followed.
2. Incubate the printed plates overnight at 4°C (see Note 10).
3. Block the plate with 150 μL/well of blocking buffer for 1 h.
4. Standard cocktail preparation: Using dilution buffer, mix each recombinant cytokine of interest into a heterogeneous cocktail. A 16-point standard curve (including the blank) that begins at

Table 2
Example layout for the multiplex array in Fig. 4

IL-1β	IL-1β	IL-1β	IL-1β	IL-1rα	IL-1rα	IL-1rα	IL-1rα
IL-2	IL-2	IL-2	IL-2	IL-4	IL-4	IL-4	IL-4
IL-5	IL-5	IL-5	IL-5	IL-6	IL-6	IL-6	IL-6
IL-10	IL-10	IL-10	IL-10	IL-12	IL-12	IL-12	IL-12
IL-13	IL-13	IL-13	IL-13	IFN-γ	IFN-γ	IFN-γ	IFN-γ
MIP-2	MIP-2	MIP-2	MIP-2	MIP-1α	MIP-1α	MIP-1α	MIP-1α
RANTES	RANTES	RANTES	RANTES	Eotaxin	Eotaxin	Eotaxin	Eotaxin
Eotaxin-2	Eotaxin-2	Eotaxin-2	Eotaxin-2	MCP-1	MCP-1	MCP-1	MCP-1
TNF-α	TNF-α	TNF-α	TNF-α	TNF-sr1	TNF-sr1	TNF-sr1	TNF-sr1
TNF-sr2	TNF-sr2	TNF-sr2	TNF-sr2	IL-17	IL-17	IL-17	IL-17
ICAM	ICAM	ICAM	ICAM				

50,000 pg/mL/cytokine followed by twofold serial dilutions is sufficient (i.e.: dilutions of 1×, 2×, 4×, *n*×., 16,384×, or, 50,000–3.05 pg/mL).

5. Incubate standards, samples, and the detection Ab cocktail as with the standard ELISA procedure. For the sake of simplicity, dilute all detection Abs into the cocktail at the same concentration.
6. After the detection cocktail incubation is complete, wash the plate and add the streptavidin dye conjugate diluted in dilution buffer, 50 μL/well, and incubate for 30 min *in the dark* (see Note 11).
7. Wash the plate and dry thoroughly by spinning upside down in a centrifuge.
8. Inspect surface of plate for fingerprints or other optical obstructions. If necessary, clean the bottom of the plate and then place into the scanner.
9. Scan and analyze as suggested by the specific system used.

3.5.2. Cross-Reactivity Determination

As with the sequential ELISA, cross-reactivity poses a potential problem in a multiplex assay and the issue should be addressed prior to analyzing a sample. An experimental approach to determine the presence of cross-reactivity is listed in Table 3.

3.5.3. Cost Comparison

An obvious advantage of the multiplex versus the standard ELISA is the greater amount of data generated from a single sample and within the same window of time. Less obvious is the money saved

Table 3
Determination of cross-reactivity for multiplex ELISA

	Capture	Standard	Detection	Assessment
1	Capture	Complete standard cocktail	Complete detection cocktail	Ensures that Ab pairs provide signal
2	Capture	Complete standard cocktail	Ab of interest absent from cocktail	Ensures that other detection Abs do not interact with Std of interest
3	Capture	Std of interest absent from cocktail	Ab of interest absent from cocktail	Ensures that other std Abs and other det. Abs do not interact with capture of interest
4	Capture	Std of interest absent from cocktail	Complete detection cocktail	Ensures that capture does not interact with any detection Abs or other stds
5	Capture	Only Std of interest	Only Ab of interest	Ensures that Ab pairs work well and multiplex assay works as well

using the multiplex approach. The comparison is not straightforward and varies from lab to lab. Without regards to specialized equipment, the multiplex is more expensive than a standard ELISA on a per plate basis. However, when compared using a data per sample basis, the multiplex becomes more cost-efficient than the standard ELISA.

For example, our lab has found that it is more cost-effective to use the multiplex when measuring seven cytokines or more. This takes into account the cost of antibodies, technician time, and overhead costs (see Table 4 for an example comparison). To be most efficient, samples that require 7+ cytokine measurements are stored until there are enough samples to fill an entire plate.

4. Notes

1. In general, high binding plates (>400 ng/cm^2) work best for ELISAs, but may result in higher background readings. If the high background cannot be resolved through titration, one should consider the use of medium binding plates (250 ng/cm^2).
2. It is acceptable, but cumbersome, to manually wash the plates with a multichannel pipette. It is not advisable to use a squirt bottle for washing because wash fluid from one well can spill into another, skewing results. Additionally, it is possible to leave the plates soaking in wash buffer for up to an hour; however, it is recommended that this is determined empirically for your ELISAs.

Table 4
Example cost comparison between multiplex and single cytokine ELISA. Costs are for analysis of 17 cytokines. Up to 40 samples/plate may be assayed

	Multiplex	Single
Antibody costs		
Capture	\$17.28	\$72.00
Standard	\$0.81	\$2.00
Detection	\$41.04	\$91.20
Consumables' costs		
Pipette tips	\$2.00	\$40.00
Clean room wipes	\$0.20	\$0.00
Plates	\$4.00	\$68.00
IR dye	\$0.27	\$0.00
Odyssey blocking buffer	\$5.72	\$0.00
Buffers	\$0.50	\$40.00
HRP	\$0.00	\$2.00
TMB	\$0.00	\$8.80
Total costs	\$71.82	\$324.00

3. Any length of time may be used for each incubation so long as the incubation times used in the checkerboarding step of development are used going forward. It is likely that shorter incubation times will require an increased concentration of reactants. For instance, our lab has a standard IL-6 ELISA assay that takes 6+ hours to complete. We also have a rapid IL-6 ELISA that can be performed in less than 90 min and is useful when a rapid value is necessary (9).
4. The optimal dilution for SA–HRP typically remains the same across different cytokine ELISAs. This needs to be determined only once.
5. There is a linear relationship between the OD_{590} and the ΔOD(10). An OD_{590} of 0.4–0.5 translates to a ΔOD of ~ 1.2–1.6 (see Note 7).
6. There are several reasons why a plate may turn uniformly blue or not change color at all. Uniform color development across the plate may be due to the following: (1) User coated with biotinylated detection Ab instead of unlabeled capture antibody. (2) The biotinylated antibody has degraded causing NSB. (3) Too much capture and/or detection antibody was used. This is especially true in the initial stages of development

(checkerboard), when the amount of antibody to be used is unknown. Lack of color development usually indicates a problem with the capture antibody or HRP conjugate. If all wells with standards turn blue too quickly, but the zero standard remains clear, repeat the ELISA with a more dilute standard range.

7. The optimal dilution of antibody may not have the highest signal-to-noise ratio for the top standard when compared to other dilutions. The signal:noise of the low and mid standard should guide your selection. This is because the OD of the top standard is likely representative of saturation of the system (either the capture or the detection Ab is not in excess to standard). What is important, however, is that the top standard has an OD around 1.5 or higher. ODs above 1.5–1.7 are unreliable for quantitative measurements in many plate readers (11), but they are a good starting point for the top standard. If the OD of the top standard does not reach this area, the range of the standard curve is decreased. If raising the concentration of standard does not increase the OD, it is likely that either the capture or detection antibodies are limiting.
8. The upper asymptote is a result of the standard being in excess to the capture or detection antibodies. It is important that this asymptote does not exceed the linear range of the plate reader being used. Decreasing capture or detection antibody concentrations decrease the maximum OD observed; however, there may be a trade-off in sensitivity of the ELISA. If sensitivity cannot be sacrificed, it is advisable to investigate the five-parameter logistic (5PL) fit model as an alternative to the 4PL. The 5PL can better handle curves with asymmetries such as those that may come from not achieving an upper or lower asymptote. Regardless of which curve model is selected, the OD of the standards must be within the linear range of the plate reader.
9. Samples can be stored overnight at 4°C to continue the sequential ELISAs the next day, since each additional plate adds 2 hours to the time devoted to the assay.
10. Printed plates may be stored for up to 6 months at 4°C. It is recommended to determine the stability of the printed plates under the conditions in your lab.
11. As with many fluorophores, exposure to ambient light may result in photobleaching. Care should be taken to limit the exposure of the IR dye to light.

References

1. Feghali, C.A., Wright, T.M. (1997) Cytokines in acute and chronic inflammation, *Front Biosci* **2**, d12–26.
2. Sokol, H., Pigneur, B., Watterlot, L. et al. (2008) *Faecalibacterium prausnitzii* is an anti-inflammatory commensal bacterium identified by gut microbiota analysis of Crohn disease patients, *Proc Natl Acad Sci USA* **105**, 16731–16736.
3. Polpitiya, A. D., McDunn, J. E., Burykin, A. et al. (2009) Using systems biology to simplify complex disease: immune cartography, *Crit Care Med* **37**, S16–21.
4. De Santo, C., Arscott, R., Booth, S. et al. (2010) Invariant NKT cells modulate the suppressive activity of IL-10-secreting neutrophils differentiated with serum amyloid A, *Nat Immunol* **11**, 1039–1046.
5. DeForge, L. E., Remick, D. G. (1991) Sandwich ELISA for detection of picogram quantities of interleukin-8, *Immunol Invest* **20**, 89–97.
6. Osuchowski, M. F., Remick, D. G. (2006) The repetitive use of samples to measure multiple cytokines: the sequential ELISA, *Methods (San Diego, Calif)* **38**, 304–311.
7. Findlay, J.W.A., Dillard, R. F. (2007) Appropriate calibration curve fitting in ligand binding assays, *AAPS Journal* **9,** 2 E260-E267. doi: 10.1208/aapsj0902029.
8. Morgan, E., Varro, R., Sepulveda, H. et al. (2004) Cytometric bead array: a multiplexed assay platform with applications in various areas of biology, *Clin Immunol* **110**, 252–266.
9. Nemzek, J. A., Siddiqui, J., Remick, D. G. (2001) Development and optimization of cytokine ELISAs using commercial antibody pairs, *J Immunol Methods* **255**, 149–157.
10. Natarajan, S., Remick, D.G. (2008) The ELISA Standard Save: Calculation of sample concentrations in assays with a failed standard curve, *J Immunol Methods* **336**, 242–245.
11. Crowther, J. R. (2000) The ELISA guidebook, *Methods in molecular biology (Clifton, NJ)* **149**, III-IV, 1–413.

Chapter 3

Flow Cytometry Analysis of Cell Cycling and Proliferation in Mouse Hematopoietic Stem and Progenitor Cells

Valérie Barbier, Bianca Nowlan, Jean-Pierre Lévesque, and Ingrid G. Winkler

Abstract

The hematopoietic system is highly proliferative in the bone marrow (BM) due to the short half-life of granulocytes and platelets in the blood. Analysis of cell cycling and cell proliferation in vivo in specific populations of the mouse BM has highlighted some key properties of adult hematopoietic stem cells (HSCs). For instance, despite their enormous proliferation and repopulation potential, most true HSC are deeply quiescent in G_0 phase of the cell cycle and divide very infrequently, while less potent lineage-restricted progenitors divide rapidly to replace the daily consumption of blood leukocytes, erythrocytes, and platelets. In response to stress, e.g., following ablative chemotherapy or irradiation, HSC must enter the cell cycle to rapidly repopulate the BM with progenitors. Due to their extreme rarity in the BM, at least five color flow cytometry for cell surface antigens has to be combined with staining for DNA content and nuclear markers of proliferation to analyze cell cycle and proliferation of HSC in vivo. In this chapter, we describe two methods to stain mouse HSC to (1) distinguish all phases of the cell cycle (G_0, G_1, S, and G_2/M) and (2) analyze the divisional history of HSC in vivo by incorporation of the thymidine analog 5-bromo-2-deoxyuridine.

Key words: Hematopoietic stem cells, Bone marrow, Bone marrow stroma, Perfusion, Stem cell niche, Flow cytometry

1. Introduction

The hematopoietic system is hierarchically organized in the bone marrow (BM) of adult mammals with a self-renewing hematopoietic stem cell (HSC) at the apex. The HSCs remain largely quiescent in adult BM. They also divide either symmetrically to self-renew or asymmetrically to generate multipotent HPCs that will in turn commit to one of the many blood cell lineages, divide and mature to replace blood leukocytes, platelets, and erythrocytes. Due to the relatively short half-life of leukocytes (from days for granulocytes

Robert B. Ashman (ed.), *Leucocytes: Methods and Protocols*, Methods in Molecular Biology, vol. 844,
DOI 10.1007/978-1-61779-527-5_3, © Springer Science+Business Media, LLC 2012

and platelets to weeks for erythrocytes or years for memory B/T lymphocytes), the BM hematopoietic tissue is highly regenerative. For instance, approximately 6×10^6 erythrocytes and 5×10^5 granulocytes per second need to be produced in healthy human adults just for their replacement to maintain their numbers within normal ranges in the blood. The bulk of proliferative cells in the BM are HPC (1). In sharp contrast, true HSC divide very infrequently in the BM despite having the highest proliferative potential of the whole hematopoietic system. In the mouse, it has been recently reported that the most primitive HSC, defined functionally as cells able to reconstitute the whole hematopoietic and immune system once transplanted serially in successive lethally irradiated hosts, are highly quiescent in G_0 phase of the cell cycle and divide very infrequently (about once every 145 days or five times in the lifespan of an adult mouse) (2, 3). These highly quiescent HSC are a genetic reserve that can be induced to divide in emergency situations, such as following cytotoxic therapy, irradiation, or systemic cytokine treatments, to rapidly repopulate the BM with all the necessary HPC to produce mature blood cells (2).

Consequently, the BM is an interesting tissue to study stem cell division in adult mammals in vivo. Cell cycling can be analyzed in two different manners (1) by cell cycle analysis based on DNA copy numbers and nuclear markers of cell cycle progression, which gives a snapshot distribution of a cell type between the G_0, G_1, S, and G_2/M phases of the cell cycle, or (2) cumulative incorporation of a nucleotide analog (such as 5-bromo-2-deoxyuridine or BrdU) in the genomic DNA during S phase to analyze the divisional history or turnover of a cell type in a tissue (1). Unfortunately, the BM is a very heterogeneous tissue and HSC represent less than 1/20,000 BM leukocytes. Even the frequency of the more abundant HPC frequency does not exceed 5% of the BM. Consequently in the BM, cell cycle analysis must be combined with at least five color flow cytometry for cell surface antigens in order to phenotypically separate HSC from highly proliferative HPC (2, 3).

This chapter provides detailed methods to analyze cell cycle or BrdU incorporation in HSC and HPC from the mouse BM. These methods can be easily adapted to measure HSPC proliferation and cycling in other tissues such as liver, spleen, and blood. For BrdU incorporation studies, mice are administered BrdU for various durations, whereas cell cycle analyses do not require BrdU administration. In both methods, BM cells are harvested and in a first step enriched for HSPC as these cells are very rare in the BM. Once enriched, live HSPC are stained for cell surface markers with fluorescent antibodies. They are then fixed and permeabilized for further staining for genomic DNA, markers of cycling (Ki-67), or newly synthesized DNA (BrdU). DNA content and Ki-67 expression or BrdU incorporation are then analyzed together with cell surface antigens by flow cytometry.

2. Materials

2.1. Animal Treatment

BrdU: 5-Bromo-2′ deoxyuridine (Sigma Aldrich). Light sensitive. Resuspend the powder at 10 mg/mL in injectable saline. Sterile filter and store frozen at –20°C (or –80°C for long-term storage).

2.2. Tissue Harvest

1. 5-mL polypropylene tubes (Greiner).
2. Dulbecco's phosphate-buffered saline (DPBS) without calcium or magnesium (Biowhittaker) supplemented with 2% heat-inactivated newborn calf serum (NCS), (GIBCO) sterile filtered.
3. Mortar and pestle.
4. Cell strainers 40-μm nylon (BD).

2.3. Bone Marrow Processing and Magnetic-Activated Cell Sorting

1. Automated hematology analyzer, such as KX-21 N (Sysmex, Kobe, Japan) or manual counting of cells, following (>1:20) dilution in white cell counting fluid using a brightfield Neubauer microscope counting chamber.
2. 10× red cell lysis buffer. 1.5 M NH_4Cl, 100 mM $NaHCO_3$, 10 mM EDTA pH 7.4. Sterile stock can be kept in fridge for many months. On day of the experiment, dilute 1 part of 10× red cell lysis buffer with 9 parts of sterile water to make 1× red cell lysis buffer.
3. Refrigerated centrifuge to rotate 1–50-mL tubes and microplates at 370×*g*.
4. 1.5-mL Eppendorf tubes.
5. 0.5 M EDTA pH8.0.
6. Tissue culture Petri dishes 35×10mm (NUNC).
7. Microscope slides rough with frosted-glass ends.
8. Anti-Kit magnetic-activated cell sorting (MACS) beads: mouse "CD117 microbeads" (Miltenyi Biotec).
9. MACS buffer: DPBS+0.5% bovine serum albumin+2 mM EDTA.
10. "autoMACS pro-separator" with "autoMACS separation columns" (Miltenyi Biotec). Manual positive MACS columns "autoMACS Separator" are also available.

2.4. Flow Cytometry Staining

1. 1.5-mL polypropylene tubes (Greiner).
2. Stain tubes: ideally 1.2-mL micro tibertube (Quality Scientific Plastics) or other polypropylene staining tubes.
3. Sterile cannulas (Unomedical).
4. Purified rat anti-mouse Fcγ receptor II/III clone 2.4 G2 (Fc Block) (BD) (see Note 1).
5. MACS buffer: DPBS+0.5% bovine serum albumin+2 mM EDTA.

6. DPBS + 2% NCS.
7. Conjugated monoclonal antibodies specific for mouse antigens:

   CD3ε-biotin clone 145-2C11 (BD), 0.5 mg/mL

   CD5-biotin clone 53–7.3 (BD), 0.5 mg/mL

   CD45R (B220)-biotin clone RA3-6B2 (BD), 0.5 mg/mL

   Gr1- biotin clone RB6-8 C5 (BD), 0.5 mg/mL

   F4/80-biotin clone BM8 (eBioscience), 0.2 mg/mL

   CD41-biotin clone MWreg30 (eBioscience), 0.5 mg/mL

   Ter119-biotin clone Ter119 (BD) 0.5 mg/mL

   Sca-1-PECY7 clone D7 (BD), 0.2 mg/mL

   Kit (CD117)-APC clone 2B8 (Biolegend) 0.2 mg/mL

   CD48-PE clone HM48-1 (BD) 0.2 mg/mL

   CD48-Pacific blue clone HM48-1 (Biolegend) 0.5 mg/mL

   CD150-PE clone TC15-12 F 12.2 (Biolegend) 0.2 mg/mL

   Streptavidin-Alexa700 (invitrogen) 0.5 mg/mL
8. Ki67-FITC Kit. BD pharmingen. Includes mouse anti-human Ki67-FITC clone B56 and IgG1κ isotype control clone MOPC-21.
9. For Hoechst33342 and Ki-67 stains only, purchase Fix & Perm Cell Permeabilization reagents, including fixation medium A and permeabilization medium B from Caltag Laboratories—Invitrogen.
10. Saponin from Sigma Chemicals for Hoechst33342 and Ki-67 stains only.
11. FITC-BrdU Flow Kit, BD Pharmingen (50 tests and 4 × 50 tests). This kit includes permeabilization and fixation buffers together with the DNAse solution.
12. mIgG1κ-FITC (BD) 0.5 mg/mL.
13. RNAse A (Sigma Chemicals) 1 mg/mL DNAse heat-inactivated 10 min at 95°C.
14. Hoescht33342 (Sigma) stock 20 mM (11.2 mg/mL) stored –20°C (see Note 2).
15. BD LSR II flow cytometer for cell cycle analysis, equipped with 350-nm ultraviolet laser (with 450/50 filter for Hoechst33342), 406-nm violet laser (with 450/50 filter for Pacific blue), 488-nm blue laser (with 530/40, 575/25, 710/30 and 787/43 filters for FITC, PE, PercPCY5.5, PECY7, respectively) and 643-nm red laser (with 665/20 and 750LP filters for APC and APCCY7/Alexa700 respectively) (see Notes 3).
16. FloJo software (Tree Star, Ashland, OR) or other for analysis of results.

3. Methods

3.1. Animal Treatment

3.1.1. In Vivo BrdU Labeling for Turnover Analysis

1. Dilute BrdU at 0.25 mg/mL in mouse drinking water. Cover the bottle with aluminum foil as BrdU is light sensitive.
2. Keep the BrdU in drinking water for the whole 3 days and nights prior to harvest (see Note 5).

3.1.2. In Vivo BrdU Labeling with Long Chase to Identify Long-Term BrdU Label-Retaining HSC

1. Dilute BrdU at 0.5 mg/mL in mouse drinking water. Cover the bottle with aluminum foil as BrdU is light sensitive.
2. Keep the BrdU in drinking water for 14 days ("Pulse") by changing the drinking solution every 3 days.
3. On day 14, remove drinking bottle containing BrdU and replace it by drinking bottle containing plain tap water without BrdU for 60–100 days ("Chase").

3.2. Bone Marrow Harvest

At all times cells should be kept on ice and in the dark, reagents and centrifuges should be kept at 4°C.

1. Before euthanasia record mouse weight.
2. Immediately euthanize mouse by cervical dislocation or any other ethically approved method.
3. Remove skin to access and collect hind limb bones (two hips, two tibias, two femurs) using sterile scissors, tweezers, and scalpel.
4. Clean the bones by removing the muscles attached to them.
5. Place bones in ice-cold DPBS + 2%NCS.
6. Gently crush bones in ice-cold mortar and pestle with 5 mL of DPBS + 2%NCS. Do not over crush.
7. Filter through 40-μm cell strainer into 50-mL tube on ice.
8. Repeat the crush three times (each crush in 5 mL of buffer). The final volume will then be around $3 \times 5 = 15$ mL.
9. Discard the carcass following ethical procedures.

3.3. Cell Preparation

1. Add 2 mM final EDTA to the BM cells collected in Subheading 3.2, step 8 to avoid cell clumping.
2. Dilute 20 μL whole BM cell suspension with 80 μL DPBS + 2% NCS (1/5 dilution) into an Eppendorf tube. Count leukocytes on automated Sysmex cell counter. Multiply by five to obtain number of leukocytes per mL.
3. Spin the 50-mL tubes at $370 \times g$ for 5 min at 4°C.
4. Aspirate the supernatant. Leave 2 mL behind. Resuspend the cell pellet.
5. These BM cells can now be used for Kit^+ cell enrichment.

3.4. Enrichment for Kit-Positive Cells by autoMACS

1. Add 0.5 μL of anti-Kit MACS beads per 10^8 mononucleated cells.
2. Incubate at 4°C for 25–40 min with gentle agitation.
3. Wash cells once in 10 mL of MACS buffer.
4. Spin the 50-mL tubes at 370×*g* for 5 min at 4°C.
5. Aspirate supernatant to leave pellet.
6. Resuspend cell pellet in 1 mL MACS buffer per 2×10^8 cells.
7. Enrich Kit⁺ cells by using an autoMACS Pro-Separator (automatic). Choose the positive selection program "POSSEL" and collect the Kit-enriched cells (positive fraction) in 5-mL tube. Note that an autoMACS separator (manual) can also be used.
8. Take a 60 μL aliquot and count leukocytes on automated Sysmex cell counter. Expect 1% recovery. Keep remaining cells on ice.

3.5. Flow Cytometry Stains

All single color controls for compensation are performed on total BM leukocytes taken prior autoMACS separation.

3.5.1. Flow Cytometry Analysis of Cell Cycling with Ki67 and Hoechst33342 Stains on BM Cells

In this method, cells are surface labeled with fluorescent antibodies specific of blood lineage markers, HPC and HSC. Cells are then fixed, permeabilized and stained with an FITC-conjugated monoclonal antibody specific for Ki67, a nuclear antigen exclusively expressed by cells entering (phase G_1), and progressing through cell cycle (Phases S, G_2, and M). Ki67 is absent in quiescent cells in phase G_0 (4). Cells are finally stained for DNA content with the fluorescent DNA intercalating agent Hoechst33342 after RNA digestion with RNAse A.

1. Transfer 10^6 Kit⁺ enriched BM cells into labeled stain tubes. Add an extra tube for control stain with non-immune mIgG1-FITC (isotype control).
2. Fill up tubes with DPBS+2%NCS.
3. Spin at 370×*g* for 5 min at 4°C.
4. Aspirate supernatant with cannula on vacuum line and leave 25 μL on cell pellet. Resuspend the cells by tapping the tubes with fingers or vigorous vortex (do not create foam).
5. Keep seven cell aliquots to make single color controls for color compensation see step 25.
6. Keep a cell aliquot to be fixed, permeabilized, and stained with Hoechst33342 alone to use as a single color control for Hoechst33342 see step 27.
7. Add 25 μL of antibody mix to each cell aliquot (final stain volume 50 μL for 10^6 cells). The antibody mix is made of Fcblock hybridoma supernatant or in DPBS+2%NCS with 2–5 μg/mL purified Fc Block antibody containing lineage-biotin (CD3, CD5, B220, Gr1, F4/80, Ter119, CD41),

Sca-1-PECY7, Kit-APC, CD48-PE antibodies in order to obtain a 1/300 final dilution for each conjugated antibody.

8. Mix then incubate on ice in the dark for 30 min with gentle rocking.
9. Wash stains with 1 mL of straight DPBS or MACS buffer. Repeat steps 3 and 4 (wash).
10. Add 25 μL of streptavidin-Alexa700 1/200 final in MACS buffer.
11. Repeat step 8 (mix and incubate).
12. Fill up tubes with DPBS.
13. Spin at 370 × *g* for 5 min at 4°C.
14. Aspirate supernatant with cannula on vacuum line and leave 20 μL on cell pellet.
15. Resuspend the cells.
16. Add 80 μL fixation medium A (Caltag), mix well.
17. Incubate for 15 min at room temperature with rocking in the dark.
18. Wash twice in 1 mL DPBS + 5% NCS and leave 20 μL of cell pellet.
19. Add 80 μL permeabilization medium B (Caltag).
20. Add 10 μL anti-Ki67-FITC or mIgG1-FITC control.
21. Incubate 30 min on ice with gentle agitation.
22. Wash with DPBS + 2% NCS.
23. Resuspend cells in 1 mL of DPBS containing 1 μg/mL RNAse A (1/1,000 of 1 mg/mL stock), 0.05% saponin, 20 μM Hoechst33342.
24. Put on shaker for 10 min before reading at the flow cytometer.
25. For the single color controls (required to set flow cytometer compensation values), add 0.2 μL of fluorochrome conjugated antibody (one antibody per stain) to 10^6 unstained control BM cells in a final volume of 25 μL. For biotinylated antibodies stained with fluorochrome-conjugated streptavidin, add 0.2 μL of both biotinylated antibody and streptavidin at the same time. After 20 min incubation on ice, wash single color controls once as in steps 2 and 3 and fix as in steps 16 and 17.
26. Analyze on a flow cytometer without washing the samples from the Hoechst33342 dye. On LSRII flow cytometer, it is essential to untick the logarithmic box on the Hoechst33342 channel (same as DAPI channel with excitation in the UV and reading through a 450/50 filter) and record peak height, area, and width to eliminate doublets and analyze DNA content on a linear scale.
27. Setup of photomultiplier voltages and compensation of the LSRII flow cytometer (see Note 3). The unstained sample is

acquired first to set photomultiplier voltages of all colors with a peak of negative fluorescence at 200 on a logarithmic scale. In second must be acquired the sample containing Hoescht33342 alone. The corresponding photomultiplier voltage is set to obtain a nice cell cycling distribution in the middle of the linear scale for signal-area. Once photomultiplier voltage for Hoechst33342 is adjusted, negative cell are appended to the file with Hoechst33342-labeled cells in order to have both negative and positive events for Hoechst33342 within the same file. This is used to compensate Hoechst33342 in all other color channels. Then, all other individual colors are acquired and compensated into all other channels one by one.

28. Acquisition of files. Once the flow cytometer is properly adjusted and compensated, acquire files at low speed in order to have maximum resolution.

29. Analyze results with FloJo software. Figure 1 is an example of the gating strategy to determine frequency of cells in phases G_0

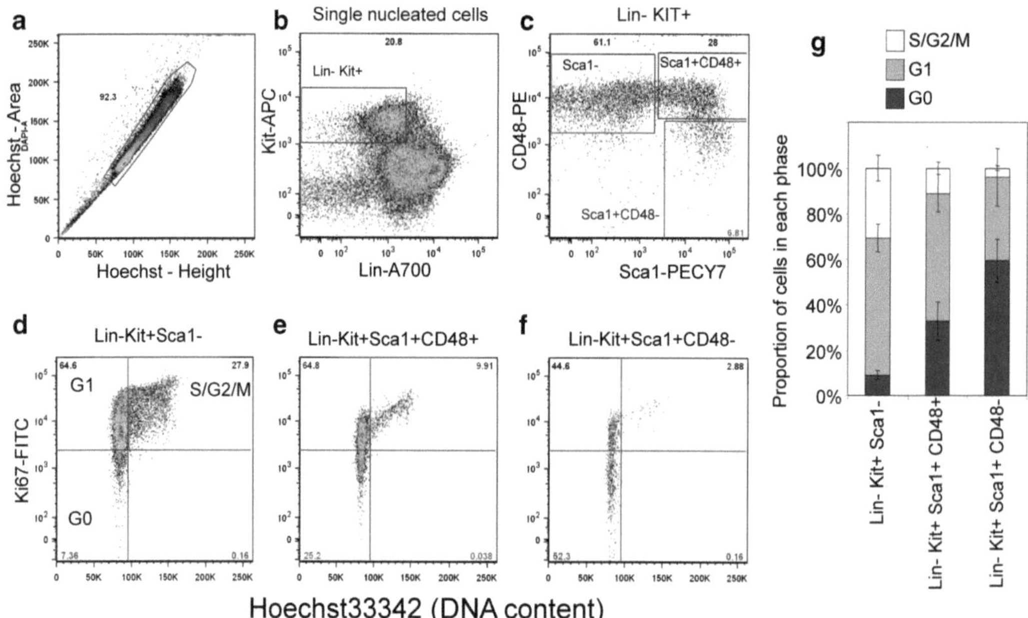

Fig. 1. Gating strategy to analyze cell cycle in HSPC from the mouse BM. BM leukocytes from C57BL/6 mice were stained for blood lineage markers, Sca1, Kit, and CD48 cell surface antigens and then for Ki67 and DNA content with Hoechst33342. (**a**) After Kit^+ cell enrichment by MACS, intact cells were gated on forward scatter versus side scatter plot (not shown), and then single nucleated cells were gated by drawing a diagonal region on Hoechst peak height versus Hoechst peak area on a linear scale. Note that Hoechst negative anucleated cells or apoptotic cells with weak Hoechst signal are gated out. (**b**) Gating of Lin^- Kit^+ cells. (**c**) Gating of Lin^- Kit^+ $Sca1^-$ myeloid progenitors, Lin^- Kit^+ $Sca1^+$ $CD48^+$ lineage-restricted HPC, and Lin^- Kit^+ $Sca1^+$ $CD48^-$ multipotent progenitors and HSC. (**d–f**) Cell cycle analysis in Lin^- Kit^+ $Sca1^-$, Lin^- Kit^+ $Sca1^+$ $CD48^+$, and Lin^- Kit^+ $Sca1^+$ $CD48^-$ cells, respectively. Note that DNA content is measured on Hoechst peak area in a linear scale, whereas all other markers are in logarithmic scales. Cells in phase G_0 are $Ki67^-$ with 2n DNA, in G_1 $Ki67^+$ with 2n DNA, and in S/G2/M $Ki67^+$ with >2n DNA. (**g**) Distribution of HPC and HSC between the different phases of the cell cycle. Note that more primitive HSPC have a higher proportion of quiescent cells in G_0. These data are mean ± SD of 3 individual mice.

(Ki67$^-$, 2n DNA), G_1 (Ki67$^+$, 2n DNA), and S/G_2/M (Ki67$^+$, >2n DNA) on lineage-negative (Lin$^-$) Sca1$^-$ Kit$^+$ myeloid HPC, Lin$^-$ Sca1$^+$ Kit$^+$ HSPC which can be further subdivided into Lin$^-$ Sca1$^+$ Kit$^+$ CD48$^+$ lineage-restricted HPC and Lin$^-$ Sca1$^+$ Kit$^+$ CD48$^-$ multipotent progenitors and HSC (see Note 4).

3.5.2. Flow Cytometry Analysis of Cell Proliferation by BrdU Stain

In this method, mice have been fed with BrdU in their drinking water from 1 to 14 days (see Note 5). Any cell dividing during this period of time will incorporate BrdU in their genomic DNA each time they progress through S phase. The more the given cell has divided during this BrdU loading period, the more the cell will be positive for BrdU. Conversely, cells that remain quiescent during this period of time will remain BrdU-negative (see Note 6).

1. Transfer 10^6 Kit$^+$ enriched BM cells into labeled stain tubes. Add an extra stain for mIgG1-FITC (isotype control).
2. Fill up tubes with 1 mL DPBS + 2%NCS.
3. Spin at 370 × g for 5 min at 4°C.
4. Aspirate supernatant with cannula on vacuum line and leave 25 μL on cell pellet. Resuspend the cells by tapping the tubes with fingers or vigorous vortex (do not create foam).
5. Add 25 μL of antibody mix to each cell aliquot (final stain volume 50 μL for 10^6 cells). The antibody mix is made of Fcblock hybridoma supernatant or in DPBS + 2% NCS + 2–5 μg/mL purified Fc Block antibody containing lineage-biotin (CD3, CD5, B220, Gr1, F4/80, Ter119, CD41), Sca-1-PECY7, Kit-APC, CD48-Pacific blue, CD150-PE antibodies in order to obtain a 1/300 final dilution for each conjugated antibody.
6. Mix then incubate on ice in the dark for 30 min with gentle rocking.
7. Wash stains with 1 mL of straight DPBS or MACS buffer. Repeat steps 3 and 4 (wash).
8. Add 25 μL of streptavidin-Alexa700 1/200 final in MACS buffer.
9. Mix then incubate on ice in the dark for 15 min with gentle rocking.
10. Wash by repeating steps 2–4 before fixation and permeabilization steps below.
11. Resuspend cells with 80 μL cytofix–permeabilization buffer per tube (provided in FITC-BrdU staining kit), mix well.
12. Incubate for 15–30 min on ice with agitation.
13. Wash cells with 800 μL permeabilization–wash buffer (provided in FITC-BrdU staining kit).
14. Spin at 370 × g for 5 min at 4°C. Aspirate the supernatant and leave 25 μL of cell pellet.

15. Add 80 μL of cytoperm-plus buffer (provided in FITC-BrdU staining kit).
16. Incubate for 10 min on ice with agitation.
17. Wash cells with 800 μL permeabilization–wash buffer.
18. Spin at 370 × *g* for 5 min at 4°C. Aspirate the supernatant and leave 25 μL of cell pellet to proceed to the second fixation–permeabilization step.
19. Add 80 μL of cytofix–permeabilization buffer.
20. Incubate 5 min on ice with agitation.
21. Wash cells with 800 μL permeabilization–wash buffer.
22. Spin at 370 × *g* for 5 min at 4°C. Aspirate the supernatant and leave 25 μL of cell pellet.
23. Per sample, make a DNAse dilution of 13 μL of DNAse (provided in FITC-BrdU staining kit) in 37 μL DPBS (see Note 7).
24. Add 50 μL of diluted DNAse (30 μg DNAse/tube) to cell pellets.
25. Incubate 1 h at 37°C with agitation in the dark.
26. Wash cells with 800 μL of permeabilization–wash buffer.
27. Spin at 370 × *g* for 5 min at 4 C. Aspirate the supernatant and leave 25 μL of cell pellet to proceed with the BrdU staining.
28. Add to each cell pellet 25 μL of permeabilization–wash buffer containing diluted (1/60) FITC-conjugated anti-BrdU antibody (provided in FITC-BrdU staining kit).
29. Incubate for 30–60 min at room temperature in the dark.
30. Then, add 800 μL of permeabilization–wash buffer and incubate a further 10 min with mixing to wash off unbound antibody.
31. Spin at 370 × *g* for 5 min at 4 C. Aspirate the supernatant and leave 25 μL of cell pellet.
32. Resuspend cells in 200 μL DPBS + 2%NCS.
33. For the single color controls (needed to set flow cytometer compensation values), add 0.2 μL of fluorochrome-conjugated antibody (one antibody per stain) to 10^6 unstained control BM cells in a final volume of 25 μL. For biotinylated antibodies stained with fluorochrome-conjugated streptavidin, 0.2 μL of both biotinylated antibody and streptavidin can be added at the same time. After 30 min incubation on ice, wash the single color controls once as in steps 2–4 and resuspend in 300 μL DPBS + 2% NCS.
34. Analyze on a flow cytometer after setting of photomultiplier voltages on unstained cells and adjustment of color compensation parameters for each individual color.

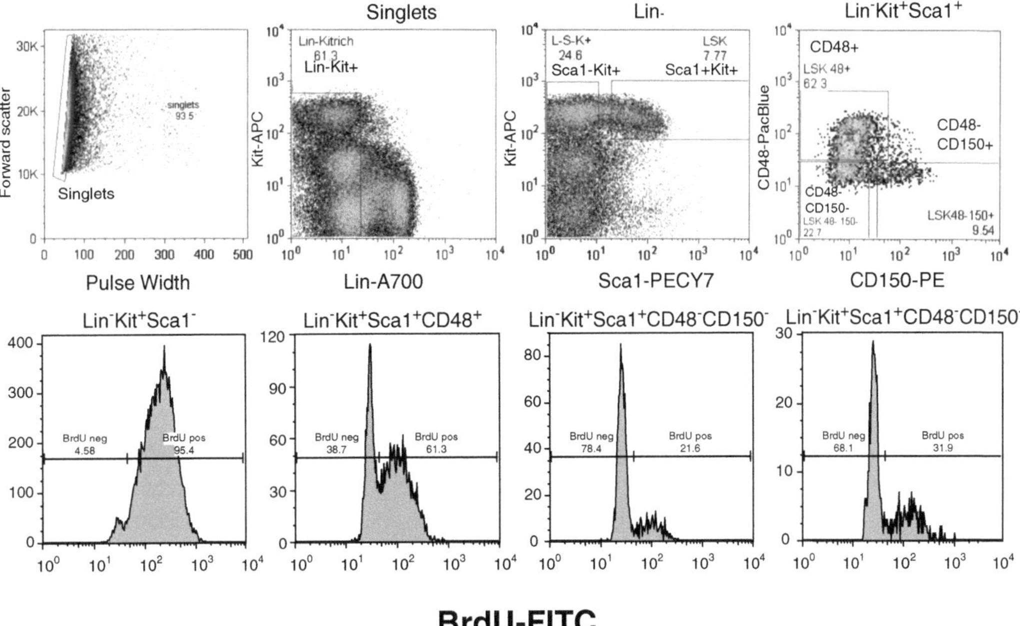

Fig. 2. Gating strategy to analyze BrdU incorporation in HSPC from mouse BM. Single cells are gated using forward scatter versus pulse width. Lineage-negative cells are then gated using dot-plots of lineage versus Kit. Lin^- cells are further gated into $Lin^-Kit^+Sca1^-$ myeloid progenitors and $Lin^-Kit^+Sca1^+$ HSPC. Finally, $Lin^-Kit^+Sca1^+$ HSPC are gated into $Lin^-Kit^+Sca1^+CD48^+$ lineage-restricted progenitors, $Lin^-Kit^+Sca1^+CD48^-CD150^-$ multipotent progenitors, and $Lin^-Kit^+Sca1^+CD48^-CD150^+$ HSC. BrdU incorporation is then measured in each of these populations. Typically, we find that following a 3-day period of continuous BrdU feeding, 96 ± 4% myeloid progenitors are $BrdU^+$, decreasing to 64 ± 1% in lineage-restricted HPC, 29 ± 10% in multipotent progenitors, and 43 ± 6% in phenotypic HSC. These data are mean ± SD of 4 adult C57BL/6 male mice.

35. Analyze results with FlowJo software. Figure 2 shows an example of gating strategy to measure BrdU incorporation in Lin^- $Sca1^-$ Kit^+ myeloid HPC, Lin^- $Sca1^+$ Kit^+ HSPC which can be further subdivided into Lin^- $Sca1^+$ Kit^+ $CD48^+$ lineage-restricted HPC, Lin^- $Sca1^+$ Kit^+ $CD48^-$ $CD150^-$ multipotent progenitors, and Lin^- $Sca1^+$ Kit^+ $CD48^-$ $CD150^+$ phenotypic HSC (5). BrdU-negative cells that remained in G_0/G_1 phase of the cell cycle for 3 days are most abundant in the HSC and multipotent progenitor fractions, whereas most HSC have divided or entered S phase and are $BrdU^+$.

4. Notes

1. To save money, blockage of Fcγ receptors II/III can be achieved with the straight hybridoma supernatant from hybridoma 2.4 G2. However, supernatants must be batch tested for their ability to block CD16/32-PE binding to mouse macrophages or mouse monocytic M1 cell line differentiated with interleukin-6.

2. If a cytometer equipped with an ultraviolet laser is not available, Hoechst33342 can be replaced by Vybrant DyeCycle Violet from Invitrogen and DNA content analyzed following excitation with a 405-nm violet laser and detection through a 450/50 filter. We do not advise to use 7-amino actinomycin D for cell cycle analysis when many other dyes excited by the blue laser are used simultaneously as PE, and PECY7 have some spectral overlap in the 7-AAD channel. Despite compensation, this broadens the width of the G_0/G_1 peak and reduces resolution of cycle phase analysis.
3. We use a BD LSR II flow cytometer (Serial number: H48200015; Build date: 2005; Software: BD FACSDiVa Version 6.1.3) to perform cell cycle analyses. This instrument is regularly checked with BD CS&T beads to maintain integrity of service and maintenance values, particularly the time delay between the four lasers. Although auto-compensation is often recommended for multicolor experiments, manual compensation using an unstained control and single stains yields superior compensation in our hands. Compensation using BD CompBeads (Anti-Rat Ig) has been used, but precise compensation requires the negative control (FBS) beads supplied with the kit. With practice the setting of the auto-fluorescent PMT voltages and performing a manual compensation can be completed within 30 min. To optimize the cell cycle results, the sheath tank should be filled prior to the start of the analysis, and any bubbles in the sheath line removed at the filter and by priming the LSR II twice. By reducing the air volume at the top of the sheath container, sheath pressure stability is increased. Cell concentration is also very important. The cells should be at a sufficient concentration to allow a reasonable flow rate past the laser while allowing the instrument to be run on the low sample speed setting. For normal analysis, the LSR II collects data most accurately at event rates of <20,000 events/s. For accurate cell cycle data, the rate should be <5–10,000 events/s. It is important to realize that cell cycle is acquired in linear mode, and it is important to set the PMT voltage for that parameter based on a stained sample and a linear scale. However, to complete compensation a log scale is required using the same PMT voltage. See Subheading 3.5.1, step 27. A clean cell cycle is dependent on single cell data. This is why area, height, and width parameters are collected. By using a bivariate histogram (dot plot) of Hoechst33342 height versus Hoechst33342 width and gating the center core of the data (essentially the diagonal population excluding outliers) then looking at a histogram of Hoechst33342 area should provide a clean clear cell cycle.
4. We find that adding CD150 to separate Lin^- $Sca1^+$ Kit^+ $CD48^-$ $CD150^-$ multipotent progenitors from Lin^- $Sca1^+$ Kit^+

CD48$^-$ CD150$^+$ HSC does not resolve better cell cycle analysis. However, if cell cycle analysis need be performed by further subgating the Lin$^-$ Sca1$^+$ Kit$^+$ CD48$^-$ population, the CD48-PE antibody can be replaced by CD48-PercPCY5.5 together with CD150-PE in order to analyze separately the Lin$^-$ Sca1$^+$ Kit$^+$ CD48$^-$ CD150$^-$ multipotent progenitors from Lin$^-$ Sca1$^+$ Kit$^+$ CD48$^-$ CD150$^+$ HSC populations.

5. If BrdU is to be pulsed for a period of less than 24 h, it is preferable to inject it intraperitoneally at 1 mg/mouse in sterile injectable saline instead of oral administration in drinking water.
6. An interesting variation of this technique is a pulse-chase experiment in which mice are fed with BrdU continulously for 2 weeks to load all HSC and HPC with BrdU (pulse) and then given BrdU-free water for up to 100 days (chase) to detect long-term BrdU label retaining cells (2, 3). These cells correspond to the most deeply quiescent HSC able to serially reconstitute successive lethally irradiated hosts (2). They reside in the least perfused areas of the BM in hypoxic niches (3).
7. As genomic DNA is tightly packed in the nucleus, BrdU epitopes incorporated into the genomic DNA are not very accessible to monoclonal antibodies because of steric hindrance. For this reason, a step of partial DNA digestion is necessary to unpack the DNA and make incorporated BrdU epitope accessible to antibodies.

Acknowledgments

JPL was supported by a Senior Research Fellowship from the Cancer Council of Queensland, IGW by a CDA fellowship form the National Health and Medical Research Council of Australia. This work was supported by project grants 434515, 543706, and 350406 from the National Health and Medical Research Council.

References

1. Bradford, G. B., Williams, B., Rossi, R. et al. (1997) Quiescence, cycling, and turnover in the primitive hematopoietic stem cell compartment, *Exp Hematol* **25**, 445–453.
2. Wilson, A., Laurenti, E., Oser, G. et al. (2008) Hematopoietic stem cells reversibly switch from dormancy to self-renewal during homeostasis and repair, *Cell* **135**, 1118–1129.
3. Winkler, I. G., Barbier, V., Wadley, R. et al. (2010) Positioning of bone marrow hematopoietic and stromal cells relative to blood flow in vivo: serially reconstituting hematopoietic stem cells reside in distinct nonperfused niches, *Blood* **116**, 375–385.
4. Lalor, P. A., Mapp, P. I., Hall, P. A. et al. (1987) Proliferative activity of cells in the synovium as demonstrated by a monoclonal antibody, Ki67, *Rheumatol Int* 7, 183–186.
5. Kiel, M. J., Yilmaz, O. H., Iwashita, T. et al. (2005) SLAM family receptors distinguish hematopoietic stem and progenitor cells and reveal endothelial niches for stem cells, *Cell* **121**, 1109–1121.

Chapter 4

Flow Cytometry Measurement of Bone Marrow Perfusion in the Mouse and Sorting of Progenitors and Stems Cells According to Position Relative to Blood Flow In Vivo

Valérie Barbier, Ingrid G. Winkler, Robert Wadley, and Jean-Pierre Lévesque

Abstract

Identification of the precise location, where hematopoietic stem cells (HSCs) reside in the bone marrow, has made a great leap forward with the advance of live time-lapse video 2-photon fluorescent microscopy. These studies have shown that HSCs preferentially resides in the endosteal region of the BM, at an average of two cell diameters from osteoblasts covering endosteal bone surfaces. However, this equipment is very sophisticated and only a very few laboratories can perform these studies. To investigate functional attributes of these niches, we have developed a flow cytometry technique in which mice are perfused with the cell-permeable fluorescent dye Hoechst33342 in vivo before bone marrow cells are collected and antibody stained. This method enables to position phenotypic HSC, multipotent and myeloid progenitors, as well as BM nonhematopoietic stromal cells relative to blood flow in vivo. This technique enables prospective isolation of HSCs based on the in vivo perfusion of the niches in which they reside.

Key words: Hematopoietic stem cells, Bone marrow, Bone marrow stroma, Perfusion, Stem cell niche, Flow cytometry

1. Introduction

Hematopoietic stem cells (HSCs) and lineage-restricted hematopoietic progenitor cells (HPCs) localize in specific microdomains termed "niches" according to their differentiation stage. These specific microenvironments play a critical role in controlling HSC and HPC fate, and regulate whether they remain quiescent, self-renew, differentiate, or apoptose (1–3).

Robert B. Ashman (ed.), *Leucocytes: Methods and Protocols*, Methods in Molecular Biology, vol. 844, DOI 10.1007/978-1-61779-527-5_4, © Springer Science+Business Media, LLC 2012

In situ microscopy to observe the location of HSC in unmanipulated mice has proven to be challenging as multicolor labeling is needed to identify HSCs. $Lin^-Sca-1^+KIT^+$ cells, which comprise HSCs and multipotent progenitors, have been observed both at the endosteum near osteoblasts in naïve nontransplanted mice (4, 5) and against sinusoid endothelial cells (5). Similarly, phenotypic long-term reconstituting HSCs ($Lin^-CD41^-CD48^-CD150^+$) have been reported on the abluminal side of BM vasculature (6).

Whether these two HSC niches are functionally different or overlap is unknown and remains hotly debated, in part because the endosteum is often in close range to endothelial sinuses (3, 7, 8). It has been proposed that vascular niches, which are perfused in nutrients and oxygen by sinusoidal blood, may represent "proliferative niches," whereas endosteal niches, poorer in blood nutrients and oxygen, could represent more "quiescent niches" (2, 9–12). The presence of two types of niches could also explain why "phenotypically homogeneous" HSCs defined as LSK $CD48^-CD34^-Flt3^-CD150^+$ contains two pools of HSCs proliferating at two different rates (13).

In order to further explore whether local blood perfusion defines functionally distinct niches for HSCs, we took advantage of in vivo perfusion of the vital fluorescent DNA intercalant Hoechst33342, which enables the measurement of blood perfusion in various normal or malignant tissues (14), including the BM (11). By combining in vivo Ho perfusion with arrays of up to six fluorescent antibodies for specific cell surface antigens, a positional hierarchy within the BM between HSC and lineage-restricted HPC relative to rapid blood flow, as well as for stromal cells, such as endothelial cells, mesenchymal stem cells (MSCs), and osteoblast lineage cells, can be established (15).

This method is based on the molecule diffusion across tissues. Once a cell-permeable fluorescent DNA intercalent (e.g., Hoechst33342) is injected intravenously, it rapidly distributes and equilibrates in the circulation. The dye then passively diffuses across the endothelial barrier into the adjacent tissue. If measurements are made within a relatively short period of time before the concentration of dye equilibrates between the blood and the adjacent tissues (typically, 10 min in the mouse BM), the further a cell will be from the blood flow bringing the dye into the tissue and lower the concentration of dye will be where the cell resides. As the amount of Hoechst33342 dye entering the cell is directly dependent on the concentration of dye in contact with the cells (15), the level of Hoechst33342 cell uptake in vivo is directly related to how close the cell is from the blood circulation or how well it is perfused. Using this method, we have found that the position of phenotypic HSC, multipotent and myeloid progenitors relative to blood flow, follows a hierarchy reflecting differentiation stage, whereas mesenchymal stromal cells are perivascular (15). Importantly, phenotypic $Lin^-Sca1^+KIT^+CD41^-CD48^-CD150^+$ HSC segregate into two groups based on the degree of blood (Hoechst33342) perfusion of their niche. HSCs capable of serial transplantation and long-term bromodeoxyuridine label retention were poorly perfused by blood in vivo with negative Hoechst33342 uptake while HSCs in more perfused niches cycled more frequently and only reconstituted a single host (15).

Importantly, this method can be applied to many other non-phagocytic cell types or malignant cells in other solid tissues easy to dissociate for flow cytometry, such as spleen and liver.

2. Materials

2.1. Animal Treatment

1. Clinical-grade, injectable-quality saline, sodium chloride injection BP, 0.9%, 50 × 10 mL sterile ampoules (Pfizer) to dilute Hoechst33342.
2. Hoescht33342 (Sigma) stock 20 mM (11.2 mg/mL) stored at −20°C. Use at ½ dilution in clinical-grade, sterile-injectable saline (10 mM) for injections. Use 0.8 mg per 25 g mouse per injection which is equivalent to 144 μL of Hoechst33342 10 mM per 25 g mouse.
3. Isoflurane inhalation anesthetic (Fortane, Abbott).
4. Insulin syringe with attached needle 27 G1/2 (Terumo, Somerset, NJ) for intravenous retro-orbital injections of Hoechst33342.

2.2. Tissue Harvest

1. Isoptin (Verapamil, a *P*-glycoprotein/Mdr1 inhibitor) 2-mL vial at 2.5 mg/mL (Abbott Laboratories cat# M075.784): Its molecular weight is 454 g per mole which is equivalent to a 5.5-mM stock. Use in all buffers at 50 μM final.
2. Reserpine (Sigma): Its molecular weight is 608 g per mole. Make 5 mM stock in DMSO (3 mg/mL), store at −20°C. For use, dilute to 5 μM (1/1000) to stop ATP-binding ABCG2 transporters which are responsible for the formation of the Hoechst33342 side population. Make fresh for each experiment.
3. Dulbecco's Phosphate Buffered Saline (DPBS) without calcium or magnesium (Biowhittaker) supplemented with 2% heat-inactivated newborn calf serum (NCS) (GIBCO), sterile filtered + 5 μM Reserpine + 50 μM Verapamil.
4. Magnetic activated cell sorting (MACS) washing buffer: Ice-cold MACS buffer (PBS + 2 mM EDTA + 0.5% BSA) + 5 μM Reserpine + 50 μM Verapamil.
5. 1-mL tuberculin syringe (Terumo) mounted with 23-G needles.
6. 5-mL polypropylene tubes (Greiner).

7. 1.5-mL Eppendorf tubes.
8. Heparin for blood collection to inactivate thrombin clots. DBL heparin sodium from porcine mucous (Hospira, Lake Forest, IL): Dilute to 1 U/μL in sterile-injectable saline. Store at 4°C.
9. Collagenase type 1 from Clostridium histilyticum (Worthington Biochemical).
10. 5-mL screw cap tubes (Sarstedt).

2.3. Cell Processing

1. Automated Hematology Analyzer KX-21 N (Sysmex) or manual counting of cells following (>1:20) dilution in white cell counting fluid using a bright-field Neubauer microscope counting chamber.
2. 50-mL polypropylene tubes (Falcon).
3. Cell strainers 40-μm nylon (BD).
4. 10× red cell lysis buffer: 1.5 M NH_4Cl, 100 mM $NaHCO_3$, 10 mM EDTA, pH 7.4. Sterile stock can be kept in fridge for many months. On the day of the experiment, dilute one part of 10× red cell lysis buffer with nine parts of sterile water to make 1× red cell lysis buffer.
5. Refrigerated centrifuge to rotate 1–50-mL tubes and microplates at 370 ×*g*.

2.4. Phenotypic Stains

1. 5-mL polypropylene tubes (Greiner).
2. Stain tubes: Ideally, 1.2-mL micro tibertube (Quality Scientific Plastics) or other polypropylene staining tubes.
3. Purified rat anti-mouse Fcγ receptor II/III clone 2.4 G2 (Fc Block) (BD) or culture supernatant from 2.4 G2 hybridoma.
4. MACS buffer: DPBS + 0.5% bovine serum albumin + 2 mM EDTA.
5. DPBS + 2% NCS.
6. Conjugated monoclonal antibodies specific for mouse antigens:

   CD3ε-biotin clone 145-2C11 (BD), 0.5 mg/mL.

   CD5-biotin clone 53–7.3 (BD), 0.5 mg/mL.

   CD45R (B220)-biotin clone RA3-6B2 (BD), 0.5 mg/mL.

   CD11b-biotin clone M1/70 (BD), 0.5 mg/mL.

   Gr1-biotin clone RB6-8C5 (BD), 0.5 mg/mL.

   CD127 (IL7Rα)-biotin clone B12.1 (BD), 0.5 mg/mL.

   CD41-biotin clone MWreg30 (eBioscience), 0.5 mg/mL.

   Ter119-biotin clone Ter119 (BD) 0.5 mg/mL.

   CD3ε-FITC clone 145-2 C11 (BD), 0.5 mg/mL.

CD5-FITC clone 53–7.3 (BD), 0.5 mg/mL.
CD11b-FITC cloneM1/70 (BD), 0.5 mg/mL.
Gr1-FITC clone RB6-8C5 (BD), 0.5 mg/mL.
Ter119-FITC clone Ter119 (BD), 0.5 mg/mL.
CD45R (B220)-FITC clone RA3-6B2 (BD), 0.5 mg/mL.
CD51-PE clone RMV-7 (BD), 0.2 mg/mL.
CD31-APC clone MEC 13.3 (BD), 0.2 mg/mL.
CD45-APCCY7 clone 30-F11 (BD), 0.2 mg/mL.
CD34-FITC clone RAM34 (BD), 0.5 mg/mL.
CD16/32-PE clone 2.4G2 (BD), 0.2 mg/mL.
Sca-1-PECY7 clone D7 (BD), 0.2 mg/mL.
Kit (CD117)-APC clone 2B8 (Biolegend), 0.2 mg/mL.
CD48-FITC clone HM 48–1 (BD), 0.5 mg/mL.
CD150-PE clone TC15-12F 12.2 (Biolegend), 0.2 mg/mL.
Streptavidin-Alexa700 (Invitrogen), 0.5 mg/mL.

7. 7-amino actinomycin D (7-AAD) for cell viability stain (Sigma), 1 mg/mL stock stored at 4°C.
8. We use an eight color ARIA cell sorter (BD) equipped with 406-nm violet laser to excite Hoechst33342 (with 450/50 filter for blue channel and 610/20 filter for the red channel), 488-nm blue laser (with 530/40, 575/25, 710/30, and 787/43 filters for FITC, PE, 7-AAD, and PECY7, respectively), and 643-nm red laser (with 665/20 and 750LP filters for APC and APCCY7/Alexa700, respectively) in order to sort cells according to their positioning relative to blood flow in vivo.
9. FloJo software (Tree Star, Ashland, OR) or other for analysis of results.

3. Method

3.1. Animal Treatment and Tissue Harvest

Hoechst33342 can be effluxed by cells by plasma membrane pumps, such as mdr1 or ABCG2 which are both expressed by HSC. It is, therefore, critical to keep bone and bone marrow cells in ice-cold buffers in the presence of 5 μM reserpine and 50 μM verapamil which block ATP-dependent transporters as rapidly as possible after mouse sacrifice. Before harvesting bones, make sure that all buffers and mortar and pestle are prechilled on ice.

1. Weigh two mice that are not going to be injected with Hoechst. These provide cells for single-color controls and unlabeled controls for flow cytometry.

2. To lightly anesthetize mouse, place some gauze in a 50-mL tube and imbibe it with 50 μL of isoflurane.
3. Gently introduce the mouse in the tube, head first. Keep the mouse nose away from the gauze.
4. When the mouse breathing slows down (after about 10 s), retrieve the mouse from the tube and collect blood by cardiac puncture. To do this, insert mounted 23-G needle into chest cavity and gently aspirate blood.
5. Collect 200–300 μL of blood in a 5-mL polypropylene tube containing 5 U of heparin. Mix well.
6. Euthanize mouse by cervical dislocation and collect the bones.
7. Remove skin to access and collect femurs, tibias, and pelvis using sterile scissors, tweezers, and scalpel.
8. Clean the bones by removing the muscles attached to them.
9. Place bones in ice-cold washing buffer (DPBS + 2%NCS + 5 μM reserpine + 50 μM verapamil).
10. Gently crush bones in ice-cold mortar and pestle with 4 mL of DPBS + 2%NCS + 5 μM reserpine + 50 μM verapamil (see Note 1). Do not overcrush as it may damage BM cells.
11. Filter through cell strainer into 50-mL tube on ice.
12. Repeat the crush three times (each crush in 4 mL of buffer). The final volume is then around $3 \times 4 = 12$ mL.
13. Discard the carcass following ethical procedures.
14. Weigh the mice that are going to be injected with Hoechst33342. Prepare the Hoechst33342 injectable solution. Make enough for two injections per mouse at 144 μL of Ho33342 10 mM per 25 g body weight.
15. To lightly anesthetize mouse, place some gauze in a 50-mL tube and imbibe it with 50 μL of isoflurane.
16. Gently introduce the mouse in the tube, head first. Keep the mouse nose away from the gauze (isoflurane is an irritant).
17. When the mouse breathing slows down (after about 10 s), pull the mouse out of the tube and inject 144 μL Hoechst33342 10 mM per 25 g body weight into the right retro-orbital sinus. Start timer.
18. Replace the mouse in the cage. The mouse wakes up within a minute.
19. Precisely 5 min later, repeat steps 15–18, but inject in the left retro-orbital sinus to avoid a hemorrhage due to weakened sinus following the first injection. This is $T = 5$ min.
20. At $T = 10$ min precisely, lightly anesthetize the mouse by putting some gauze in a 50-mL tube and imbibe it with 50 μL of isoflurane.

21. Gently introduce the mouse in the tube, head first. Keep the mouse nose away from the gauze.
22. When the mouse breathing slows down (after about 10 s), retrieve the mouse from the tube and collect blood by cardiac puncture. To do this, insert mounted 23-G needle into chest cavity and gently aspirate blood.
23. Collect 200–300 μL of blood in a 5-mL polypropylene tube containing 5 U of heparin. Mix well.
24. Immediately euthanize mouse by cervical dislocation and collect the bones. At this stage, the timing is crucial. Work as quickly as possible. BM cells' isolation should be completed within 3 min of death (see Note 2).
25. Remove skin to access and collect femurs, tibias, and pelvis using sterile scissors, tweezers, and scalpel.
26. Clean the bones by removing the muscles attached to them. Note that using a paper towel is a quick and efficient way to clean the bones.
27. Place bones in mortar and pestle with about 4 mL of ice-cold washing buffer DPBS + 2%NCS + 5 μM reserpine + 50 μM verapamil.
28. Gently crush bones. Do not overcrush.
29. Filter through cell strainer into 50-mL tube on ice.
30. Repeat the crush three times (each crush in 4 mL of buffer). The final volume is then around 3 × 4 = 12 mL.
31. Keep the bone fragments in a 5-mL screw-cap tube in the dark. They will be processed to isolate endosteal cells as described in Subheading 3.2.3.
32. Discard the carcass following ethical procedures.
33. Repeat steps 14–32 for each treated mouse.

3.2. Cell Processing

3.2.1. Blood

200–300 μL has been collected in heparin + verapamil + reserpine on ice.

1. Add 600 μL of ice-cold 1× red cell lysis buffer containing 5 μM reserpine + 50 μM verapamil per 200 μL of blood collected by cardiac puncture and incubate for 10 min on ice with gentle mixing (on a rotator in the dark).
2. Spin tubes at 370 × *g* for 5 min at 4°C.
3. Aspirate supernatant (leave 500 μL as the cell pellet can rarely be seen at this stage). Mix well to resuspend cells, then immediately add 4 mL of DPBS + 2%NCS + verapamil + reserpine, and mix.
4. Spin tubes at 370 × *g* for 5 min at 4°C.
5. Aspirate supernatant and leave 50 μL of cell pellet. Mix well to resuspend cells.

6. Resuspend the pellet in 500 μL final DPBS+2%NCS+verapamil+reserpine. Keep on ice until ready for staining (cf Subheading 3.3).

3.2.2. Bone Marrow

Dilute 20 μL whole bone marrow cell suspension with 80 μL DPBS+2% NCS (1/5 dilution) into an Eppendorf tube. Count leukocytes on automated Sysmex cell counter or by manual (microscopy) count of cells loaded on Neubauer counting chamber. Multiply by 5 to obtain the number of leukocytes per milliliter.

3.2.3. Endosteal Cells

1- Wash the crushed bone fragments three times with washing buffer DPBS+2%NCS+5 μM reserpine+50 μM verapamil to remove central BM cells.

2- Cover the bone fragments with 3 mg/mL collagenase type 1 in DPBS containing 5 μM reserpine+50 μM verapamil and vortex strongly.

3- Incubate for 30 min at 37°C in the dark. Use an incubator with shaker.

4- Filter through cell filter into 50-mL tube on ice.

5- Wash bone fragments twice with 1 mL of ice-cold washing buffer+verapamil+reserpine. Vortex and filter each time and pool with cells extracted from the bones with collagenase. This represents the endosteal fraction.

6- Dilute 20 μL endosteal cell suspension with 80 μL DPBS+2% NCS (1/5 dilution) into an Eppendorf tube. Count leukocytes. Multiply by 5 to obtain the number of leukocytes per milliliter.

3.3. Flow Cytometry Stains

3.3.1. Single-Color Control on Blood Leukocytes and Adjustment of the Flow Cytometer

1. Transfer 20 μL of control unstained blood cells (see Subheading 3.2.1, step 6, from a mouse that did not get injected with Hoechst33342) into one stain tube per color control and add 0.2 μL of each single-color control antibody to compensate spectral overlap on flow cytometer. Mix and then incubate on ice in the dark for 30 min with gentle rocking.
2. Transfer 20 μL of control unstained blood cells into a stain tube and add 2 μL of 10× 7-AAD stock solution (2 μg/mL final). This is the 7-AAD single-color control to compensate flow cytometer. Mix and then incubate at room temperature in the dark for 30 min with gentle rocking. Do not wash.
3. Pool 50 μL from each control-unstained mouse into a stain tube. This is the unstained control that is used for FACS compensation.
4. Take an aliquot of blood leukocytes from mice injected with Hoechst33342 and do not stain them further to make single Hoechst33342 color stain to compensate in other fluorescence channels.

5. Using unlabeled cells and single-color stains, adjust photomultiplier voltage for each individual fluorochrome. For Hoechst33342, acquire fluorescence following excitation with the violet laser in both the blue and red channels. Adjust the corresponding photomultiplier voltages of these two channels in order to have population in the right top corner of the dot plot (Fig. 1a). Compensate each individual dye into all other fluorescence channels. Do not compensate Hoechst blue fluorescence with its red fluorescence as it comes from the same dye molecule.

3.3.2. Fully Perfused Blood Leukocyte-Positive Controls

In order to assess that the correct amount of Hoechst33342 dye has been injected intravenously in each individual mouse, it is important to check that blood leukocytes from each mouse have taken up similar amounts of dye. This is measured by flow cytometry in viable CD45$^+$ blood leukocytes. The variability of Hoechst33342 fluorescence between blood leukocytes from each mouse should be less than 10% from the group mean. If a mouse has leukocytes with Hoechst mean fluorescence intensity below 90% of the group average, the concentration of Hoechst33342 in the blood is below anticipated likely due to suboptimal intravenous delivery. If this happens, samples from this mouse should be discarded.

1. Spin blood leukocytes kept on ice from Subheading 3.2.1, step 6, at 370 × *g* for 5 min at 4°C.

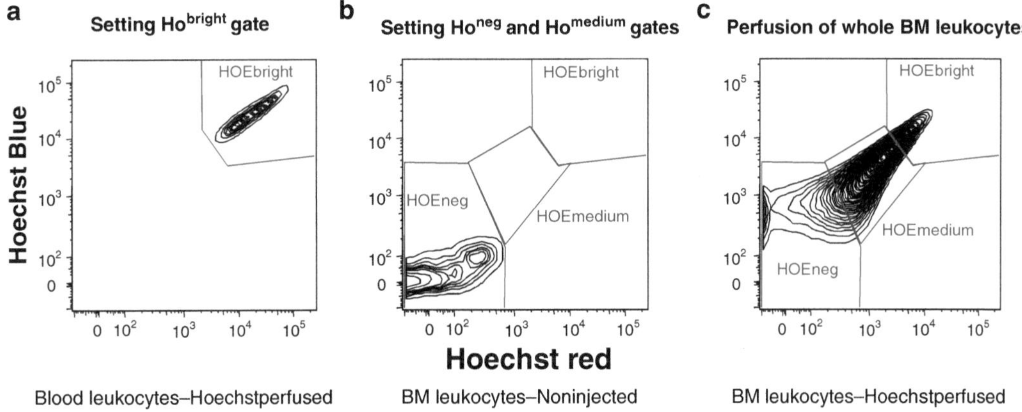

Fig. 1. Delineation of Ho^{neg}, Ho^{med}, and Ho^{bright} gates. For each individual experiment, the Ho^{neg}, Ho^{med}, and Ho^{bright} gates were defined by analyzing Hoechst33342 *blue* and *red* fluorescences on a logarithmic scale before running antibody-stained samples. Viable 7-AAD-negative blood CD45$^+$ leukocytes from a mouse injected retro-orbitally with Hoechst33342 10 and 5 min before sacrifice (panel **a**) were used to define the Ho^{bright} gate, and viable BM leukocytes from a control mouse that was not injected with Ho (panel **b**) were used to define the Ho^{neg} gate. The Ho^{med} gate was drawn between the Ho^{neg} and Ho^{bright} gates. Panel **c** is an example of Hoechst33342 staining obtained from BM leukocytes from a Hoechst33342-injected mouse. Note in panel **c** that unlike blood leukocytes (A) Hoechst33342 uptake in BM leukocytes spreads over the three Hoechst33342 gates.

2. Aspirate supernatant with cannula on vacuum line and leave 25 μL on cell pellet. Resuspend the cells by tapping the tubes with fingers or vigorous vortex (do not create foam).
3. Add 25 μL of CD45-APCCy7 1/200 final to each cell aliquot diluted in Fcblock hybridoma supernatant + 5 μM reserpine + 50 μM verapamil or in DPBS + 2%NCS + 2 to 5 μg/mL purified Fc Block antibody + 5 μM reserpine + 50 μM verapamil.
4. Mix and then incubate on ice in the dark for 30 min with gentle rocking.
5. Wash stains by adding 1 mL of DPBS + 2%NCS + reserpine + verapamil to each tube.
6. Spin at 370 × *g* for 5 min at 4°C.
7. Aspirate supernatant. Mix well and resuspend cells in 200 μL of DPBS + 2%NCS + verapamil + reserpine.
8. Analyze on a flow cytometer. Add 7-AAD to the cells 10 min before analyzing samples at a final concentration of 2 μg/mL to gate viable 7-AAD-negative cells.

3.3.3. Hematopoietic Stem Cell Staining

1. From the crushed BM (see Subheading 3.1, step 30), transfer 10^7 cells into labeled stain tubes.
2. Fill up tubes with DPBS + 2%NCS + verapamil + reserpine.
3. Spin at 370 × *g* for 5 min at 4°C.
4. Aspirate supernatant with cannula on vacuum line and leave 25 μL on cell pellet. Resuspend the cells by tapping the tubes with fingers or vigorous vortex (do not create foam).
5. Add 75 μL of antibody mix to each cell aliquot (final stain volume 100 μL for 10×10^6 cells). The antibody mix is made of lineage biotin (CD3, CD5, B220, Gr1, CD11b, CD41, Ter119) all at 1/300 dilution final, Sca-1-PECY7 1/300 final, Kit-APC 1/200 final, CD48-FITC 1/200 final, and CD150-PE 1/300 final diluted in Fcblock hybridoma supernatant + verapamil + reserpine or in DPBS + 2%NCS + 2 to 5 μg/mL purified Fc Block antibody + verapamil + reserpine.
6. Mix and then incubate on ice in the dark for 30 min with gentle rocking.
7. Wash stains with 1 mL of straight DPBS + verapamil + reserpine or MACS buffer + verapamil + reserpine.
8. Spin at 370 × *g* for 5 min at 4°C.
9. Aspirate supernatant with cannula on vacuum line and leave 25 μL on cell pellet. Resuspend the cells by tapping the tubes with fingers or vigorous vortex (do not create foam).
10. Add 75 μL of streptavidin-Alexa700 1/200 final in MACS buffer + verapamil + reserpine.

11. Mix and then incubate on ice in the dark for 30 min with gentle rocking.
12. Repeat steps 2–3 (wash).
13. Resuspend the cells in 300 μL of DPBS + 2%NCS + verapamil + reserpine.
14. Analyze on the cell sorter. Add 7-AAD to the cells 10 min before analyzing samples at a final concentration of 2 μg/mL to gate viable 7-AAD-negative cells.
15. Once the cell sorter has been adjusted for photomultiplier voltages in each channel, and compensation values (see Subheading 3.3.1), establish gates to identify the region of poor blood perfusion (no Hoechst33342 cell uptake), intermediate perfusion (medium Hoechst33342 uptake), and high blood perfusion (high Hoechst33342 uptake). This is achieved by using BM leukocyte sample from mouse noninjected with Hoechst33342 to define the Hoechst^{negative} region on dot plot representing blue and red fluorescence of Hoechst33342 (Fig. 1b). Using blood leukocyte sample from mice injected with Hoechst33342 and stained with CD45-APCCY7 and 7-AAD, establish the Hoechst^{bright} region (Fig. 1a). The Hoechst^{medium} region is defined as the region between the Hoechst^{negative} and Hoechst^{bright} regions.
16. Analyze all blood samples from Hoechst-injected mice to make sure that all mice have similar concentration of Hoechst33342 in their blood (samples prepared in Subheading 3.3.2). This is done by measuring the mean fluorescence intensity for Hoechst33342 in all CD45^{+} 7-AAD^{-}-viable blood leukocytes. All samples should have a mean fluorescence intensity within 10% of the average of all mice. Discard samples from any mouse which does not fall within this range as it failed to receive the correct concentration of Hoechst33342 in the blood circulation.
17. Analyze and sort (optional) all BM samples from Hoechst33342-injected mice. Define gates for Lin^{-} Sca1^{-} Kit^{+} myeloid HPC, Lin^{-} Sca1^{+} Kit^{+} HSPC, which can be further subdivided into Lin^{-} Sca1^{+} Kit^{+} CD48^{+} lineage-restricted HPC, Lin^{-} Sca1^{+} Kit^{+} CD48^{-} CD150^{-} multipotent progenitors, and Lin^{-} Sca1^{+} Kit^{+} CD48^{-} CD150^{+} phenotypic HSC (6). An example is provided in Fig. 2. Analyze Hoechst33342 uptake in each of these population. Sort if required.

3.3.4. Myeloid Progenitors Staining

1. From the bone marrow crush (see Subheading 3.1, step 30), transfer 10^6 cells into labeled stain tubes.
2. Fill up tubes with DPBS + 2%NCS + verapamil + reserpine.
3. Spin at 370 × *g* for 5 min at 4°C.

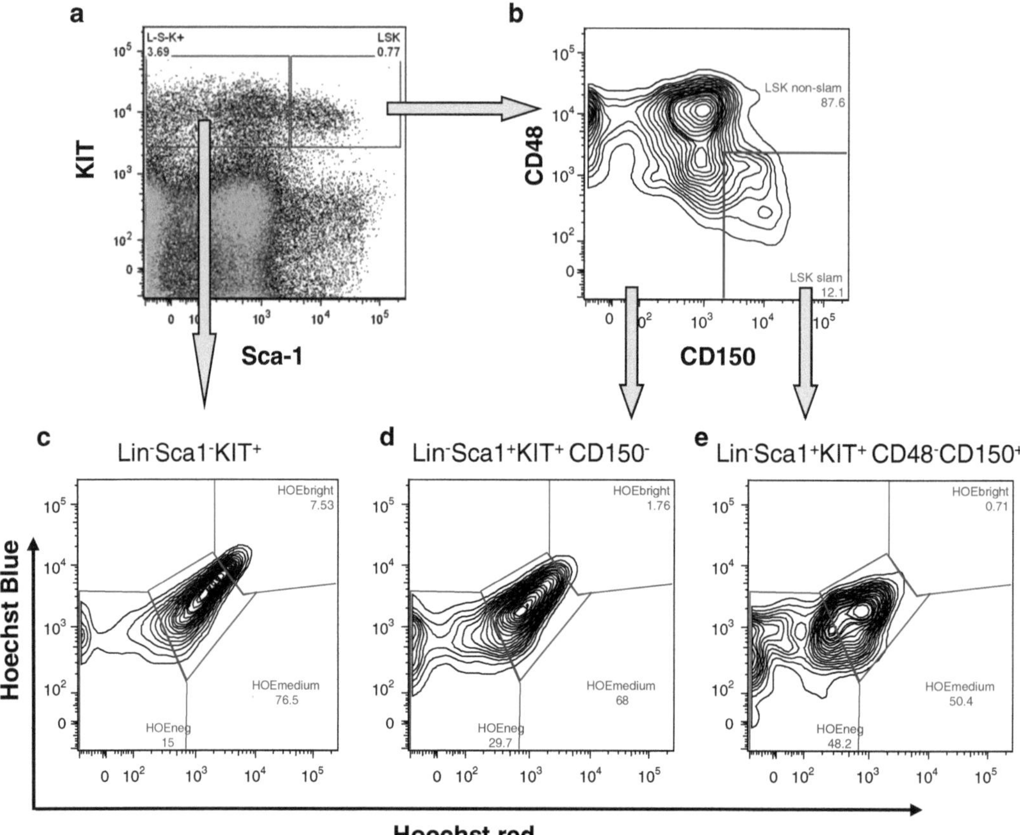

Fig. 2. In vivo Hoechst33342 uptake by BM HSC, multipotent, and lineage-restricted progenitor cells. C57BL/6 mice were perfused with Hoechst33342 dye intravenously 10 and 5 min prior sacrifice, BM cells harvested on ice in the presence of verapamil and reserpine to block ATP-dependent transporters, and stained for lineage, CD41, Sca-1, KIT, CD48, and CD150 surface antigens. Panel **a** is a representative dot plot of Sca-1 versus KIT expression on viable 7-ADD$^-$ Lin$^-$ CD41$^-$-gated BM cells. Panel **b** is dot plot of CD48 versus CD150 expression on Lin$^-$ CD41$^-$ Sca-1$^+$ KIT$^+$ cells gated in panel **a**. Panels **c–e** are representative dot plots of Hoechst33342 blue fluorescence versus red fluorescence of gated viable Lin$^-$ CD41$^-$ Sca-1$^-$ KIT$^+$ lineage-restricted progenitors, Lin$^-$ CD41$^-$ Sca-1$^+$ KIT$^+$ CD150$^-$ short-term reconstituting multipotent progenitors, and Lin$^-$ CD41$^-$ Sca-1$^+$ KIT$^+$ CD48$^-$ CD150$^+$ HSC, respectively.

4. Aspirate supernatant with cannula on vacuum line and leave 25 μL on cell pellet. Resuspend the cells by tapping the tubes with fingers or vigorous vortex (do not create foam).
5. Add 25 μL of antibody mix to each cell aliquot (final stain volume 50 μL for 10^6 cells). The antibody mix is made of lineage biotin (CD3, CD5, B220, Gr1, Ter119) all at 1/300 dilution final, IL7Rα-biotin 1/200 final, CD34-FITC 1/75 final, CD16/32-PE 1/200 final, Kit-APC 1/300 final, and Sca1-PECy7 1/400 diluted in DPBS + 2%NCS + verapamil + reserpine.
6. Mix and then incubate on ice in the dark for 30 min with gentle rocking.

7. Wash stains with 1 mL of straight DPBS + verapamil + reserpine or MACS buffer + verapamil + reserpine.
8. Spin at 370 × *g* for 5 min at 4°C.
9. Aspirate supernatant with cannula on vacuum line and leave 25 μL on cell pellet. Resuspend the cells by tapping the tubes with fingers or vigorous vortex (do not create foam).
10. Add 25 μL of streptavidin-A700 at 1/150 final in MACS buffer + verapamil + reserpine.
11. Mix and then incubate on ice in the dark for 30 min with gentle rocking.
12. Repeat steps 2–3 (wash).
13. Resuspend the cells in 200 μL of DPBS + 2%NCS + verapamil + reserpine.
14. Analyze on a flow cytometer. Add 7-AAD to the cells 10 min before analyzing samples at a final concentration of 2 μg/mL.
15. Adjust photomultiplier voltages and compensation values, and gates for low, medium, and bright Hoechst33342 uptake as defined in Subheading 3.3.3.
16. Analyze and sort (optional) all BM samples from Hoechst33342-injected mice. Define gates for different Lin^- $Sca1^-$ Kit^+ myeloid progenitors as follows: Lin^- $Sca1^-$ Kit^+ $CD16/32^-$ $CD34^+$ common myeloid progenitors (CMPs), Lin^- $Sca1^-$ Kit^+ $CD16/32^-$ $CD34^-$ megakaryocyte erythroid progenitors (MEPs), and Lin^- $Sca1^-$ Kit^+ $CD16/32^+$ granulocyte–monocyte progenitors (GMPs) (16). An example is provided in Fig. 3. Analyze Hoechst33342 uptake in each of these population. Sort if required.

3.3.5. Endothelial Cells, MSCs, and Osteoblasts

1. From the bone fragments collagenase treated (Subheading 3.2.3), transfer 10^6 cells into labeled stain tubes.
2. Fill up tubes with DPBS + 2%NCS + verapamil + reserpine.
3. Spin at 370 × *g* for 5 min at 4°C.
4. Aspirate supernatant with cannula on vacuum line and leave 25 μL on cell pellet. Resuspend the cells by tapping the tubes with fingers or vigorous vortex (do not create foam).
5. Add 25 μL of antibody mix to each cell aliquot (final stain volume 50 μL for 10^6 cells). The antibody mix is made of FITC-conjugated lineage antibodies (CD3, CD5, B220, CD11b, Gr1, Ter119) all at 1/300 dilution final, CD45-APCCy7 1/200 final, CD31–APC 1/300 final, Sca1-PECy7 1/300 final, and CD51-PE 1/200 final diluted in Fcblock hybridoma supernatant + verapamil + reserpine or in DPBS + 2%NCS + 2 to 5 μg/mL purified Fc Block antibody + verapamil + reserpine. Mix and then incubate on ice in the dark for 30 min with gentle rocking.

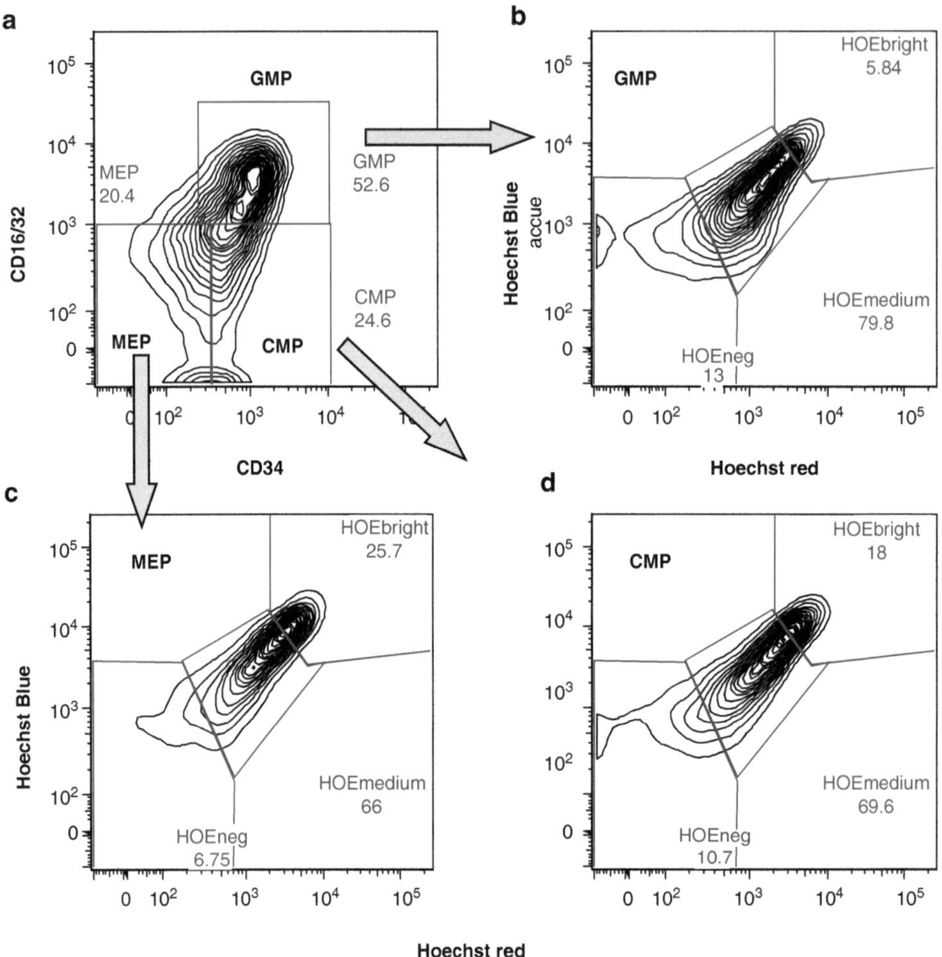

Fig. 3. In vivo Ho uptake by BM myeloid progenitors. C57BL/6 mice were perfused with Hoechst33342 dye intravenously 10 and 5 min prior sacrifice, BM cells harvested on ice in the presence of verapamil and reserpine to block ATP-dependent transporters, and stained for lineage IL7Rα, Sca-1, KIT, CD16/32, and CD34 surface antigens. Panel **a** is a representative dot plot of CD16/32 versus CD34 expression on viable 7-ADD⁻ Lin⁻ IL7Rα⁻ Sca-1⁻ KIT⁺-gated BM cells and shows the gates representing CMP, GMP, and MEP. Panels **b–d** are representative dot plots of Ho blue fluorescence versus Ho red fluorescence of gated viable GMP, MEP, and CMP, respectively.

6. Wash stains with 1 mL of straight DPBS + verapamil + reserpine or MACS buffer + verapamil + reserpine.
7. Spin at 370 × *g* for 5 min at 4°C.
8. Aspirate supernatant with cannula on vacuum line and leave 25 μL on cell pellet. Resuspend the cells by tapping the tubes with fingers or vigorous vortex (do not create foam).
9. Repeat steps 2–3 (wash).
10. Resuspend the cells in 200 μL of DPBS + 2%NCS + verapamil + reserpine.

11. Adjust photomultiplier voltages and compensation values, and gates for low, medium, and bright Hoechst33342 uptake as defined in Subheading 3.3.3.

12. Analyze and sort (optional) all BM samples from Hoechst33342-injected mice. Define gates for different nonhematopoietic stromal cells as follows: Lin^- $CD45^-$ $CD31^+$ endothelial cells, Lin^- $CD45^-$ $CD31^-$ $Sca1^+$ multipotent stromal cells (MSCs), and Lin^- $CD45^-$ $CD31^-$ $Sca1^-$ $CD51^+$ osteoblastic cells. An example is provided in Fig. 4. Analyze Hoechst33342 uptake in each of these populations. Note that endothelial cells are maximally perfused with maximal Hoechst33342 uptake similar to blood leukocytes, MSCs are about half less fluorescent for

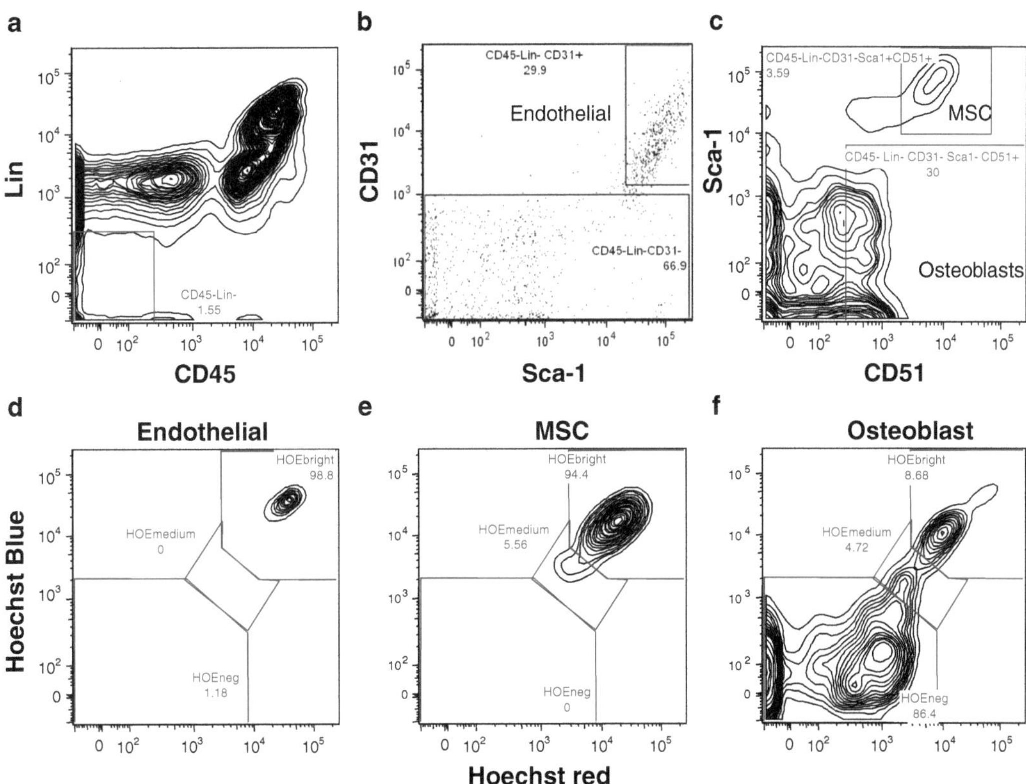

Fig. 4. In vivo Ho uptake by phenotypic BM endothelial cells, MSC, and osteoblastic cells. 129SvJ mice were perfused with Hoechst33342 dye intravenously 10 and 5 min prior sacrifice, hind limb bones were taken and crushed on ice in the presence of verapamil and reserpine, and BM cells were removed by several washes. Endosteal cells were then isolated by incubating crushed bones with collagenase in the presence of verapamil and reserpine. Cells were then stained with CD45, lineage, CD31, Sca-1, and CD51 antibodies. Panel **a** shows gating of $CD45^-$ Lin^- nonhematopoietic cells. Panel **b** shows gating of $CD31^{bright}$ endothelial cells and $CD31^-$ cells from the $CD45^-$ Lin^- gate in panel **a**. Panel **c** is the gating of $Sca\text{-}1^+$ $CD51^+$ MSC and $Sca\text{-}1^-$ $CD51^+$ osteoblast-lineage cells from the $CD45^-$ Lin^- $CD31^-$ gate defined in panel **b**. Panels **d–f** are representative dot plots of Hoechst33342 blue fluorescence versus Hoechst33342 red fluorescence of gated viable $CD45^-$ Lin^- $CD31^{bright}$ $Sca\text{-}1^{bright}$ BM endothelial cells, $CD45^-$ Lin^- $CD31^-$ $Sca\text{-}1^{bright}$ $CD51^+$ MSC, and $CD45^-$ Lin^- $CD31^-$ $Sca\text{-}1^{bright}$ $CD51^+$ osteoblast-lineage cells, respectively.

hoechst33342 indicative of their perivascular location, whereas most osteoblastic cells are very poorly perfused with 15% highly perfused.

4. Notes

1. HSCs express ATP-dependant pumps that actively efflux drugs and dyes, such as Hoechst33342. The property to actively efflux Hoechst33342 in vitro at 37°C is used to identify a "side population" which contains most of the stem cell activity in the BM and other tissues (17). In order to truly represent the level of perfusion in which HSCs reside in vivo, these transporters must be pharmacologically inhibited to prevent Hoechst efflux. In this way, the level of Hoechst fluorescence is not dependant anymore on cellular efflux but on the concentration of Hoechst33342 dye around the cells. Verapamil and reserpine are effective inhibitors of active Hoechst33342 efflux by stem cells. However, these inhibitors cannot be injected in vivo in mice to stop efflux during the perfusion period as they both depress heart beat rates and blood pressure. This would defeat the purpose of the whole experiment. By incubating BM leukocytes from precisely 10 min at 37°C (the duration of the in vivo perfusion before mouse sacrifice and tissue harvest) with increasing concentrations of Hoechst33342, we have found that the amount of Hoechst incorporated by HSPC was the same in the presence or absence of verapamil and reserpine during the 10-min incubation period at 37°C. Furthermore, the level of fluorescence was directly dependant on the concentration of Hoechst33342 present around the cells (Fig. 5). Therefore, Hoechst33342 does not get the time to be actively effluxed by transporter within the 10-min period of perfusion.

Fig. 5. Absence of detectable Ho efflux from BM HSC and multipotent progenitor cells stained with Ho ex vivo for 10 min. BM cells from nonperfused C57BL/6 mice were isolated and stained ex vivo for 10 min at 37°C with increasing Ho concentrations in the presence or absence of verapamil and reserpine as indicated on the top row. Cells were then immediately washed and stained on ice in the presence of verapamil and reserpine. Cells were gated for viable Lin$^-$ CD41$^-$ Sca-1$^-$ KIT$^+$ lineage-restricted progenitors, Lin$^-$ CD41$^-$ Sca-1$^+$ KIT$^+$ CD150$^-$ short-term reconstituting multipotent progenitors and Lin$^-$ CD41$^-$ Sca-1$^+$ KIT$^+$ CD48$^-$ CD150$^+$ HSC as described in Fig. 2 and Ho uptake was measured on dot plots of Ho blue fluorescence versus Ho red fluorescence at each Ho concentration in the absence or presence of pump inhibitors. Ho fluorescence dot plots for Lin$^-$ CD41$^-$ Sca-1$^+$ KIT$^+$ CD48$^-$ CD150$^+$ phenotypic HSC labeled in vitro with increasing concentrations of Ho in the presence or absence of verapamil and reserpine are shown. Values of MFI for Hoechst were then plotted as a function Ho concentration and linear regression calculated for the three above-mentioned populations. Differences in regression slopes were not significant demonstrating that these cells do not efflux Hoechst33342 dye during a 10-min period at 37°C.

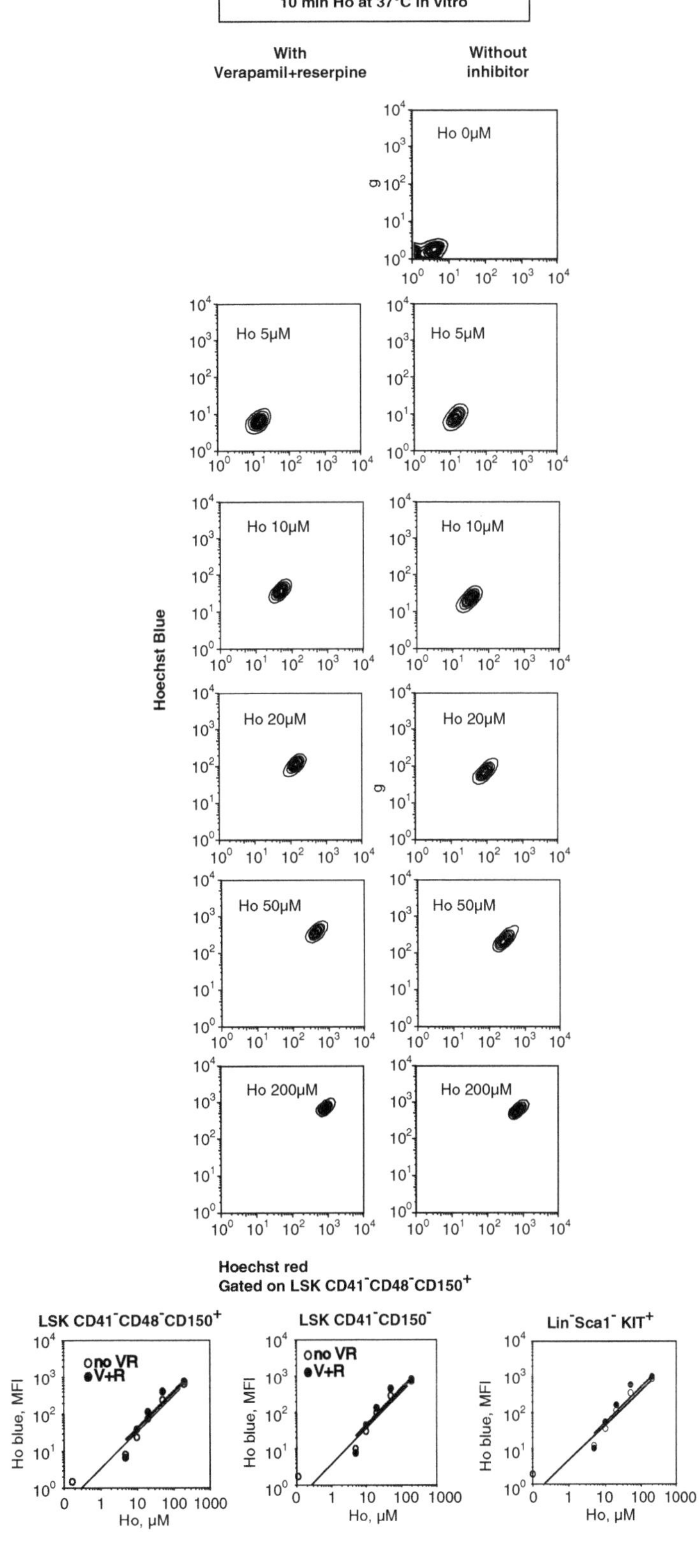
10 min Ho at 37°C in vitro
With Verapamil+reserpine
Without inhibitor
Ho 0μM
Ho 5μM
Ho 10μM
Ho 20μM
Ho 50μM
Ho 200μM
Hoechst Blue
Hoechst red
Gated on LSK CD41⁻CD48⁻CD150⁺
LSK CD41⁻CD48⁻CD150⁺
LSK CD41⁻CD150⁻
Lin⁻Sca1⁻ KIT⁺
no VR
V+R
Ho blue, MFI
Ho, μM

Moreover, the level of cell fluorescence was independent of the maturation stage of HSPC as HSC, multipotent progenitors, and myeloid progenitors were equally fluorescent.

2. The rapidity of tissue harvesting is critical to minimize the possibility of Hoechst33342 dye efflux by stem cells at 37°C. Typically, we take less than 3 min between death by cervical dislocation and crushing clean bones into ice-cold mortar with reserpine and verapamil. To achieve this, we have a team of three persons each performing an individual task. Person one: Injecting dye, bleeding mouse, and cervical dislocation; person two: dissecting the bones; person three: crushing the bones in mortar. When cells are in ice-cold buffer with reserpine and verapamil, they can be left for hours on ice without detectable Hoechst33342 efflux.

Acknowledgments

JPL is supported by a Senior Research Fellowship from the Cancer Council of Queensland and IGW by a CDA fellowship from the National Health and Medical Research Council of Australia. This work was supported by project grants 434515, 543706, and 350406 from the National Health and Medical Research Council.

References

1. Schofield, R. (1978) The relationship between the spleen colony-forming cell and the haemopoietic stem cell, *Blood Cells* **4**, 7–25.
2. Wilson, A., Trumpp, A. (2006) Bone-marrow haematopoietic-stem-cell niches. *Nat Rev Immunol* **6**, 93–106.
3. Kiel, M.J., Morrison, S.J. (2008) Uncertainty in the niches that maintain haematopoietic stem cells, *Nat Rev Immunol* **8**, 290–301.
4. Arai, F., Hirao, A., Ohmura, M et al. (2004) Tie2/angiopoietin-1 signaling regulates hematopoietic stem cell quiescence in the bone marrow niche, *Cell* **118**, 149–161.
5. Sugiyama, T., Kohara, H., Noda, M. et al. (2006) Maintenance of the hematopoietic stem cell pool by CXCL12-CXCR4 chemokine signaling in bone marrow stromal cell niches, *Immunity* **25**, 977–988.
6. Kiel, M.J., Yilmaz, O.H., Iwashita, T. et al. (2005) SLAM family receptors distinguish hematopoietic stem and progenitor cells and reveal endothelial niches for stem cells, *Cell* **121**, 1109–1121.
7. Xie, Y., Yin, T., Wiegraebe, W. et al. (2009) Detection of functional haematopoietic stem cell niche using real-time imaging, *Nature* **457**, 97–101.
8. Lo Celso, C., Fleming, H.E., Wu, J.W. et al. (2009) Live-animal tracking of individual haematopoietic stem/progenitor cells in their niche, *Nature* **457**, 92–97.
9. Wilson, A., Murphy, M.J., Oskarsson, T. et al. (2004) c-Myc controls the balance between hematopoietic stem cell self-renewal and differentiation, *Genes Dev.* **18**, 2747–2763.
10. Lévesque, J.-P., Winkler, I.G., Hendy, J. et al. (2007) Hematopoietic progenitor cell mobilization results in hypoxia with increased hypoxia-inducible transcription factor-1α and vascular endothelial growth factor A in bone marrow, *Stem Cells* **25**, 1954–1965.
11. Parmar, K., Mauch, P., Vergilio, J.A. et al. (2007) Distribution of hematopoietic stem cells in the bone marrow according to regional hypoxia, *Proc Natl Acad Sci USA* **104**, 5431–5436.

12. Trumpp, A., Essers, M., Wilson, A. (2010) Awakening dormant haematopoietic stem cells, *Nat Rev Immunol* **10**, 201–209.

13. Wilson, A., Laurenti, E., Oser, G. et al. (2008) Hematopoietic stem cells reversibly switch from dormancy to self-renewal during homeostasis and repair, *Cell* **135**, 1118–1129.

14. van Laarhoven, H. W., Bussink, J., Lok, J. et al. (2004) Effects of nicotinamide and carbogen in different murine colon carcinomas: immunohistochemical analysis of vascular architecture and microenvironmental parameters, *Int J Radiat Oncol Biol Phys* **60**, 310–321.

15. Winkler, I. G., Barbier, V., Wadley, R. et al. (2010) Positioning of bone marrow hematopoietic and stromal cells relative to blood flow in vivo: serially reconstituting hematopoietic stem cells reside in distinct nonperfused niches, *Blood* **116**, 375–385.

16. Akashi, K., Traver, D., Miyamoto, T. et al. (2000) A clonogenic common myeloid progenitor that gives rise to all myeloid lineages, *Nature* **404**, 193–197.

17. Scharenberg, C.W., Harkey, M.A., Torok-Storb, B. (2002) The ABCG2 transporter is an efficient Hoechst 33342 efflux pump and is preferentially expressed by immature human hematopoietic progenitor, *Blood* **99**, 507–512.

Chapter 5

Analyzing Cell Death Events in Cultured Leukocytes

Karin Christenson, Fredrik B. Thorén, and Johan Bylund

Abstract

Cell death is of utmost importance in immunity, in part as a way to control the development and activity of leukocytes, but also as a strategy employed by leukocytes to rid the body of unwanted cells. Apoptosis is the classic type of programmed cell death involving an ordered sequence of cellular events, resulting in morphological changes that include cleavage/fragmentation of DNA, condensation of nuclei, cell shrinkage, and alterations of the plasma membrane. The apoptotic cell is a nonfunctional, but structurally intact, entity with preserved membrane integrity that is engulfed by surrounding cells (a process known as clearance) in an immunologically silent manner. In contrast, necrotic cells, i.e., nonfunctional cells that have lost membrane integrity, are freely permeable and leak intracellular constituents that may shift immunological homeostasis. Thus, membrane integrity of dead leukocytes is very important from an immunological point of view.

For the analysis of leukocyte cell death, a wide variety of assays are available to monitor different events along the cell death pathway; a combination of different methods is advantageous in order to gain a more complete understanding of this dynamic process. In this chapter, we describe several in vitro methods for evaluating leukocyte cell death, mainly focusing on apoptosis in human neutrophils and lymphocytes. Special emphasis is given to assessment of membrane integrity of the cultured cells. Furthermore, a protocol for monitoring clearance of apoptotic neutrophils by monocyte-derived macrophages is provided.

Key words: Cell death, Apoptosis, Necrosis, Leukocytes, Neutrophils, Phagocytosis, Inflammation

1. Introduction

Programmed cell death is a process of crucial importance for the survival of all multicellular organisms, a necessity for development and life. Programmed cell death during development has been thoroughly investigated in lower multicellular organisms, like *Caenorhabditis elegans* (1), but various kinds of programmed cell death are central in life forms ranging from slime molds (2) and plants (3) to mammals. Apoptosis, the most common type of

Robert B. Ashman (ed.), *Leucocytes: Methods and Protocols*, Methods in Molecular Biology, vol. 844,
DOI 10.1007/978-1-61779-527-5_5, © Springer Science+Business Media, LLC 2012

programmed cell death, is an ordered sequence of cellular events leading to a peaceful death and is dependent on a family of enzymes known as *c*ysteine *asp*artyl prote*ases*, caspases. Apoptosis is the end of the cell as a functional entity with classic morphological changes, including cleavage/fragmentation of DNA, condensation of nuclei, cell shrinkage, and alterations of the plasma membrane (4). With regards to the latter, exposure of the membrane phospholipid phosphatidylserine (PS) on the outside of the cell is a prominent characteristic of an apoptotic cell. Despite alteration of the membranes, apoptotic cells still preserve membrane integrity for some time, giving neighboring cells the opportunity to clear away the dead cells from the system with minimal leakage of intracellular components. This is in contrast to necrotic cell death, where membranes are no longer intact and intracellular constituents leak out in an uncontrolled fashion (5). Other forms of programmed cell death, such as oxidant-induced cell death, share various characteristics of apoptosis, but this form of cell death proceeds independently of caspases (6). Pyroptosis (7) and autophagic cell death (8) are additional modes of cell death, but are not covered in this chapter.

For leukocytes in the immune system, apoptosis is important in two different ways. Not only is apoptosis induced in target cells by leukocytes as a weapon to rid the body of infected or malignant cells (9), but leukocytes are also susceptible to apoptosis themselves which serves to modulate immune reactivity. For example, when T cells develop in the thymus, their specificity is generated randomly, and many T cells display specificity for self-antigens. These T cells undergo apoptosis at an early stage to protect the body from autoreactivity (10). The fate of lymphocytes is also regulated by myeloid cells; myeloid-derived suppressor cells utilize various mechanisms, including reactive oxygen species (ROS), peroxynitrite, arginase activity, etc., to limit lymphocyte activity and induce lymphocyte cell death (11). Neutrophil granulocytes are short lived, professional phagocytes belonging to the innate immune system and central effector cells of acute inflammation. A multitude of neutrophils arrive early at an inflammatory site to engulf and kill microbes using toxic substances and degrading enzymes stored in their richly granulated cytoplasm (12). Given the abundance of these cells and their rapid and efficient transmigration to inflammatory sites, apoptosis is a means to prevent overly massive accumulation of neutrophils in the tissues and to regain homeostasis after the threat has been eliminated (13). The plethora of toxic substances with which the neutrophils are filled constitutes a potential danger to surrounding tissues, should they escape to the extracellular milieu. It is, therefore, crucial that neutrophils enter apoptosis, which keeps the membranes intact, after their missions are accomplished. Neutrophils may also die by necrosis, either secondary to apoptosis if not removed or directly due to general abuse (14, 15). This more violent end point is characterized by freely permeable cell membranes and allows for the toxic innards of neutrophils to cause collateral damage on surrounding cells and tissues.

Apoptotic cells are in general rapidly engulfed and digested by macrophages, a process often referred to as clearance. Macrophage recognition of the apoptotic prey is a very complex process involving a wide variety of receptor–ligand interactions (16–18). One central molecule is PS that, when exposed on the outside of the apoptotic cell, functions as an "eat-me" signal that facilitates engulfment. Clearance of neutrophils is a pivotal process in terminating acute inflammation and the recognition of apoptotic cells causes macrophages to produce anti-inflammatory cytokines (19) that may be important for the resolution of inflammation.

Although neutrophils spontaneously enter apoptosis, the process can be modulated, delayed, or accelerated (20, 21). These potent but short-lived phagocytes need time to act at the inflammatory site, and therefore their viability may be prolonged by danger signals typically encountered in tissues at the beginning of acute inflammation. In culture, inhibition (or rather delay) of neutrophil apoptosis is triggered by a variety of endogenous or microbial components, such as proinflammatory cytokines or TLR agonists (22). Factors shown to enhance (or accelerate) apoptosis of cultured neutrophils are UV radiation (23), phagocytosis of certain microbes (14), phorbol ester treatment (24), or endogenous molecules, e.g., FasL which mediates apoptosis by cross-linking Fas (CD95), a member of the TNF receptor family (25).

We have mainly been studying cell death events in human leukocytes, primarily neutrophils and different subsets of lymphocytes; described in this chapter are several methods that can be employed in vitro to monitor different events along the cell death pathways in these cells. Most of the protocols provided here can probably be used for other leukocytes (or cells from other species) with minor adjustments. In addition, we describe a protocol for the assessment of the clearance process in which monocyte-derived macrophages (MDMs) are allowed to engulf apoptotic neutrophils. It has to be kept in mind that cell death is a dynamic process and most detection methods are restricted to identifying cells that are currently undergoing apoptosis. Thus, these methods may fail to identify cells that have already undergone apoptosis, lost plasma membrane integrity (indistinguishable from cells dying as a result of a necrotic process), and subsequently disintegrated or cells that are destined to die but have not yet obtained any signs of apoptosis. If an individual cell-based (e.g., flow cytometry) assay is used, it is of particular importance to make sure that a significant proportion of cells have not been lost (by necrosis and subsequent disintegration or by adherence to plastic at some stage) during culture. Since such losses often occurs in an asymmetrical manner (e.g., only viable cells adhere to plastic and only dead cells disintegrate), this may significantly skew the analysis (see Note 1).

When investigating cell death, it is thus advantageous to combine bulk assays with individual cell-based assays at multiple time points and to count cells in samples before and after culture.

2. Materials

Neutrophils and peripheral blood mononuclear cells (PBMCs) can be separated from peripheral human blood using a standard technique (26, 27) involving Dextran sedimentation and Ficoll-Paque gradient centrifugation. Neutrophils are isolated from the pellet after hypotonic lysis of remaining erythrocytes and repeated washes, resuspended in buffer or medium of choice, and stored on ice. Lymphocytes can be isolated by incubating PBMCs in complete medium in a Petri dish. Monocytes adhere to the polystyrene surface while lymphocytes remain in suspension. Specific lymphocyte subsets can be isolated using commercially available magnetic separation kits. Macrophages can be obtained using various protocols; for the clearance protocol described below, we use human monocytes matured for 7 days in complete medium with addition of M-CSF (28).

2.1. Cell Culture

1. RPMI 1640 medium (see Note 2).
2. Fetal calf serum (FCS).
3. Penicillin/streptomycin (PEST) (see Note 3).
4. Polypropylene tubes (5-ml Round-bottom Tube, Becton-Dickinson; see Note 4).
5. Proapoptotic stimulation: e.g., α-CD95 (FAS) monoclonal antibody (eBioscience) (see Note 5).
6. Antiapoptotic stimulation: e.g., rhGM-CSF (Sigma–Aldrich).

2.2. Depolarization of the Mitochondrial Transmembrane Potential

1. JC-1 (Invitrogen): Reconstitute in dimethylsulfoxide (DMSO; 5 mg/ml), aliquot in 5-μl portions, and store at −20°C.
2. Mitotracker Deep Red 633 (Invitrogen): Dissolve one vial (50 μg) in 92 μl DMSO to yield a 1 mM solution. Aliquot in 2-μl portions and store at −20°C.

2.3. Activation of Caspases

2.3.1. FLICA Poly Caspases Assay Kit

1. Green fluorochrome-labeled inhibitors of caspase (FLICA) Poly Caspases Assay Kit (ImmunoChemistry Technologies) includes the FLICA reagent to be reconstituted in 50 μl DMSO per vial, aliquoted in 5-μl portions (150× stock solutions), and stored at −20°C and a 10× wash buffer.

2.3.2. Caspase-Glo 3/7 Assay

1. Caspase-Glo 3/7 Assay Systems (Promega) contains one vial of lyophilized Caspase-Glo 3/7 substrate and one vial of Caspase-Glo 3/7 buffer which are mixed and stored at –20°C (see Note 6).
2. 96-Well white flat-bottom polystyrene plate (Corning Life Sciences).
3. Krebs-Ringer phosphate buffer (KRG) (120 mM NaCl, 4.9 mM KCl, 1.7 mM KH_2PO_4, 8.3 mM Na_2HPO_4, 1.2 mM $MgSO_4$, 10 mM glucose, and 1 mM $CaCl_2$, in dH_2O, pH 7.3).
4. General caspase inhibition: Z-VAD-FMK (Calbiochem).

2.4. PS Exposure

1. Annexin V-FLUOS (Roche Diagnostics): Aliquot in 10-μl portions and store at –20°C.
2. 1× Annexin binding buffer: 10 mM HEPES, 140 mM NaCl, 2.5 mM $CaCl_2$ in dH_2O, pH 7.4. Prepare as sterile 10× stock solution.

2.5. Plasma Membrane Permeabilization

1. Sytox Green (Molecular Probes, Invitrogen), 5 mM solution.
2. 7-Amino-actinomycin D (7-AAD, BD Biosciences), 50 μg/ml solution.
3. To-Pro-3 (Invitrogen), 1 mM solution. Aliquot in 10-μl portions and store at –20°C.
4. Live/Dead fixable Violet Dead cell stain kit (ViViD, Invitrogen): Add 50 μl of DMSO to one vial, aliquot in 2-μl portions, and store at –20°C.
5. Live/Dead fixable Far Red Dead cell stain kit (FarViD, Invitrogen): Add 50 μl of DMSO to one vial, aliquot in 2-μl portions, and store at –20°C.

2.6. Measurement of Lactate Dehydrogenase

1. Cytotoxicity Detection kit (lactate dehydrogenase [LDH]; Roche Diagnostics) contains one vial of dye solution and one vial of catalyst.
2. Lysis: Triton X-100, 10% stock solution.
3. Optically clear 96-well microplates (flat bottomed).

2.7. Morphology

1. Staining: Giemsa and May Grünwald solution (Sigma–Aldrich).
2. Buffers: PBS (see Note 7).
3. Cytospin: Shandon EZ Double Cytofunnel (Anatomical Pathology).

2.8. Confocal Microscopy

1. Glass slides with coverslips or Petri dishes (MatTek Corporation).
2. Annexin V-APC (Invitrogen).
3. 7-AAD (BD Biosciences).
4. Buffer: KRG.

2.9. Clearance of Apoptotic Neutrophils by Monocyte-Derived Macrophages

1. RPMI 1640 supplemented with FCS (10%) and PEST (1%).
2. RPMI 1640 supplemented with FCS (20%).
3. rhM-CSF (R&D Systems); a stock solution of 15 μg/ml (diluted in PBS with 0.1% BSA) is prepared, aliquoted in 10 μl/vial, and stored at −20°C.
4. 24-well nontreated polystyrene plate (Nunc).
5. Staining of neutrophils: Vybrant CFDA-SE Cell tracer kit (Invitrogen) contains vials of CFDA-SE and one vial of DMSO. A stock solution of 10 mM is prepared and stored at 4°C. When needed, CFDA-SE is further diluted in PBS to working concentrations of 0.5–25 μM.
6. Proapoptotic stimulation: α-CD95 (FAS) monoclonal antibody (eBioscience; see Note 5).
7. Detachment of cells: PBS with EDTA (0.02%) supplemented with Lidocaine hydrochloride monohydrate (4 mg/ml, Sigma–Aldrich).
8. Fixation with paraformaldehyde (PFA) 4% solution. Store at −20°C and avoid repeated freeze thawing.
9. Labeling of macophages: α-CD14-PE/Cy7 antibody (eBioscience) diluted in PBS.

3. Methods

3.1. Cell Culture

3.1.1. Neutrophil Culture

Cells can be cultured in RPMI 1640 supplemented with 10% FCS (see Note 8) and 1% PEST and incubated overnight (37°C with 5% CO_2), with or without addition of pro- or antiapoptotic stimulation, for use with the protocols described below. We routinely use a density of 5×10^6 neutrophils/ml in round-bottom polypropylene tubes (see Note 4) and culture in the presence or absence of proapoptotic stimulation (α-CD95 monoclonal antibody at 10 μg/ml) or antiapoptotic factors (e.g., GM-CSF at 100 ng/ml).

3.1.2. Lymphocyte Culture

Lymphocytes can be cultured in RPMI 1640 supplemented with 10% FCS and 1% PEST (37°C with 5% CO_2). Cell death can be induced by adding 50–200 μM of hydrogen peroxide (see Note 9). We routinely use a density of 10^6 lymphocytes/ml in 96-well round-bottomed polystyrene plates.

3.2. Depolarization of the Mitochondrial Transmembrane Potential

In viable healthy cells, there is a membrane potential over the inner mitochondrial membrane ($\Delta\Psi_m$). Since the inner side is electronegative, lipophilic cations accumulate inside mitochondria. If such lipophilic cations are fluorescent, they can be used to label intact mitochondria and measure the $\Delta\Psi_m$. Mitochondrial outer membrane permeabilization (MOMP) is a point of no return in many cell death processes (29). MOMP is in many cases paralleled by a depolarization of $\Delta\Psi_m$. Several reagents can be used to monitor this event including derivates of rhodamine-123, carbocyanine dyes, such as 5,5′,6,6′-tetrachloro-1,1′,3,3′-tetraethylbenzimidazolcarbocyanine iodide (JC-1), and Mitotracker Deep Red. In cells with an intact $\Delta\Psi_m$, JC-1 accumulates in the mitochondria and form aggregates that can be detected as orange fluorescence. In cells with altered $\Delta\Psi_m$, the reagent is predominantly in monomeric form, which emits light in the green part of the spectrum. Leukocytes with altered $\Delta\Psi_m$ can, thus, be identified as cells with increased green fluorescence and decreased orange fluorescence using flow cytometry. For simultaneous evaluation of necrotic/permeabilized cells, a membrane-impermeable dye, e.g., To-Pro-3 (see Subheading 3.4) can be included in the assay (Fig. 1).

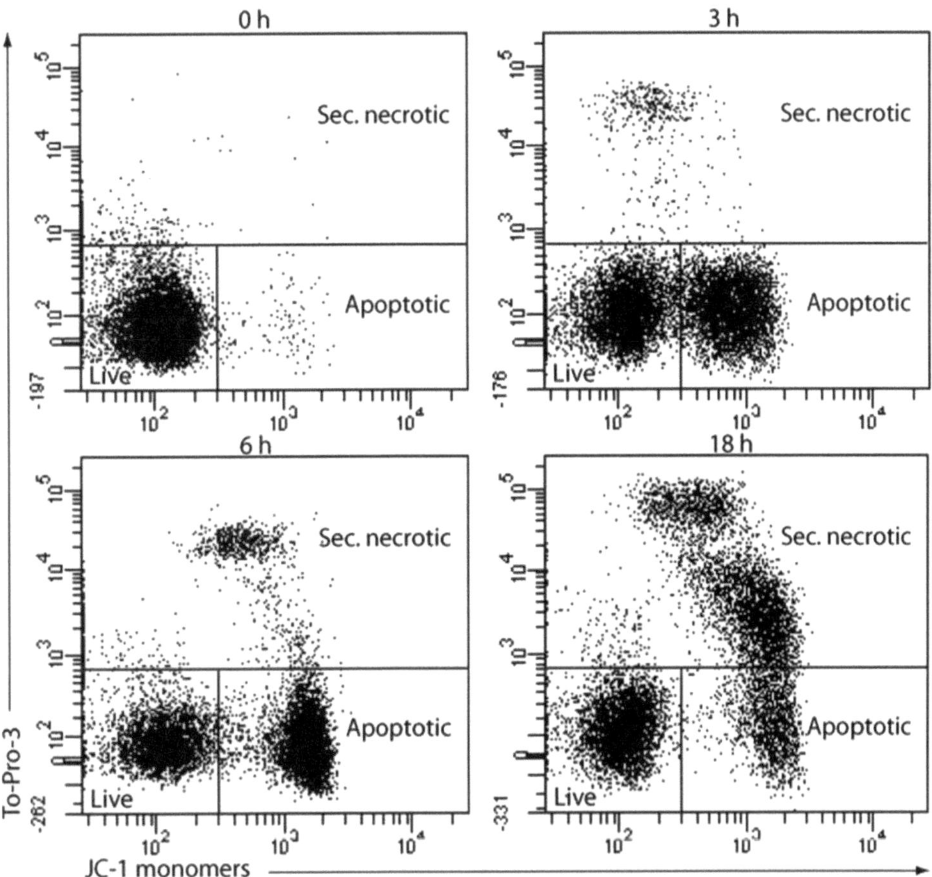

Fig. 1. Lymphocytes were exposed to hydrogen peroxide and assayed for altered $\Delta\Psi_m$ and plasma membrane integrity at various time points. Depolarization of the $\Delta\Psi_m$ is seen as an increase in green fluorescence (JC-1 monomers). After 3 h, a large fraction of lymphocytes displayed altered $\Delta\Psi_m$ (apoptotic). With time, more cells obtained depolarized mitochondrial membranes and then became increasingly stained by the membrane-impermeant stain To-Pro-3 (secondary necrotic).

3.2.1. JC-1 Staining

1. Incubate 2×10^5 leukocytes at 10^6 cells/ml in RPMI medium with 10% FCS with or without apoptosis-inducing agents.
2. Centrifuge cells at $300 \times g$ for 5 min and discard the supernatant.
3. Carefully resuspend the cell pellet in 250 μl of 2.5 μg/ml JC-1 solution (dilute the stock solution 1:2,000 in complete medium or PBS before use).
4. Incubate at 37°C for 15 min.
5. Fill up tubes with PBS, centrifuge cells at $300 \times g$ for 5 min, and discard the supernatant.
6. Resuspend the cells in 200 μl PBS. To-Pro-3 (or another permeability dye; see Subheading 3.4) can also be added here; include To-Pro-3 (0.5 μM; see Note 10) in the PBS.
7. Analyze on a flow cytometer (see Note 11).

3.2.2. Mitotracker Deep Red Staining

Mitotracker Deep Red is one of many mitochondrial dyes from Molecular Probes. One advantage with this reagent is that it is fixation compatible and its emission is in the infrared part of the spectrum, which leaves fluorescein and phycoerythrin channels available in multicolor experiments.

1. Incubate 2×10^5 leukocytes of choice at 10^6 cells/ml in RPMI medium with 10% FCS with or without apoptosis-inducing agents.
2. Centrifuge cells at $300 \times g$ for 5 min and discard the supernatant.
3. Carefully resuspend the cell pellet in 180 μl of PBS. Add 20 μl of 0.5 μM Mitotracker Deep Red (dilute the stock solution 1:2,000 in PBS before use) to yield a final concentration of 50 nM and incubate at 37°C for 20 min (see Note 12).
4. Fill up tubes with PBS, centrifuge cells at $300 \times g$ for 5 min, and discard the supernatant.
5. Resuspend the cells in 200 μl PBS and analyze on a flow cytometer.

3.3. Activation of Caspases

The mammalian proteins involved in apoptosis are known as caspases (30). In man, there are at least seven caspases that are involved in cell death (31). These caspases are divided into two main groups: the initiator caspases (−2, −8, −9, and −10) and the executioner (or effector) caspases (−3, −6, and −7). Caspases are transcribed as inactive zymogens, and are either activated by proteolytic cleavage or upon interaction with activating proteins (32, 33). Caspase activation can be monitored using inhibitors of or substrates for caspases both in individual cells (e.g., FLICA assay) and in bulk assays using cell lysates (e.g., Caspase-Glo 3/7 assay). If caspases

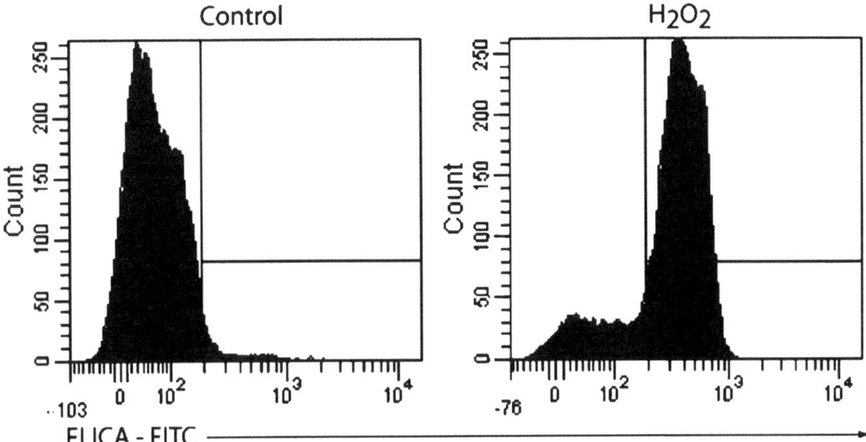

Fig. 2. A mixed lymphocyte population was cultured overnight in the absence or presence of 200 μM hydrogen peroxide, and assayed for caspase activation.

are activated, the reagent is cleaved to yield a fluorescent molecule, and total fluorescence can be assessed using a fluorometer (bulk assay) or the percentage of cells with active caspases can be determined using flow cytometry.

3.3.1. FLICA Assay

The FLICA reagent binds to the active site of caspases. Since the active site is not accessible in zymogens, only cells with active caspases are fluorescently labeled (see Note 13; Fig. 2).

1. Incubate 2×10^5 leukocytes of choice at 10^6 cells/ml in RPMI medium with 10% FCS with or without apoptosis-inducing agents.
2. One hour before the end of culture, dilute the FLICA stock solution (150×) 1:30 in culture medium to yield a 5× solution. Then, add the diluted FLICA reagent to your samples at a 1:5 ratio (i.e., 50 μl to a 200 μl cell suspension).
3. After 1 h, transfer cell suspensions to tubes and add 2 ml of wash buffer supplied with the kit (1×).
4. Centrifuge cells at $300 \times g$ for 5 min and discard the supernatant. Repeat the washing procedure once.
5. Resuspend cells in 200 μl wash buffer and analyze on a flow cytometer (see Note 14); cells with active caspases display increased green fluorescence (Fig. 2).

3.3.2. Caspase-Glo 3/7 Assay

The Caspase-Glo 3/7 kit provides a bulk assay (see Note 15) that measures the activity of the effector caspases-3 and -7. Addition of the reagent leads to cell lysis followed by caspase cleavage of the substrate and generation of a luminescent signal, proportional to the amount of caspase activity in the cells.

1. Dilute neutrophils (see Note 16) in KRG to 10^6 cells/ml and add 100 μl to a white 96-well plate.
2. Add 100 μl of caspase 3/7 reagent and incubate according to manufacturer's instructions (see Note 17).
3. Luminescence is evaluated in a plate reader.

3.4. Extracellular Exposure of Phosphatidylserine, Altered Light Scatter, Plasma Membrane Permeabilization, and Secondary Necrosis

In viable cells, phospholipids are asymmetrically distributed between the two leaflets of the plasma membrane. This distribution is maintained by an enzyme known as the aminophospholipid translocase (34, 35), which is inactivated in apoptotic cells, resulting in PS exposure on the surface (36, 37). Annexin V specifically binds to PS, and fluorochrome-conjugated Annexin V can be utilized to identify apoptotic cells. The binding of Annexin V to PS is strictly calcium dependent, so all buffers introduced after addition of Annexin V must contain calcium (see Notes 18 and 19).

Late in the apoptotic process, dying leukocytes start to display altered light scatter properties as determined by flow cytometry. Thus, necrotic/permeabilized leukocytes can be identified as cells displaying a reduced forward scatter and a slightly increased side scatter (Fig. 3). The change in scattering properties often coincides with loss of structural integrity of the plasma membrane. Late apoptotic cells with leaky membranes are often referred to as secondary necrotic. Membrane-impermeable dyes with different spectral properties can be used to monitor cell integrity. Many of these dyes are DNA-intercalating reagents, such as Sytox green (excitation, 488 nm; emission, 523 nm), Propidium iodide (488 nm; 615 nm), 7-AAD (488 nm; 650 nm), and To-Pro-3 (633 nm; 670 nm), and they can be used interchangeably depending on detector availability (see Notes 20–22) (Fig. 4). The availability of

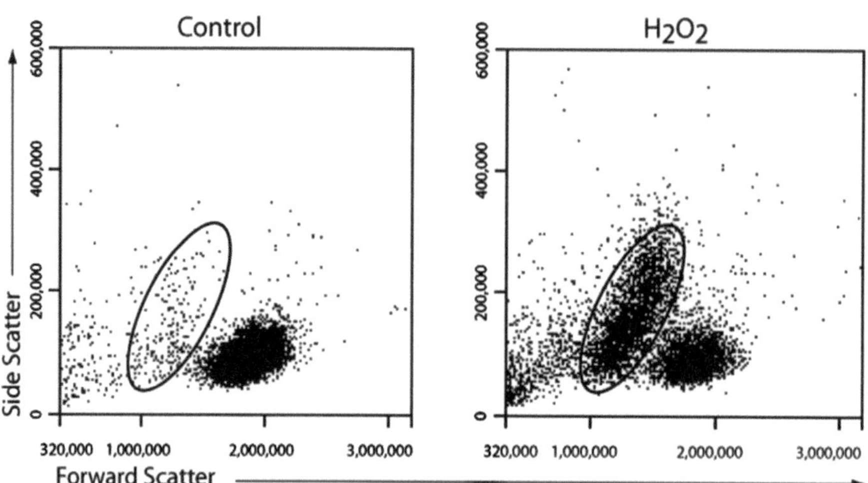

Fig. 3. Lymphocytes incubated overnight in the absence or presence of hydrogen peroxide. Secondary necrotic lymphocytes display reduced forward scatter and elevated side scatter (region).

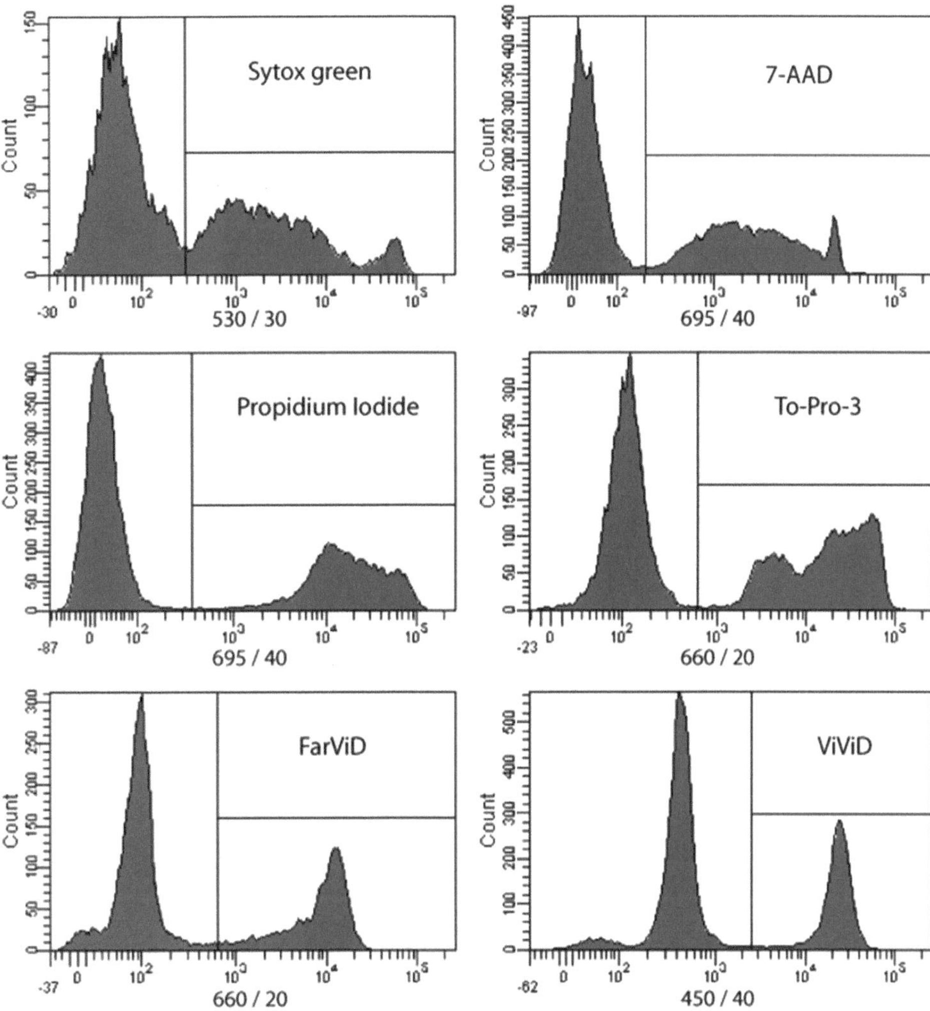

Fig. 4. Lymphocytes incubated overnight in the presence of hydrogen peroxide were stained with indicated membrane-impermeable dyes and fluorescence was measured in the indicated channels.

multiple different permeability dyes as well as Annexin V coupled to different fluorochromes makes it possible to choose a combination that enables further staining and minimize the need for electronic compensation of the flow cytometer (see Note 23). These permeability dyes are also useful in combination with non-Annexin-based techniques, e.g., JC-1 staining (Fig. 1).

3.4.1. Annexin V and 7-AAD Staining

1. Wash 10^6 cells in 2 ml Annexin binding buffer, $190 \times g$ for 10 min.
2. Mix 2 μl Annexin V-FLUOS and 5 μl 7-AAD with 100 μl Annexin binding buffer and add it to the pelleted cells.
3. Resolve the pellet carefully.

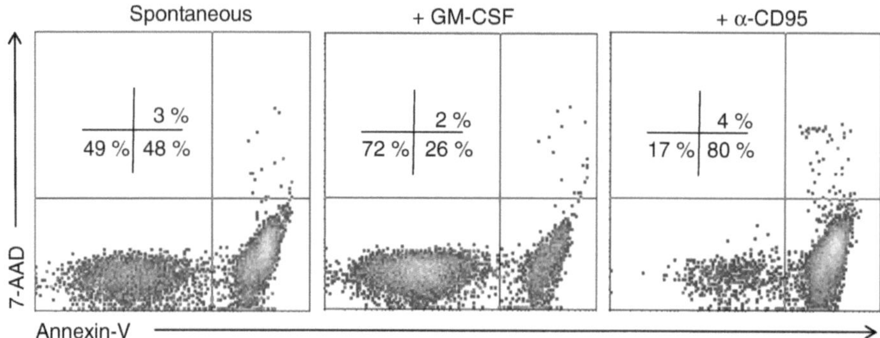

Fig. 5. Cell death of human neutrophils evaluated by flow cytometry after a 20-h incubation as described in text. Apoptotic cells labeled with Annexin V are visible in the LR quadrant while necrotic cells positive for both Annexin V and 7-AAD are seen in the UR quadrant. Spontaneous apoptosis (*left*) can be modulated by stimulation with antiapoptotic factors, e.g., GM-CSF (100 ng/ml; middle) that delays cell death, or by proapoptotic factors, e.g., FAS (anti-CD95 antibody, 10 μg/ml; *right*) that enhances cell death.

4. Incubate the cells in the dark for 10 min in room temperature.
5. Add 400 μl Annexin binding buffer to the cells before analysis by flow cytometry.
6. Apoptotic cells can be evaluated in FL-1 and permeabilized cells in FL-3 (see Notes 24 and 25) (Fig. 5).

3.4.2. Amine-Reactive Dyes to Distinguish Permeable Membranes

Live/Dead fixable stains are amine-reactive dyes that take advantage of the fact that more free amines are accessible for binding in a permeabilized cell. Thus, cells with compromised plasma membranes are more intensely stained than healthy cells with intact plasma membranes (Fig. 4). These dyes are also compatible with fixation and permeabilization and thus allow subsequent staining of intracellular antigens (38).

1. Incubate 2×10^5 leukocytes of choice at 10^6 cells/ml in RPMI medium with 10% FCS with or without apoptosis-inducing agents.
2. Transfer cells to 4-ml tubes and add 2 ml of PBS. Centrifuge cells at $300 \times g$ for 5 min and discard the supernatant.
3. Dilute the stock solution of amine-reactive dye 1:1,000 in PBS (see Note 26).
4. Resuspend cells in 100 μl of the diluted staining solution and incubate at 4°C for 20 min.
5. Add 2 ml of PBS, centrifuge cells at $300 \times g$ for 5 min, and discard the supernatant.
6. Repeat the washing procedure once (see Note 27).
7. Resuspend the cells in 200 μl PBS and analyze by flow cytometry.

3.5. Measurement of Lactate Dehydrogenase

LDH is an enzyme present at high concentrations in the cytoplasm of leukocytes. After disruption of the plasma membrane, LDH is released to the extracellular milieu and is therefore a useful bulk assay for permeabilization during in vitro culture.

1. Prepare LDH standards using freshly prepared leukocytes of the same density (and from the same donor) as samples to be tested, resuspended in identical medium.
   (a) Background: 250 μl cell culture medium
   (b) 0% LDH release: 250 μl cell-free supernatant from untreated cells
   (c) 100% LDH: Lyse cells with 1% (final concentration) Triton X-100 and thorough vortexing
2. Also prepare (a)–(c) with addition of the (cytotoxic) compounds to be tested to make sure that these compounds do not affect the assay per se or cause immediate lysis of the cells (see Note 28).
3. From cultured samples, remove approximately 300 μl supernatant and centrifuge at $300 \times g$ for 1 min.
4. Carefully aspirate 250 μl of the cell-free supernatant without disturbing possible cell pellet.
5. Prepare triplicate samples of each sample (including standards) by pipetting 75 μl into each well.
6. Add 75 μl reaction mixture from the LDH kit to each sample well.
7. Incubate the 96-well plate at room temperature, protected from light, for up to 30 min.
8. Measure the absorbance of the samples at 490 nm (see Note 29).
9. First, subtract the background readings from all values and then make sure that none of the compounds tested affect the assay per se (change the background or 100% LDH readings) or is immediately cytotoxic to cells (increase the 0% control). Next, calculate necrosis/lysis for each sample by relating the values obtained to the 100% lysis control.

3.6. Morphology

Morphological changes, such as a reduction of the cellular and nuclear volume, are characteristic for apoptotic cells. Apoptotic neutrophils, with condensed round nuclei, are easily separated from their viable counterparts with multilobulated nuclei (Fig. 6). Giemsa and May Grünwald are often used to stain peripheral blood cells (both stains contain methanol which permeabilizes cell membranes, a necessity for intracellular staining) and are good tools to evaluate apoptotic morphology.

1. Attach 2×10^5 cells (suspended in KRG) to glass slides by cytospin and leave them to dry.

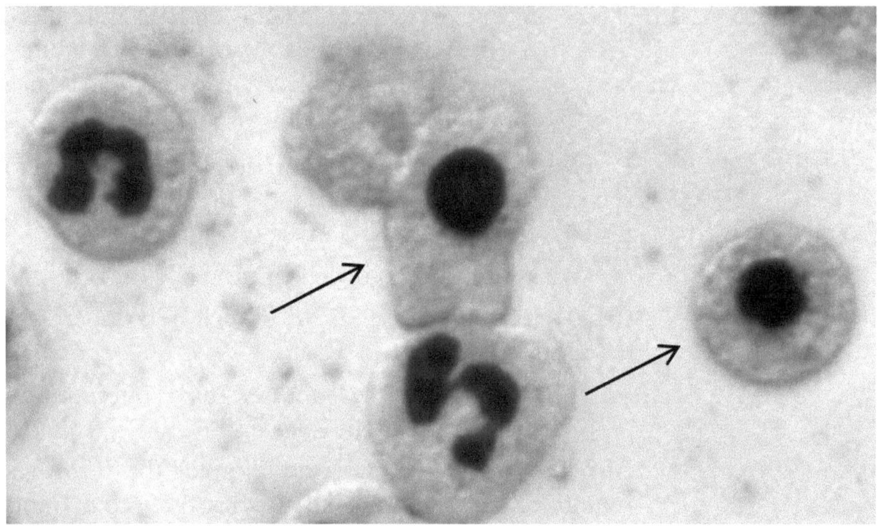

Fig. 6. Morphological assessment of neutrophils attached to microscopic slides by cytospin and stained with Giemsa and May Grünwald. The multilobulated nuclei make the two viable neutrophils easy to distinguish from the two apoptotic neutrophils with rounder, condensed nuclei (*arrows*).

2. Put the glass slides in May Grünwald solution for 5 min and then in PBS for 1.5 min.
3. Cover the slides with Giemsa (diluted 1:20) for 20 min, wash them with dH_2O, and leave them to dry overnight.
4. Mount and inspect by light microscopy. Typical morphologies of viable and apoptotic neutrophils are shown in Fig. 6.

3.7. Confocal Microscopy

1. Wash 10^6 cells (concentration of 5×10^6/ml) in 2 ml Annexin binding buffer, $190 \times g$ for 10 min.
2. Mix 2 μl Annexin V-APC and 5 μl 7-AAD with 100 μl Annexin binding buffer and add it to the pelleted cells.
3. Resolve the pellet carefully.
4. Incubate the cells in the dark for 10 min at room temperature.
5. Dilute the cells with 200 μl of KRG and place sample on slides.
6. By confocal microscopy, apoptotic cells can be seen as halo-shaped circles as Annexin V binds to PS on the outer membranes. Permeable cells also bind Annexin V, but in addition 7-AAD staining can be seen in the middle of these cells.

3.8. Clearance of Apoptotic Neutrophils by MDMs

Neutrophils are the predominating cell type in an acute inflammatory situation and removal of the apoptotic cells is necessary for the termination of the acute inflammation. Described below is a flow cytometric method for quantifying the clearance process using

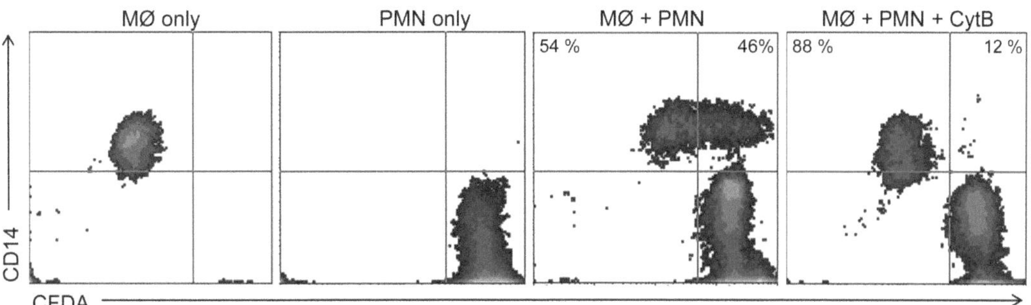

Fig. 7. Macrophage (MØ) phagocytosis of CFDA-labeled apototic neutrophils (PMN), evaluated by flow cytometry. Percentages reflect the proportion of CD14-positive events negative (UL quadrants) or positive (UR quadrants) for CFDA. Addition of cytochalasin B (10 μg/ml) prevents the uptake but not adherence of apoptotic PMN.

fluorescently labeled apoptotic neutrophils and monocyte-derived macrophages (MDMs), but the protocol can easily be adjusted to use with other macrophage types and/or apoptotic cells from other sources (see Note 30). Unlike neutrophils (viable or apoptotic), MDMs have a high expression of CD14 on the surface and can therefore easily be distinguished from neutrophils. The MDMs that have engulfed one or more CFDA-labeled apoptotic neutrophil are seen as double-positive events, whereas the nonphagocytosing MDMs are negative for CFDA (Fig. 7).

Day 1

3.8.1. Preparation of CFDA-Stained Neutrophils

1. Add 0.5 μM CFDA-SE (see Note 31) to fresh viable neutrophils at a density of 5×10^6 cells/ml resolved in prewarmed (37°C) PBS.
2. Incubate the cells at 37°C for 15 min.
3. Wash the cells with warm PBS, resuspend them in RPMI 1640 supplemented with 10% FCS and 1% PEST, and culture overnight (see Note 32).

3.8.2. Preparation of MDMs

1. Add 0.5 ml detached MDMs suspended in RPMI 1640 supplemented with 20% FCS (to 6×10^5 cells/ml) to each well in a 24-well plate.
2. Add additional 0.5 ml RPMI 1640 with 20% FCS to each well together with 15 ng/ml M-CSF.
3. Incubate the plate overnight at 37°C and with 5% CO_2.

Day 2

3.8.3. MDM Phagocytosis of Apoptotic Neutrophils

1. Prewarm RPMI 1640 (without FCS; see Note 33) to 37°C.
2. Wash and count the CFDA-labeled apoptotic neutrophils, centrifuge for 10 min at $190 \times g$, and dilute with the prewarmed medium to 10^7 cells/ml.
3. Remove the medium from the plate with macrophages and add 500 μl of the prewarmed medium and 500 μl of the

apoptotic neutrophils to each well (see Note 34). Incubate the plate at 37°C with 5% CO_2 for 90 min.

4. Remove the medium (containing the majority of nonattached, nonengulfed neutrophils) and detach the MDMs by adding 1 ml cold PBS, with EDTA (0.02%) and lidocaine (4 mg/ml), to each well and put the plate on a rocking platform shaker for 15 min at 4°C (see Note 35).
5. Pipette the cells up and down in the wells and move them to premarked tubes on ice containing 2 ml of ice-cold PBS.
6. Repeat steps 4 and 5 one more time to make sure that all MDMs are detached (see Note 36).
7. Centrifuge the tubes at 335 × *g* for 10 min at 4°C.
8. Resuspend the cell pellets in 200 μl ice-cold PBS, fix the cells by adding 200 μl PFA (final concentration of 2%), and keep the tubes cold for 10 min.
9. Centrifuge at 335 × *g* for 10 min at 4°C and carefully remove all supernatants.
10. Resuspend the cells in 100 μl of premixed PBS with α-CD14 antibody (diluted 1:50, 0.1 μg per sample) and leave in the dark (room temperature) for 1 h.
11. Centrifuge at 335 × *g* for 10 min at room temperature and resuspend in 300 μl PBS before evaluating clearance by flow cytometry.
12. CFDA-labeled neutrophils can be seen in FL-1- and CD14-positive macrophages in FL-3. Macrophages containing neutrophils are double positive for CD14 and CFDA (see Note 37) (Fig. 7).

4. Notes

1. One example of where cell losses affect the interpretation of data is the effect on neutrophil viability by the human cathelicidin LL-37 (39). This peptide induces a very rapid and selective permeabilization of apoptotic neutrophils leaving viable cells unaffected (40) (Fig. 8). If neutrophils are permeabilized immediately upon PS exposure, many of these dead cells will be disintegrated during culture and thus lost from subsequent individual cell-based, end-point assays that would mostly show viable cells. In this manner, stimuli that induce secondary necrosis could be mistaken for having antiapoptotic effects.
2. RPMI 1640 should be interchangeable with similar cell culturing media, e.g., Dulbecco's modified Eagle's medium.

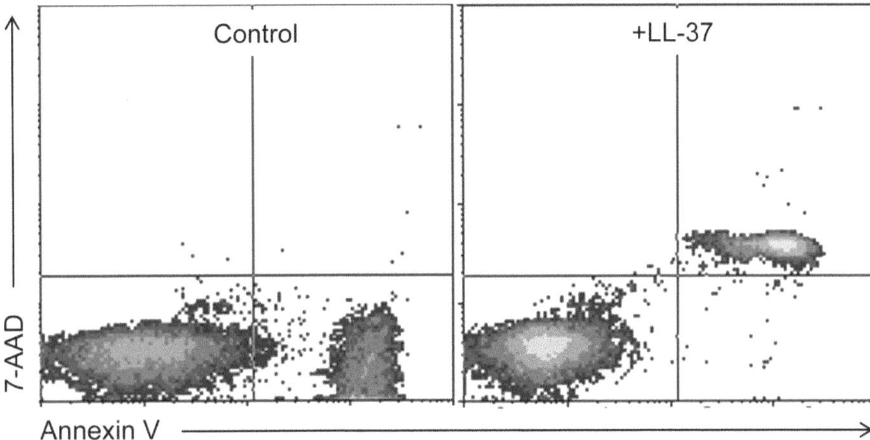

Fig. 8. LL-37 permeabilizes apoptotic neutrophils. Overnight-cultured human neutrophils were analyzed before (control) and 5 min after addition of LL-37 (50 μg/ml). As seen, the peptide swiftly permeabilized only the apoptotic cells, leaving viable cells largely unaffected.

3. PEST can be added as a precaution to avoid microbial contamination.
4. It is of importance to choose the right plastic in apoptosis assays if the cells of study are prone to adhere. Viable and activated phagocytes (in particular neutrophils) are inclined to attach to several plastic materials (even sometimes when stored on ice!), and are thus easily missed in the subsequent analyses. Polypropylene tubes are preferred for neutrophil culture, but culture plates (in other plastic materials) can also be used if the neutrophils are to be lysed after the incubation (bulk assays).
5. It is of importance to use nonfluorescent α-CD95 antibody of functional grade (azide free), since azide can affect the apoptotic process.
6. Mixed Caspase-Glo 3/7 (substrate and buffer) can be stored protected from light at 4°C for a shorter time (1–2 weeks), but the luminescent signal will be weakened if the solution is not stored properly.
7. The specific phosphate buffer suggested by the protocol accompanying the Giemsa and May Grünwald stains can be replaced by regular PBS.
8. Different batches of FCS affect cell viability in different ways. Hence, the amount of FCS could be adjusted so that the spontaneous neutrophil apoptosis is around 50% after a 20-h incubation, making it easy to study both pro- and antiapoptotic effects. Neutrophils are also very sensitive to endotoxin contamination which could result in low levels of apoptosis after culture. It is, therefore, useful to ensure minimal levels of endotoxin in all reagents used.

9. Different batches of FCS have different contents of ROS scavengers and effective concentrations of hydrogen peroxide may thus vary.
10. If more than 10 min pass between analysis of the first and the last sample, the latter sample may display increased background staining as To-Pro-3 enters also into cells with intact membranes given enough time. If many samples are to be analyzed, it could be of value to stain the samples sequentially to avoid this problem.
11. One sample can be pretreated for 5 min with 50 μM carbonyl cyanide 3-chlorophenylhydrazone (CCCP)—a disrupter of the mitochondrial membrane potential—before JC-1 staining. These cells can be used to set gates to discriminate between cells with intact and disrupted mitochondria.
12. Increase the concentration of Mitotracker Deep Red to 200 nM if cells are to be fixed and permeabilized after staining.
13. In cells incubated with FLICA, the apoptotic process may be discontinued since the reagent is also an inhibitor of caspase activity.
14. At this stage, the cells could also be fixed and analyzed later.
15. In all bulk assays, it is very important to have equal amounts of cells in all samples and the cell concentrations should, therefore, be determined before starting the assay to get comparable values.
16. A 4-h preincubation of neutrophils at 37°C is sufficient time to see differences in caspase activity after stimulation with pro- or antiapoptotic factors. Z-VAD-FMK is a cell-permeable, general caspase inhibitor that binds irreversibly to the catalytic site of caspases and can be used as a negative control.
17. According to manufacturer's instructions, the plate should be incubated at 22°C for 1 h after addition of sample and reagent, but this time can be shortened by incubation at 37°C for 20–30 min.
18. In some experimental setups, high calcium concentrations can influence cell viability (41), so it is important not to leave cells longer than necessary in calcium-rich buffers.
19. A sterile stock solution of 10× Annexin binding buffer can be stored at 4°C for at least 6 months.
20. In the protocol describing Annexin V and 7-AAD staining (Subheading 3.4.1), 7-AAD is interchangeable with Sytox Green (10 nM), To-Pro-3 (0.5 μM), or Propidium iodide (2.5 μg/ml) if required.
21. One advantage with 7-AAD as compared to To-Pro-3 is that cells can be kept in Annexin binding buffer with 7-AAD for up to 30 min without the dye leaking into viable cells (see Note 10 above). It is advised that careful time-titration experiments are

performed when setting up systems, including permeability dyes, to ensure that viable cells are not stained under the conditions used.

22. Propidium iodide has two major disadvantages: it has a broad emission spectrum and it has a propensity to stick to plastic. Although propidium iodide is commonly included in many commercial apoptosis kits, many flow cytometry laboratories have replaced propidium iodide with other dyes.
23. When assessing necrosis by flow cytometry, it is advised that the analysis is combined with a bulk assay, e.g., measurement of LDH (see Subheading 3.5), to assure that cells are not missed from the flow cytometric analysis by disintegration (see Note 1 above) or adherence to plastic (see Note 4 above).
24. For the sake of simplicity, we use the terms FL-1, FL-2, etc. for the most common fluorescence channels. In many flow cytometers, FL-1 is the channel used to detect fluorescein (FITC) fluorescence (~515 nm); FL-2 is used to detect R-phycoerythrin (PE) fluorescence (~575 nm); FL-3 is used to detect PerCP or similar fluorescence (~675 nm); and FL-4 is used to detect allophycocyanin (APC) fluorescence (~660 nm).
25. A disadvantage with the combination of Annexin V-FLUOS and 7-AAD is that since the same laser is used for both FL-1 and FL-3 it is necessary with an electronic compensation to correct for leakage between the two channels. Another combination of stains using different lasers, e.g., Annexin V-FLUOS and To-Pro-3, can be used instead to avoid this problem.
26. Since the reagent reacts with free amine groups, it is important that the dilution is carried out in protein-free buffers.
27. The second round of washing is important for staining with the Far Red Dead Cell stain kit (Invitrogen), but can be omitted for the Violet Dead Cell stain kit (Invitrogen).
28. Cell-free supernatants can be stored at 4°C at least for 24 h without affecting the LDH analysis, and it is thus suggested that all controls (a–c) are prepared from freshly isolated cells, even if the samples to be tested are to be cultured overnight.
29. 492 nm could also be used if the ELISA reader is only equipped with such a filter.
30. It is necessary to find a suitable marker with specificity for the macrophages used. In this protocol, we have used CD14 as a marker for human MDMs, but we have also used the protocol with murine bone marrow-derived macrophages with F4/80 as a macrophage marker (42). Setting the macrophage gate based on forward and/or side scatter alone is not advisable, since these parameters may change considerably during phagocytosis.
31. Staining neutrophils with CFDA-SE does not affect the apoptotic process in the cells.

32. To obtain a population of 80–90% apoptotic cells with minimal necrosis, we culture the neutrophils for 20 h at 37°C and 5% CO_2 in the presence of 10 μg/ml α-CD95 after CFDA-SE labeling (Fig. 5, right panel).
33. In a reductionist approach, we routinely omit serum from the clearance assay, whereas serum is used to prepare the cells prior to coculture.
34. A PMN: Macrophage ratio of 4:1 leads to approximately 50% clearance in the setup described (Fig. 7). This is a good baseline that gives a possibility to investigate both increases and reductions in clearance after addition of different factors.
35. Addition of lidocaine makes the detachment of cells faster and easier and it also decreases the amount of cell debris generated by manual detachment with pipettes or cell scrapers. In addition, lidocaine treatment does not affect expression of surface markers (e.g., CD14) and is thus a better choice than, e.g., trypsin-containing detachment buffers (43).
36. Complete detachment of the MDMs can be verified microscopically.
37. Neutrophils that are attached to the surface of the macrophages and not actually engulfed also give rise to double-positive events. The cytoskeleton-disrupting agent cytochalasin B can be used to prevent the uptake without affecting adherence. Addition of this reagent (10 μg/ml) to the MDMs before addition of apoptotic neutrophils can, therefore, be used as a control for actual engulfment (Fig. 7).

Acknowledgments

This work was supported by the Swedish Research Council, the King Gustav V Memorial Foundation, Ingabritt and Arne Lundgren's Research Foundation, Gunvor and Ivan Svensson's Foundation, and the Swedish state under the LUA/ALF agreement. F.B. Thorén was supported by EMBO and the European commission (Marie Curie Intra-European Fellowship).

References

1. Sulston, J. E., Horvitz, H. R. (1977) Post-embryonic cell lineages of the nematode, *Caenorhabditis elegans*, *Dev Biol* **56**, 110–156.
2. Hu, F. S., Clark, J., Lott, T. (1985) Recurrent senescence in axenic cultures of *Physarum polycephalum*, *J Gen Microbiol* **131**, 811–815.
3. Drew, M. C., He, C. J., Morgan, P. W. (2000) Programmed cell death and aerenchyma formation in roots, *Trends Plant Sci* **5**, 123–127.
4. Kerr, J. F., Wyllie, A. H., Currie, A. R. (1972) Apoptosis: a basic biological phenomenon with wide-ranging implications in tissue kinetics, *Br J Cancer* **26**, 239–257.

5. Majno, G., Joris, I. (1995) Apoptosis, oncosis, and necrosis. An overview of cell death, *Am J Pathol* **146**, 3–15.
6. Thoren, F. B., Romero, A. I., Hellstrand, K. (2006) Oxygen radicals induce poly(ADP-ribose) polymerase-dependent cell death in cytotoxic lymphocytes, *J Immunol* **176**, 7301–7307.
7. Kepp, O., Galluzzi, L., Zitvogel, L. et al. (2010) Pyroptosis - a cell death modality of its kind?, *Eur J Immunol* **40**, 627–630.
8. Yu, L., Lenardo, M. J., Baehrecke, E. H. (2004) Autophagy and caspases: a new cell death program, *Cell Cycle* **3**, 1124–1126.
9. Vujanovic, N. L., Nagashima, S., Herberman, R. B. et al. (1996) Nonsecretory apoptotic killing by human NK cells, *J Immunol* **157**, 1117–1126.
10. Werlen, G., Hausmann, B., Naeher, D. et al. (2003) Signaling life and death in the thymus: timing is everything, *Science* **299**, 1859–1863.
11. Gabrilovich, D. I., Nagaraj, S. (2009) Myeloid-derived suppressor cells as regulators of the immune system, *Nat Rev Immunol* **9**, 162–174.
12. Hager, M., Cowland, J. B., Borregaard, N. (2010) Neutrophil granules in health and disease, *J Intern Med* **268**, 25–34.
13. Fox, S., Leitch, A. E., Duffin, R. et al. (2010) Neutrophil apoptosis: relevance to the innate immune response and inflammatory disease, *J Innate Immun* **2**, 216–227.
14. Bylund, J., Campsall, P. A., Ma, R. C. et al. (2005) *Burkholderia cenocepacia* induces neutrophil necrosis in chronic granulomatous disease, *J Immunol* **174**, 3562–3569.
15. Silva, M. T., do Vale, A., dos Santos, N. M. (2008) Secondary necrosis in multicellular animals: an outcome of apoptosis with pathogenic implications, *Apoptosis* **13**, 463–482.
16. Erwig, L. P., Henson, P. M. (2008) Clearance of apoptotic cells by phagocytes, *Cell Death Differ* **15**, 243–250.
17. Ravichandran, K. S., and Lorenz, U. (2007) Engulfment of apoptotic cells: signals for a good meal, *Nat Rev Immunol* **7**, 964–974.
18. Savill, J. S., Wyllie, A. H., Henson, J. E. et al. (1989) Macrophage phagocytosis of aging neutrophils in inflammation. Programmed cell death in the neutrophil leads to its recognition by macrophages, *J Clin Invest* **83**, 865–875.
19. Fadok, V. A., Bratton, D. L., Konowal, A. et al. (1998) Macrophages that have ingested apoptotic cells in vitro inhibit proinflammatory cytokine production through autocrine/paracrine mechanisms involving TGF-beta, PGE2, and PAF, *J Clin Invest* **101**, 890–898.
20. Akgul, C., Moulding, D. A., and Edwards, S. W. (2001) Molecular control of neutrophil apoptosis, *FEBS letters* **487**, 318–322.
21. Liles, W. C., Kiener, P. A., Ledbetter, J. A. et al. (1996) Differential expression of Fas (CD95) and Fas ligand on normal human phagocytes: implications for the regulation of apoptosis in neutrophils, *J Exp Med* **184**, 429–440.
22. Colotta, F., Re, F., Polentarutti, N. et al. (1992) Modulation of granulocyte survival and programmed cell death by cytokines and bacterial products, *Blood* **80**, 2012–2020.
23. Sweeney, J. F., Nguyen, P. K., Omann, G. M. et al. (1997) Ultraviolet irradiation accelerates apoptosis in human polymorphonuclear leukocytes: protection by LPS and GM-CSF, *J Leukoc Biol* **62**, 517–523.
24. Suzuki, K., and Namiki, H. (1998) Phorbol 12-myristate 13-acetate induced cell death of porcine peripheral blood polymorphonuclear leucocytes, *Cell Struct Funct* **23**, 367–372.
25. Akgul, C., Edwards, S. W. (2003) Regulation of neutrophil apoptosis via death receptors, *Cell Mol Life Sci* **60**, 2402–2408.
26. Boyum, A. (1968) Isolation of mononuclear cells and granulocytes from human blood. Isolation of monuclear cells by one centrifugation, and of granulocytes by combining centrifugation and sedimentation at 1 g, *Scand J Clin Lab Investig* **97**, 77–89.
27. Boyum, A., Lovhaug, D., Tresland, L. et al. (1991) Separation of leucocytes: improved cell purity by fine adjustments of gradient medium density and osmolality, *Scand J Immunol* **34**, 697–712.
28. Karlsson, A., Christenson, K., Matlak, M. et al. (2009) Galectin-3 functions as an opsonin and enhances the macrophage clearance of apoptotic neutrophils, *Glycobiology* **19**, 16–20.
29. Green, D. R., Kroemer, G. (2004) The pathophysiology of mitochondrial cell death, *Science* **305**, 626–629.
30. Thornberry, N. A., Lazebnik, Y. (1998) Caspases: enemies within, *Science* **281**, 1312–1316.
31. Riedl, S. J., and Shi, Y. (2004) Molecular mechanisms of caspase regulation during apoptosis, *Nat Rev Mol Cell Biol* **5**, 897–907.
32. Li, P., Nijhawan, D., Budihardjo, I. et al. (1997) Cytochrome c and dATP-dependent formation of Apaf-1/caspase-9 complex initiates an apoptotic protease cascade, *Cell* **91**, 479–489.
33. Martins, L. M., Kottke, T., Mesner, P. W. et al. (1997) Activation of multiple interleukin-1beta converting enzyme homologues in cytosol and nuclei of HL-60 cells during etoposide-induced apoptosis, *J Biol Chem* **272**, 7421–7430.

34. Seigneuret, M., Devaux, P. F. (1984) ATP-dependent asymmetric distribution of spin-labeled phospholipids in the erythrocyte membrane: relation to shape changes, *Proc Natl Acad Sci USA* **81**, 3751–3755.
35. Zachowski, A., Herrmann, A., Paraf, A. et al. (1987) Phospholipid outside-inside translocation in lymphocyte plasma membranes is a protein-mediated phenomenon, *Biochim Biophys Acta* **897**, 197–200.
36. Bratton, D. L., Fadok, V. A., Richter, D. A. et al. (1997) Appearance of phosphatidylserine on apoptotic cells requires calcium-mediated nonspecific flip-flop and is enhanced by loss of the aminophospholipid translocase, *J Biol Chem* **272**, 26159–26165.
37. Verhoven, B., Schlegel, R. A., and Williamson, P. (1995) Mechanisms of phosphatidylserine exposure, a phagocyte recognition signal, on apoptotic T lymphocytes, *J Exp Med* **182**, 1597–1601.
38. Perfetto, S. P., Chattopadhyay, P. K., Lamoreaux, L. et al. (2006) Amine reactive dyes: an effective tool to discriminate live and dead cells in polychromatic flow cytometry, *J Immunol Meth* **313**, 199–208.
39. Björstad, Å., Brown, K., Forsman, H. et al. (2008) Antimicrobial Host Defence Peptides of Human Neutrophils - Roles in Innate Immunity, *Anti-Infective Agents in Medicinal Chemistry* 7, 155–168.
40. Björstad, A., Askarieh, G., Brown, K. L. et al . (2009) The host defense peptide LL-37 selectively permeabilizes apoptotic leukocytes, *Antimicrob Agents Chemother* 53, 1027–1038.
41. Trahtemberg, U., Atallah, M., Krispin, A. et al. (2007) Calcium, leukocyte cell death and the use of annexin V: fatal encounters, *Apoptosis* 12, 1769–1780.
42. Brown, K. L., Christenson, K., Karlsson, A. et al. (2009) Divergent effects on phagocytosis by macrophage-derived oxygen radicals, *J Innate Immun* 1, 592–598.
43. Davies, J. Q., Gordon, S. (2005) Isolation and culture of murine macrophages, *Methods Mol Biol* 290, 91–103.

Chapter 6

Towards a Four-Dimensional View of Neutrophils

Ben A. Croker, Andrew W. Roberts, and Nicos A. Nicola

Abstract

Neutrophils are constitutively produced throughout adult life and are essential for host responses to many types of pathogen. Neutropenia has long been associated with poor prognosis in the clinic, yet we have an incomplete understanding of their life cycle, not only during homeostasis but also during infection and chronic inflammation. Here, we review recent advances that provide insight into the genetic and biochemical regulators of neutrophil production, function, and survival.

Key words: Neutrophils, Apoptosis, NETs, Cytokines, Retrograde chemotaxis, NADPH oxidase, Proteases

1. Introduction

Neutrophils are classically thought to represent the first line of defense against infection and tissue injury. They are produced in the bone marrow, circulate in the blood, and are attracted to sites of infection by chemotactic molecules. They are thought to be short lived with a circulating half-life of 6–8 h, so large numbers (10^{11}–10^{12}) are required to be produced daily. While neutrophils are essential for host defense, they also contribute to inflammation and tissue damage, so their numbers and activity need to be tightly controlled.

Key aspects of neutrophil biology include the signals mediating neutrophil production, release from the bone marrow into the blood, adherence to endothelium and movement into the tissues, recruitment of other innate and adaptive immune cells, and the battery of effector functions used to fight infections as well as the signals for resolving the effector phase (1–7).

Robert B. Ashman (ed.), *Leucocytes: Methods and Protocols*, Methods in Molecular Biology, vol. 844,
DOI 10.1007/978-1-61779-527-5_6, © Springer Science+Business Media, LLC 2012

The recent descriptions of retrograde chemotaxis, neutrophil extracellular traps (NETs) and potent immunomodulatory actions of neutrophils have rejuvenated interest in the roles played by neutrophils during immune responses. The one-dimensional view of the neutrophil as a short-lived, frontline defender against infection now requires careful reevaluation in light of data from multiple investigators that collectively challenge this dogma. This review examines recent highlights from the literature and examine the role of neutrophils in shaping host immune and inflammatory responses.

2. Neutropenia and Severe Infection

Neutrophils play key roles in host defense against a range of pathogens but also contribute to acute and chronic inflammatory diseases. In 1966, a strict correlation was noted between neutropenia and severe infection in patients with acute leukemia (8). More recently, a correlation between compromised neutrophil reserves and mortality has been described for mice challenged with *Listeria monocytogenes*, *Staphylococcus aureus*, *Salmonella typhimurium*, *Streptococcus pyogenes*, or heat-killed bacteria (9). Drug-induced neutropenia is a serious complication of chemotherapy regimens, contributing to bacterial and fungal sepsis. Depletion of neutrophils during influenza infection increases viral titers, edema, respiratory disease, and mortality of mice (10). But the opposite has also been described. Neutrophil depletion increases survival of mice in response to herpes simplex virus 2 (HSV-2) and mouse cytomegalovirus and reduces hepatoxicity (11). Neutrophils are essential to prevent and recover from bacterial and fungal infections; however, their roles in viral infection appear more complicated.

3. Positive and Negative Regulation of Neutrophil Production

Much of our understanding of the production of mature neutrophils stems from the pioneering work of Bradley et al, and Ichikawa et al, who developed in vitro culture systems for the study of hematopoietic cell production (12, 13). The studies that followed led to the purification and identification of G-CSF, the primary extracellular regulator of neutrophil production and mobilization during homeostatic conditions and emergency granulopoiesis (14, 15). In its absence, neutrophil numbers are reduced by 75% and, in response to infection, neutrophil production and mobilization into the periphery are severely compromised (15). G-CSF itself is rapidly induced in response to infection and is likely to be an attractive therapeutic target to modulate neutrophilic inflammatory disease (16). The production of G-CSF is regulated in part by IL-17, a cytokine produced by a variety of cells including γδ T cells, αβ T cells, and neutrophils (17–20). The phagocytosis of apoptotic neutrophils by macrophages induces IL-23, a cytokine that stimulates IL-17 production and thereby contributes to the homeostatic regulation of neutrophil numbers (21).

IL-6 also contributes to neutrophil production in vivo, and mice deficient in both G-CSF and IL-6 are profoundly neutropenic, confirming that the ability of IL-6 to stimulate neutrophil production in vitro is also relevant in vivo (22).

The binding of G-CSF to the G-CSF receptor bring associated Janus kinases (JAKs) into close proximity, resulting in cross-phosphorylation and activation. Activated JAKs phosphorylate tyrosine residues on the receptor providing docking sites for the Signal Transducers and Activators of Transcription (STATs), which are themselves phosphorylated. Phosphorylated STATs dimerize and translocate to the nucleus to induce gene transcription. One of the genes that is induced by G-CSF is the Suppressor of Cytokine Signaling-3 (SOCS3) (23, 24). SOCS3 negatively regulates G-CSF signaling by binding to the G-CSF receptor and to JAKs, and inhibiting the activity of the receptor signaling complex as well as inducing its proteasomal degradation (25–27). SOCS3 not only controls the magnitude and duration of G-CSF and IL-6 signaling in myeloid cells but also controls the specificity of the biological responses to these cytokines (28–32). G-CSF induces pathological neutrophil infiltration into the liver, lung, muscle, and spinal tissue in mice lacking SOCS3 in hematopoietic and endothelial cells (29). These mice also develop severe antigen-induced arthritis and display dramatic increases in IL-6, G-CSF, and neutrophil production (33). Adenoviral expression of SOCS3 reduced proliferation, cytokine production, and tissue pathology in mouse models of arthritis (34). Therapeutic targeting of IL-6 and G-CSF, the dominant regulators of neutrophil production, is likely to play an important role in the treatment of acute and chronic neutrophilic inflammatory disease.

4. Multifunctional Roles for Neutrophil Granule Proteins and NADPH Oxidase

For many years it was believed that reactive oxygen species, generated by the NADPH oxidase complex, were directly responsible and sufficient for microbe killing. Mutations in NADPH oxidase (the cytochrome b245 subunit) cause chronic granulomatous disease (CGD) and immunodeficiency, but early reports also described defective neutrophil degranulation in this disease (35–38).

Divergent theories on the relative microbicidal roles of the NADPH oxidase complex and neutrophil granule proteins originate from various studies of gene targeted mice lacking proteases including neutrophil elastase and cathepsin G. Mice deficient in neutrophil elastase and cathepsin G fail to kill *A. fumigatus* despite normal production of reactive oxygen species and phagocytic capacity, demonstrating that the production of reactive oxygen species is not sufficient to kill microbes (39, 40). Cathepsin G deficiency does not affect survival of mice challenged with *S. aureus*, *Klebsiella pneumoniae*, and *Escherichia coli* (41), whereas neutrophil elastase is dispensable for responses to *S. aureus* (Gram positive bacteria) but not *K. pneumoniae* or *E. coli* (Gram negative bacteria). However, the activity of neutrophil proteases is dependent on the NADPH oxidase complex, which is required to establish an electrochemical gradient in the phagolysosome to control pH (3, 42). The movement of H^+ ions by the NADPH oxidase complex is also proposed to control the protonation of superoxide anions to H_2O_2 (42). The oxidation of cellular substrates by H_2O_2 is inhibited by the actions of myeloperoxidase, which converts H_2O_2 to hypochlorous acid. Myeloperoxidase may also enhance neutrophil elastase activity by inactivating α-1-antitrypsin (43, 44). However, it is difficult to reconcile the above data with that generated using neutrophils from patients with Papillon–Lefevre syndrome (PLS), a disease caused by mutations in cathepsin C. Despite severe defects in cathepsin G, neutrophil elastase, and proteinase 3 activity (due to the role of cathepsin C in processing the proenzymes) the microbicidal activity of neutrophils from PLS patients was normal in response to *E. coli* and *S. aureus* (45). Presumably, residual protease activity is sufficient for effective neutrophil responses to bacteria in the majority of PLS patients.

Neutrophil proteases are key contributors to the development of noninfective inflammatory diseases. Both proteinase 3 and neutrophil elastase can process the biologically inactive pro-IL-1β to its bioactive 17 kDa form, thereby driving inflammatory responses (46, 47). The loss of proteinase 3 reduces the severity of acute arthritis, typified by neutrophil accumulation (48). Proteinase 3 and neutrophil elastase are thought to contribute to the Arthus reaction (edema and hemorrhage in response to immune complexes) in humans by inactivating progranulin, which is an inhibitor of immune complex-induced superoxide production, thereby enhancing the oxidative burst induced by immune complexes (49). Neutrophil recruitment to antigen–antibody–complement complexes was impaired in mice lacking proteinase 3 and neutrophil elastase, due to an accumulation of progranulin (49). Neutrophil proteases play fundamental roles in antimicrobial defense and examination of their regulation will help unravel the discrepancies in the literature and their contribution to inflammatory disease.

5. Neutrophil Extracellular Traps

The ability of neutrophils to form extracellular traps (NETs) composed of chromatin, neutrophil proteases, and antimicrobial peptides to facilitate the capture and inactivation of microbes is a stunning biological phenomenon. The observation has profound implications for the contribution of neutrophils to infectious disease and also to autoinflammatory and autoimmune disease. NET formation can be triggered by a diverse array of stimuli including LPS, FasL, PMA, hydrogen peroxide, *C. albicans*, *A. nidulans*, *C. neoformans*, *S. aureus*, *S. pneumoniae*, *S. typhimurium*, and *S. flexneri* (50–53). The process of NET formation is generally promoted by NADPH oxidase activity, inhibited by catalases, endonucleases, and serum, and independent of caspase activity (53, 54). In response to IL-8, NET formation was shown to be dependent on the chemokine receptor CXCR2, Src family kinases, and ERK MAP kinases but independent of phosphatidylinositol 3-kinase (PI3K) and NADPH oxidase (55). Proteomics analysis of NETs from PMA-stimulated neutrophils revealed a diverse range of granule proteins, histones, cytoplasmic calcium-binding proteins, cytoskeletal proteins, and glycolytic enzymes (52). Using DNase to dismantle NETs, it was demonstrated that the antibacterial activity of neutrophils can be attributed to phagocytosis at early time points and to NET formation at later time points (53).

The phenomenon of NET formation may contribute to sepsis (56), small vessel vasculitis (SVV), cystic fibrosis sputum, experimental dysentery, and appendicitis (50) and has been proposed as a biomarker of septic arthritis (57). The high levels of circulating free DNA complexed with neutrophil peptides found in these diseases may impair circulation, damage tissues via the actions of the associated proteases, and promote inflammation. However, it is less clear whether the association of anionic nucleic acids with cationic peptides in vivo is a direct consequence of NET formation. Local application of CXCR2 inhibitors to the airways in mouse models of cystic fibrosis ameliorated disease and inhibited NET formation but had no effect on neutrophil recruitment, respiratory burst, or phagocytosis.

Small vessel vasculitis is an autoimmune condition characterized by antineutrophil cytoplasm autoantibodies (ANCAs) and chronic inflammation of small blood vessels. The ANCAs target proteinase 3 in Wegener's granulomatosis and myeloperoxidase in microscopic polyangiitis (58, 59). It is tempting to speculate that these conditions may be exacerbated by ANCA-activated neutrophils that release NETs containing proteinase 3 and myeloperoxidase. Complexes of myeloperoxidase and DNA were detected in the serum of patients with small vessel vasculitis, suggesting that ANCA-triggered NET formation supplies autoreactive B cells with antigenic complexes of DNA, proteinase 3, and myeloperoxidase. Disease progression in SVV patients may be compounded by their sensitivity to *S. aureus* infection, another potent activator of NET formation. Plasmacytoid dendritic cells can be induced to release interferon α (IFN-α) by LL37, an antimicrobial peptide found in NETs. LL37 colocalizes in the kidneys with neutrophils, histones, neutrophil granule proteins, and type I IFN-inducible proteins (49), suggesting it may contribute to the activation of plasmacytoid dendritic cells and the development of autoimmune disease.

The phenomenon of NET formation is not without controversy. Some studies suggest that mitochondrial DNA, but not nuclear DNA, is released by viable GM-CSF-primed neutrophils after stimulation with LPS or complement factor 5a (60). The high levels of mitochondrial DNA found in the plasma of trauma patients may activate Toll-like receptor 9 (TLR9) and formyl peptide receptor-1 (FPR1) to induce a systemic inflammatory response as a consequence of NET formation (61). The activation of platelet TLR4 can facilitate an interaction between platelets and neutrophils and provide a stimulus for NET production by viable neutrophils, and has been shown to reduce sinusoid perfusion and increase liver damage. The interaction with and phagocytosis of activated platelets by neutrophils is dependent on phosphatidylserine, P-selectin and β2-integrin (62), but it is not known if these interactions contribute to NET formation. Numerous factors have been described that affect NET formation, but not all are consistent and the frequency at which NET formation occurs in vivo remains unresolved.

6. Retrograde Neutrophil Chemotaxis and Antigen Presentation

The demonstration of retrograde neutrophil chemotaxis will be an important consideration when investigating the roles of neutrophils in antigen presentation and in the initiation and progression of acute and chronic neutrophilic inflammatory diseases (63). Only recently has it been appreciated that neutrophils can be rapidly recruited to sites of inflammation before returning to the microvasculature and to peripheral locations in the body. The bidirectional migration of neutrophils from wound sites suggest that this may be a novel means of resolving inflammatory responses but may promote relocation of activated neutrophils to peripheral tissues and organs. It is not known whether the organs (spleen, liver, and bone marrow) that mediate neutrophil clearance under homeostatic conditions (64) are also involved in the clearance of neutrophils that have undergone retrograde chemotaxis from inflamed tissues to the periphery.

Much has been learnt about lymphocyte chemotaxis within lymphoid tissues during immune responses, yet analysis of neutrophil chemotaxis within these same tissues is in its infancy. Using 2 photon scanning laser microscopy, neutrophil swarms were described in lymph nodes of Toxoplasma-infected mice, coinciding with the loss of CD169$^+$ macrophages in the subcapsular sinus. The formation of neutrophil swarms is thought to be mediated by a chemokine concentration threshold laid out by early migrating neutrophils (65). One possible role for neutrophils entering lymph nodes is to present antigen to T cells. Neutrophils are capable of presenting antigen in vitro, although not as efficiently as dendritic cells (66). Antigen-specific immunity can be induced by CD11b$^+$Gr-1$^+$ cells found in the tumor microenvironment of epithelial ovarian carcinomas (67). Because CD11b$^+$Gr-1$^+$ populations contain a number of myeloid populations, and are not exclusively mature neutrophils, the contribution of contaminating dendritic cells or macrophages to antigen presentation is not clear.

7. How to Subdivide the Neutrophil Population?

The multi-dimensional view of neutrophil biology now gaining favor envisages significant roles for neutrophils in the modulation of immune responses by virtue of cytokine production, antigen presentation, interactions with other hematopoietic and nonhematopoietic cells, and retrograde chemotaxis to other sites within the host. With an increasing appreciation of the role of neutrophils in immunomodulation, attempts to address functional specialization will need to be made. But how should this be done and around what parameters should this be built? Do neutrophil subsets exist that respond to specific Toll-like receptor ligands, for example, and do they produce selective responses, such as cytokine secretion, superoxide production or degranulation? Because cytokine stimulation rapidly alters the proportion of neutrophils responding to TLR ligands and the magnitude of cytokine production (68), it is sometimes difficult to distinguish distinct neutrophil subsets from altered states of a single subset.

At an elementary level, neutrophils might be subdivided on the basis of their ability to produce particular cytokines, such as TNF-α, IL-10, IL-8, IL-1α, and IL-1β (68–71). But brief exposure of neutrophils to inflammatory cytokines such as GM-CSF, IFN-γ, or G CSF can dramatically alter the proportion of neutrophils that respond to TLR ligands, and their magnitude of TNF-α production (68). A similar trend can be seen for fMLF-stimulated neutrophils producing reactive oxygen species following priming with GM-CSF (Croker et al., unpublished data). Tumor-associated neutrophils (TANs) were recently classified based on their ability to support tumor growth. TGF-β induces the expression of arginase and the chemokines Ccl2 and Ccl5 in neutrophils, promoting an immunosuppressive neutrophil phenotype that supports tumor growth. Neutrophil depletion can decrease tumor growth and enhance CD8$^+$ T cell responses (69). TGF-β inhibition blocks tumor development, but neutrophils must be present for this inhibition to occur. Increased levels of TNF-α, ICAM1, and Fas expression on neutrophils drive this antitumor neutrophil phenotype (69). The formation of immunosuppressive IL-10-producing neutrophils that were capable of suppressing the proliferation of alloreactive T cells was promoted by serum amyloid A1 (SAA-1), a factor released during inflammatory responses, and also by melanoma cells (70). The neutrophil population is therefore highly responsive to changes in the environment and can rapidly alter its behavior to regulate the nature of the inflammatory response.

8. Understanding Neutrophil Persistence During Inflammation

Considerable progress has been made in defining the regulators of neutrophil production and function. A less comprehensive framework exists to understand neutrophil survival during inflammation. During homeostasis, approximately 100 billion neutrophils are produced daily in humans, and these are typically silently removed from the circulation by macrophages in the bone marrow, spleen, liver, and lung (64). In contrast to previous studies using extrinsically labeled neutrophils, recent endogenous labeling studies have estimated the half-life of human neutrophils to be 5 days, and this may be considerably longer in the presence of inflammatory cytokines (71). During inflammatory responses, neutrophil progenitors respond rapidly to cytokine stimulation to increase neutrophil production. Defects in the removal of neutrophils during the resolution phase of immune responses will drive inflammatory disease. Numerous studies report correlations between defects in neutrophil apoptosis and human diseases such as sepsis (72), chronic obstructive pulmonary disease (73), antineutrophil cytoplasmic autoantibody-associated vasculitis (58, 74), acute lung injury (75), rheumatoid arthritis, cytomegalovirus, influenza, *L. monocytogenes*, respiratory syncytial virus, and *Leishmania major* infections (76–81). Levels of G-CSF and GM-CSF, inflammatory cytokines that strongly promote neutrophil survival, are elevated in bronchoalveolar lavage supernatant from patients with neutrophilic lung inflammation such as cystic fibrosis, pneumonia, and acute allergic alveolitis (82). Despite a wealth of studies demonstrating correlations between disease severity and neutrophil survival in humans, limited studies using genetically modified mice have interrogated the role of prosurvival and proapoptotic proteins during infection and inflammatory responses.

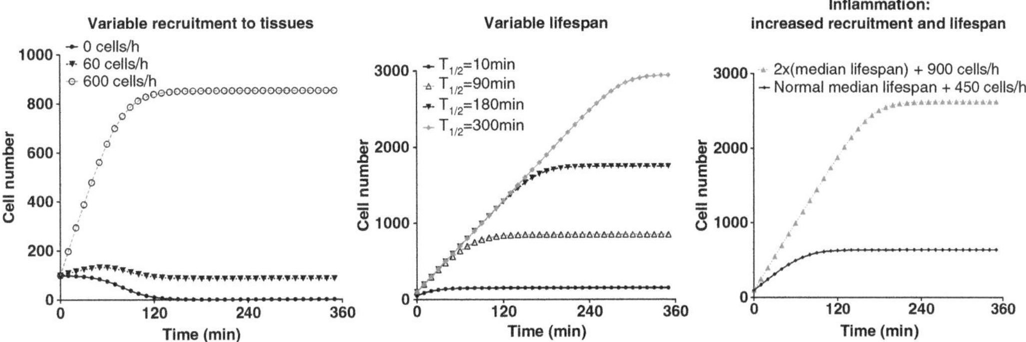

Fig. 1. Effect of recruitment and life span on neutrophil persistence during inflammation. The *left panel* illustrates the effect of neutrophil recruitment on total neutrophil numbers in tissues when the median life span of the neutrophil is fixed. As an example, 100 cells are shown at 0 min. The *middle panel* illustrates the effect of neutrophil median life span on total neutrophil numbers in tissues when the number of incoming neutrophils is fixed (recruitment number fixed to 100 cells/h for illustrative purposes). The *right panel* illustrates the combined effect of a doubling of neutrophil median life span and a twofold increase in neutrophil recruitment to total neutrophil numbers in tissues. This hypothetical model is based on neutrophils entering tissues and not returning to the circulation.

The Fas-activated and the Bcl-2 family-regulated apoptosis pathways are powerful modulators of neutrophil survival. Under homeostatic conditions, the roles of prosurvival proteins such as Mcl-1, Bcl-2, and A1 and proapoptotic proteins such as Bim, Bak, and Bax in regulating neutrophil production and survival in gene-targeted mice are well appreciated (83–87). By contrast, the specific contribution of these proteins to neutrophil survival during inflammatory responses is poorly defined. Neutrophils are highly sensitive to stimulation with FasL (88), a TNF family member that activates CD95/Fas/Apo-1 to stimulate formation of a death inducing signaling complex containing caspase 8 and FADD, leading to the activation of Bid, caspase 3, and caspase 7 (89). Probing the contributions of these cell death pathways specifically in neutrophils during infection and inflammation will be technically challenging. Understanding the relative roles played by intrinsic and extrinsic death pathways during resolution phases of neutrophilic inflammatory responses, particularly when FasL-expressing T cells are present, will be key to unraveling the mechanisms governing neutrophil persistence in tissues (Fig. 1).

We have made significant progress in our characterization of the neutrophil, particularly of human neutrophils. As a consequence, a multidimensional view of neutrophils during inflammatory responses is beginning to emerge. As we define the parameters controlling neutrophil production, mobilization, recruitment, function, survival, and clearance, we will likely reveal novel approaches to treating inflammatory disease. The adoption of protocols for the rapid purification of neutrophils from model systems and the analysis of fluorescently tagged neutrophils in vivo should reveal new dimensions in the life of the neutrophil.

Acknowledgments

B.A. Croker was supported by an Australian Research Council QEII Fellowship (DP1094854). A.W. Roberts was supported by a National Health Medical Research Council (NHMRC) Australia Practitioner Fellowship (356213) and a Victorian Cancer Agency Fellowship. N.A. Nicola was supported by a NHRMC Fellowship (637300). The authors are supported by a grant from the National Institutes of Health, USA (CA022556), Australian NHMRC Grants (461219, 508905, and 637367), the Victorian State Government Operational Infrastructure Support Grant and the NHMRC Independent Research Institutes Infrastructure Support Scheme (361646).

References

1. Dale, D.C., Boxer, L., Liles, W.C. (2008) The phagocytes: neutrophils and monocytes, *Blood* **112,** 935–45.
2. Nathan, C. (2006) Neutrophils and immunity: challenges and opportunities, *Nat Rev Immunol* **6,** 173–82.
3. Segal, A.W. (2005) How neutrophils kill microbes, *Annu Rev Immunol* **23,** 197–223.
4. Summers, C., Rankin, S.M., Condliffe, A.M. et al. (2010) Neutrophil kinetics in health and disease, *Trends Immunol* **31,** 318–24.
5. Soehnlein, O., Lindbom, L. (2010) Phagocyte partnership during the onset and resolution of inflammation, *Nat Rev Immunol* **10,** 427–39.
6. Hager, M., Cowland, J.B., Borregaard, N. (2010) Neutrophil granules in health and disease, *J Intern Med* **268,** 25–34.
7. Woodfin, A., Voisin, M.B., Nourshargh, S. (2010) Recent developments and complexities in neutrophil transmigration, *Curr Opin Hematol* **17,** 9–17.
8. Bodey, G.P., Buckley, M., Sathe, Y.S. et al. (1966) Quantitative relationships between circulating leukocytes and infection in patients with acute leukemia, *Ann Intern Med* **64,** 328–40.
9. Navarini, A.A., Lang, K.S., Verschoor, A. et al. (2009) Innate immune-induced depletion of bone marrow neutrophils aggravates systemic bacterial infections, *Proc Natl Acad Sci USA* **106,** 7107–12.
10. Tate, M.D., Deng, Y.M., Jones, J.E. et al. (2009) Neutrophils ameliorate lung injury and the development of severe disease during influenza infection, *J Immunol* **183,** 7441–50.
11. Stout-Delgado, H.W., Du, W., Shirali, A.C. et al. (2009) Aging promotes neutrophil-induced mortality by augmenting IL-17 production during viral infection, *Cell Host Microbe* **6,** 446–56.
12. Bradley, T.R. and Metcalf, D. (1966) The growth of mouse bone marrow cells in vitro. *Aust J Exp Biol Med Sci* **44,** 287–99.
13. Ichikawa, Y., Pluznik, D.H. and Sachs, L. (1966) In vitro control of the development of macrophage and granulocyte colonies, *Proc Natl Acad Sci USA* **56,** 488–95.
14. Nicola, N.A., Metcalf, D., Matsumoto, M. et al. (1983) Purification of a factor inducing differentiation in murine myelomonocytic leukemia cells. Identification as granulocyte colony-stimulating factor, *J Biol Chem* **258,** 9017–23.
15. Lieschke, G.J., Grail, D., Hodgson, G. et al. (1994) Mice lacking granulocyte colony-stimulating factor have chronic neutropenia, granulocyte and macrophage progenitor cell deficiency, and impaired neutrophil mobilization, *Blood* **84,** 1737–46.
16. Eyles, J.L., Hickey, M.J., Norman, M.U. et al. (2008) A key role for G-CSF-induced neutrophil production and trafficking during inflammatory arthritis, *Blood* **112,** 5193–201.
17. Ferretti, S., Bonneau, O., Dubois, G.R. et al. (2003) IL-17, produced by lymphocytes and neutrophils, is necessary for lipopolysaccharide-induced airway neutrophilia: IL-15 as a possible trigger, *J Immunol* **170,** 2106–12.
18. Li, L., Huang, L., Vergis, A.L. et al. (2010) IL-17 produced by neutrophils regulates IFN-gamma-mediated neutrophil migration in mouse kidney ischemia-reperfusion injury, *J Clin Invest* **120,** 331–42.
19. Hoshino, A., Nagao, T., Nagi-Miura, N. et al. (2008) MPO-ANCA induces IL-17 production

by activated neutrophils in vitro via classical complement pathway-dependent manner, *J Autoimmun* **31,** 79–89.
20. Cua, D.J., and Tato C.M. (2010) Innate IL-17-producing cells: the sentinels of the immune system, *Nat Rev Immunol* **10,** 479–89.
21. Stark, M.A., Huo, Y., Burcin, T.L. et al. (2005) Phagocytosis of apoptotic neutrophils regulates granulopoiesis via IL-23 and IL-17, *Immunity* **22,** 285–94.
22. Liu, F., Poursine-Laurent, J., Wu, H.Y. et al. (1997) Interleukin-6 and the granulocyte colony-stimulating factor receptor are major independent regulators of granulopoiesis in vivo but are not required for lineage commitment or terminal differentiation, *Blood* **90,** 2583–90.
23. Lee, C., Raz, R., Gimeno, R. et al. (2002) STAT3 is a negative regulator of granulopoiesis but is not required for G-CSF-dependent differentiation, *Immunity* **17,** 63–72.
24. Hermans, M.H., Van De Geijn, G.J., Antonissen, C. et al. (2002) Signaling mechanisms coupled to tyrosines in the granulocyte colony-stimulating factor receptor orchestrate G-CSF-induced expansion of myeloid progenitor cells, *Blood* **101,** 2584–90.
25. Schmitz, J., Weissenbach, M., Haan, S. et al. (2000) SOCS3 exerts its inhibitory function on interleukin-6 signal transduction through the SHP2 recruitment site of gp130, *J Biol Chem* **275,** 12848–56.
26. Nicholson, S.E., De Souza, D., Fabri, L.J. et al. (2000) Suppressor of cytokine signaling-3 preferentially binds to the SHP-2-binding site on the shared cytokine receptor subunit gp130, *Proc Natl Acad Sci USA* **97,** 6493–8.
27. Zhang, J.G., Farley, A., Nicholson, S.E. et al. (1999) The conserved SOCS box motif in suppressors of cytokine signaling binds to elongins B and C and may couple bound proteins to proteasomal degradation, *Proc Natl Acad Sci USA* **96,** 2071–6.
28. Croker, B.A., Mielke, L.A., Wormald, S. et al. (2008) Socs3 maintains the specificity of biological responses to cytokine signals during granulocyte and macrophage differentiation, *Exp Hematol* **36,** 786–98.
29. Croker, B.A., Metcalf, D., Robb, L. et al. (2004) SOCS3 is a critical physiological negative regulator of G-CSF signaling and emergency granulopoiesis, *Immunity* **20,** 153–65.
30. Croker, B.A., Krebs, D.L., Zhang, J.G. et al. (2003) SOCS3 negatively regulates IL-6 signaling in vivo, *Nat Immunol* **4,** 540–5.
31. Yasukawa, H., Ohishi, M., Mori, H. et al. (2003) IL-6 induces an anti-inflammatory response in the absence of SOCS3 in macrophages, *Nat Immunol* **4,** 551–6.
32. Lang, R., Pauleau, A.L., Parganas, E. et al. (2003) SOCS3 regulates the plasticity of gp130 signaling, *Nat Immunol* **4,** 546–50.
33. Wong, P.K., Egan, P.J., Croker, B.A. et al. (2006) SOCS-3 negatively regulates innate and adaptive immune mechanisms in acute IL-1-dependent inflammatory arthritis, *J Clin Invest 116*, 1571–81.
34. Shouda, T., Yoshida, T., Hanada, T. et al. (2001) Induction of the cytokine signal regulator SOCS3/CIS3 as a therapeutic strategy for treating inflammatory arthritis, *J Clin Invest* **108,** 1781–8.
35. Jackson, S.H., Gallin, J.I., Holland, S.M. (1995) The p47phox mouse knock-out model of chronic granulomatous disease, *J Exp Med* **182,** 751–8.
36. Dinauer, M.C., Orkin, S.H., Brown, R. et al. (1987) The glycoprotein encoded by the X-linked chronic granulomatous disease locus is a component of the neutrophil cytochrome b complex, *Nature* **327,** 717–20.
37. Teahan, C., Rowe, P., Parker, P. et al. (1987) The X-linked chronic granulomatous disease gene codes for the beta-chain of cytochrome b-245, *Nature* **327,** 720–1.
38. Quie, P.G., White, J.G., Holmes, B. et al. (1967) In vitro bactericidal capacity of human polymorphonuclear leukocytes: diminished activity in chronic granulomatous disease of childhood, *J Clin Invest* **46,** 668–79.
39. Belaaouaj, A., McCarthy, R., Baumann, M. et al. (1998) Mice lacking neutrophil elastase reveal impaired host defense against gram negative bacterial sepsis, *Nat Med* **4,** 615–8.
40. Tkalcevic, J., Novelli, M., Phylactides, M. et al. (2000) Impaired immunity and enhanced resistance to endotoxin in the absence of neutrophil elastase and cathepsin G, *Immunity* **12,** 201–10.
41. MacIvor, D.M., Shapiro, S.D., Pham, C.T. et al. (1999) Normal neutrophil function in cathepsin G-deficient mice, *Blood* **94,** 4282–93.
42. Reeves, E.P., Lu, H., Jacobs, H.L. et al. (2002) Killing activity of neutrophils is mediated through activation of proteases by K+flux, *Nature* **416,** 291–7.
43. Clark, R.A., Stone, P.J., El Hag, A. et al. (1981) Myeloperoxidase-catalyzed inactivation of alpha 1-protease inhibitor by human neutrophils, *J Biol Chem* **256,** 3348–53.
44. Klebanoff, S.J., Kinsella, M.G., Wight, T.N. (1993) Degradation of endothelial cell matrix heparan sulfate proteoglycan by elastase and

the myeloperoxidase-H2O2-chloride system, *Am J Pathol* **143,** 907–17.

45. Pham, C.T., Ivanovich, J.L., Raptis, S.Z. et al. (2004) Papillon-Lefevre syndrome: correlating the molecular, cellular, and clinical consequences of cathepsin C/dipeptidyl peptidase I deficiency in humans, *J Immunol* **173,** 7277–81.
46. Coeshott, C., Ohnemus, C., Pilyavskaya, A. et al. (1999) Converting enzyme-independent release of tumor necrosis factor alpha and IL-1beta from a stimulated human monocytic cell line in the presence of activated neutrophils or purified proteinase 3, *Proc Natl Acad Sci USA* **96,** 6261–6.
47. Greten, F.R., Arkan, M.C., Bollrath, J. et al. (2007) NF-kappaB is a negative regulator of IL-1beta secretion as revealed by genetic and pharmacological inhibition of IKKbeta, *Cell* **130,** 918–31.
48. Joosten, L.A., Netea, M.G., Fantuzzi, G. et al. (2009) Inflammatory arthritis in caspase 1 gene-deficient mice: contribution of proteinase 3 to caspase 1-independent production of bioactive interleukin-1beta, *Arthritis Rheum* **60,** 3651–62.
49. Kessenbrock, K., Frohlich, L., Sixt, M. et al. (2008) Proteinase 3 and neutrophil elastase enhance inflammation in mice by inactivating antiinflammatory progranulin, *J Clin Invest* **118,** 2438–47.
50. Brinkmann, V., Reichard, U., Goosmann, C. et al. (2004) Neutrophil extracellular traps kill bacteria, *Science* **303,** 1532–5.
51. Urban, C.F., Reichard, U., Brinkmann, V. et al. (2006) Neutrophil extracellular traps capture and kill Candida albicans yeast and hyphal forms, *Cell Microbiol* **8,** 668–76.
52. Urban, C.F., Ermert, D., Schmid, M. et al. (2009) Neutrophil extracellular traps contain calprotectin, a cytosolic protein complex involved in host defense against *Candida albicans. PLoS Pathog* **5:** e1000639.
53. Fuchs, T.A., Abed, U., Goosmann, C. et al. (2007) Novel cell death program leads to neutrophil extracellular traps, *J Cell Biol* **176,** 231–41.
54. Beiter, K., Wartha, F., Albiger, B. et al. (2006) An endonuclease allows *Streptococcus pneumoniae* to escape from neutrophil extracellular traps, *Curr Biol* **16,** 401–7.
55. Marcos, V., Zhou, Z., Yildirim, A.O. et al. (2010) CXCR2 mediates NADPH oxidase-independent neutrophil extracellular trap formation in cystic fibrosis airway inflammation, *Nat Med* **16,** 1018–23.
56. Margraf, S., Logters, T., Reipen, J. et al. (2008) Neutrophil-derived circulating free DNA (cf-DNA/NETs): a potential prognostic marker for posttraumatic development of inflammatory second hit and sepsis, *Shock* **30,** 352–8.
57. Logters, T., Paunel-Gorgulu, A., Zilkens, C. et al. (2009) Diagnostic accuracy of neutrophil-derived circulating free DNA (cf-DNA/NETs) for septic arthritis, *J Orthop Res* **27,** 1401–7.
58. Gilligan, H.M., Bredy, B., Brady, H.R. et al. (1996) Antineutrophil cytoplasmic autoantibodies interact with primary granule constituents on the surface of apoptotic neutrophils in the absence of neutrophil priming, *J Exp Med* **184,** 2231–41.
59. Chen, M., Kallenberg, C.G. (2010) ANCA-associated vasculitides-advances in pathogenesis and treatment, *Nat Rev Rheumatol* **6,** 653–4.
60. Yousefi, S., Mihalache, C., Kozlowski, E. et al. (2009) Viable neutrophils release mitochondrial DNA to form neutrophil extracellular traps, *Cell Death Differ* **16,** 1438–44.
61. Zhang, Q., Raoof, M., Chen, Y. et al. (2010) Circulating mitochondrial DAMPs cause inflammatory responses to injury, *Nature* **464,** 104–7.
62. Maugeri, N., Rovere-Querini, P., Evangelista, V. et al. (2009) Neutrophils phagocytose activated platelets in vivo: a phosphatidylserine, P-selectin, and {beta}2 integrin-dependent cell clearance program, *Blood* **113,** 5254–65.
63. Mathias, J.R., Perrin, B.J., Liu, T.X. et al. (2006) Resolution of inflammation by retrograde chemotaxis of neutrophils in transgenic zebrafish, *J Leukoc Biol* **80,** 1281–8.
64. Furze, R.C., Rankin, S.M. (2008) The role of the bone marrow in neutrophil clearance under homeostatic conditions in the mouse, *FASEB J* **22,** 3111–9.
65. Chtanova, T., Schaeffer, M., Han, S.J. et al. (2008) Dynamics of neutrophil migration in lymph nodes during infection, *Immunity* **29,** 487–96.
66. Beauvillain, C., Delneste, Y., Scotet, M. et al. (2007) Neutrophils efficiently cross-prime naive T cells in vivo, *Blood* **110,** 2965–73.
67. Tomihara, K., Guo, M., Shin, T. et al. (2010) Antigen-specific immunity and cross-priming by epithelial ovarian carcinoma-induced CD11b(+)Gr-1(+) cells, *J Immunol* **184,** 6151–60.
68. Croker, B.A., Lewis, R.S., Babon, J.J. et al. (2010) Neutrophils require SHP1 to regulate IL-1{beta} production and prevent inflammatory skin disease, *J Immunol* **186,** 1131–9.
69. Fridlender, Z.G., Sun, J., Kim, S. et al. (2009) Polarization of tumor-associated neutrophil phenotype by TGF-beta: "N1" versus "N2" TAN, *Cancer Cell* **16,** 183–94.

70. De Santo, C., Arscott, R., Booth, S. et al. (2010) Invariant NKT cells modulate the suppressive activity of IL-10-secreting neutrophils differentiated with serum amyloid A, *Nat Immunol* **11,** 1039–46.
71. Pillay, J., den Braber, I., Vrisekoop, N. et al. (2010) In vivo labeling with 2H2O reveals a human neutrophil lifespan of 5.4 days, *Blood* **116,** 625–7.
72. Paunel-Gorgulu, A., Zornig, M., Logters, T. et al. (2009) Mcl-1-mediated impairment of the intrinsic apoptosis pathway in circulating neutrophils from critically ill patients can be overcome by Fas stimulation, *J Immunol* **183,** 6198–206.
73. Hogg, J.C., Chu, F., Utokaparch, S. et al. (2004) The nature of small-airway obstruction in chronic obstructive pulmonary disease, *N Engl J Med* **350,** 2645–53.
74. Harper, L., Cockwell, P., Adu, D. et al. (2001) Neutrophil priming and apoptosis in anti-neutrophil cytoplasmic autoantibody-associated vasculitis, *Kidney Int* **59,** 1729–38.
75. Matute-Bello, G., Martin, T.R. (2003) Science review: apoptosis in acute lung injury, *Crit Care* 7, 355–8.
76. Elbim, C., Katsikis, P.D., Estaquier, J. (2009) Neutrophil apoptosis during viral infections, *Open Virol J* **3,** 52–9.
77. Colamussi, M.L., White, M.R., Crouch, E. et al. (1999) Influenza A virus accelerates neutrophil apoptosis and markedly potentiates apoptotic effects of bacteria, *Blood* **93,** 2395–403.
78. Engelich, G., White, M., Hartshorn, K.L. (2001) Neutrophil survival is markedly reduced by incubation with influenza virus and *Streptococcus pneumoniae*: role of respiratory burst, *J Leukoc Biol* **69,** 50–6.
79. Saez-Lopez, C., Ngambe-Tourere, E., Rosenzwajg, M. et al. (2005) Immediate-early antigen expression and modulation of apoptosis after in vitro infection of polymorphonuclear leukocytes by human cytomegalovirus, *Microbes Infect* 7, 1139–49.
80. Engelich, G., White, M., Hartshorn, K.L. (2002) Role of the respiratory burst in co-operative reduction in neutrophil survival by influenza A virus and *Escherichia coli*, *J Med Microbiol* **51,** 484–90.
81. Lindemans, C.A., Coffer, P.J., Schellens, I.M. et al. (2006) Respiratory syncytial virus inhibits granulocyte apoptosis through a phosphatidylinositol 3-kinase and NF-kappaB-dependent mechanism, *J Immunol* **176,** 5529–37.
82. Dibbert, B., Weber, M., Nikolaizik, W.H. et al. (1999) Cytokine-mediated Bax deficiency and consequent delayed neutrophil apoptosis: a general mechanism to accumulate effector cells in inflammation, *Proc Natl Acad Sci USA* **96,** 13330–5.
83. Hamasaki, A., Sendo, F., Nakayama, K. et al. (1998) Accelerated neutrophil apoptosis in mice lacking A1-a, a subtype of the bcl-2-related A1 gene, *J Exp Med* **188,** 1985–92.
84. Steimer, D.A., Boyd, K., Takeuchi, O. et al. (2009) Selective roles for antiapoptotic MCL-1 during granulocyte development and macrophage effector function, *Blood* **113,** 2805–15.
85. Dzhagalov, I., St John, A., He, Y.W. (2007) The antiapoptotic protein Mcl-1 is essential for the survival of neutrophils but not macrophages, *Blood* **109,** 1620–6.
86. Villunger, A., Scott, C., Bouillet, P. et al. (2003) Essential role for the BH3-only protein Bim but redundant roles for Bax, Bcl-2, and Bcl-w in the control of granulocyte survival, *Blood* **101,** 2393–400.
87. Villunger, A., O'Reilly, L.A., Holler, N. et al. (2000) Fas ligand, Bcl-2, granulocyte colony-stimulating factor, and p38 mitogen-activated protein kinase: Regulators of distinct cell death and survival pathways in granulocytes, *J Exp Med* **192,** 647–58.
88. Liles, W.C., Kiener, P.A., Ledbetter, J.A. et al. (1996) Differential expression of Fas (CD95) and Fas ligand on normal human phagocytes: implications for the regulation of apoptosis in neutrophils, *J Exp Med* **184,** 429–40.
89. Strasser, A., Jost, P.J., Nagata, S. (2009) The many roles of FAS receptor signaling in the immune system, *Immunity* **30,** 180–92.

Chapter 7

Isolation of Human and Mouse Neutrophils Ex Vivo and In Vitro

Yan Hu

Abstract

Neutrophils are one of main cellular elements of innate immune system that act as the first line of host defense against invasion by microorganisms. Neutrophils phagocytose and kill microbes through production of toxins such as hydrogen peroxide, superoxide anion, and nitric oxide. Recent studies have demonstrated a new strategy—so-called neutrophil extracellular traps (NETs) that are able to kill bacteria and fungi in vivo and in vitro (Brinkmann et al., Science 303:1532–1535, 2004; Wartha and Henriques-Normark, Sci Signal 1:pe25, 2008). Neutrophils are, therefore, always a major focus of investigation for scientists all over the world. Isolation of neutrophils from either human beings or animals such as mice is a very common first step for researchers to start their investigations on innate immunity to microbes. Fortunately, there are nowadays many methods available to obtain neutrophils from peripheral blood, bone marrow, and cell cultures. However, researchers still encounter technical problems in terms of purification, viability, and recovery. In this chapter, different approaches to the methodology of the isolation of neutrophils are described.

Key words: Neutrophils, Isolation, Gradient, Centrifugation, Percoll

1. Introduction

Neutrophils are generally referred to as either polymorphonuclear neutrophils (PMNs) or neutrophil granulocytes and the most abundant type of white blood cells in mammals, and form an essential part of the innate immune system (3, 4). They are normally found in the blood stream. During the very beginning phase of invasion of microbes (particularly bacteria and fungi) causing inflammation, and some cancers (5, 6), neutrophils are one of the first-responders of inflammatory cells to migrate toward the site of inflammation through the blood vessels, attracting by chemoattractant such as

Robert B. Ashman (ed.), *Leucocytes: Methods and Protocols*, Methods in Molecular Biology, vol. 844, DOI 10.1007/978-1-61779-527-5_7, © Springer Science+Business Media, LLC 2012

Interleukin-8(IL-8) and C5a in a process called chemotaxis as a hallmark of acute inflammation.

Neutrophils have an average diameter of 12–15 μm and a 12-h life span in circulation in nonactivated condition. After activated and migrating into tissues, they are able to survive for 1–2 days. After digestion of pathogens (bacteria, fungi, and viruses), they often are phagocytosed by macrophages. Neutrophils have three strategies to attack invaded microorganisms: phagocytosis, release of soluble antimicrobials (including granule proteins) and generation of neutrophil extracellular traps (NETs) (1, 7).

As the importance for a host against microbe infection, the studies on neutrophils such as mechanisms of NETs generation are always attractive to investigators today (2). However, how to obtain the pure population of neutrophils is still a key issue for investigators although there are many methods available nowadays. In this chapter, the different processes to isolate neutrophils from human and mouse will be described in details in order to present broad choice for investigators.

2. Materials

1. 70% Alcohol.
2. 1–10-mL Syringes.
3. 15–50-mL Centrifuge tubes.
4. 18–26-Gauge needles.
5. Bovine serum albumin (BSA).
6. Cell counter.
7. Dextran.
8. Ethylenediaminetetraacetic acid (EDTA).
9. Fetal bovine serum (FBS).
10. Fetal calf serum (FCS).
11. Fine scissors.
12. Forceps.
13. Hank's balanced salt solution (HBSS).
14. HBSS free of Ca^{2+} or Mg^{2+}(HBSS-CMF).
15. 4-(2-Hydroxyethyl)-1-piperazineethanesulfonic acid (HEPES).
16. Krebs–Ginger phosphate buffer plus glucose (KRPG).
17. 0.9% NaCl.
18. Pasteur pipette.
19. PBS.
20. Petri dishes.

21. RPMI-1640.
22. Temperature adjustable centrifuge.
23. Acid citrate dextrose formula A (ACD).
24. Bone marrow culture medium (BMM): For 1 L:

Glutamine	0.292 g
Pyruvic acid	0.11 g
HEPES	2.383 g
$NaHCO_3$	2 g
Monothioglycerol	10 μL
Hydrocortisone 21 hemisuccinate	0.462 M
FCS (fetal calf serum)	150 mL

Add RPMI-1640 w/o l-glutamine and $NaHCO_3$ to 1 L.

25. Mouse-osmolality phosphate-buffered saline (MPBS):

Na_2HPO_4	0.92 g
KH_2PO_4	0.2 g
KCl	0.2 g
NaCl	9.83 g

Dissolved in 1 L distilled water and filtered with 0.2-μm filter

26. BS separation column (Miltenyi Biotech, Bisley, UK).
27. Cell preparation tube (CPT) from BD Biosciences.
28. Goat anti-rat IgG microbeads (Bisley, UK).
29. Mouse-osmolality phosphate-buffered saline (MPBS).
30. Percoll.
31. Polymorphprep™.
32. Rat anti-mouse antibodies to CD2, CD5, and CD45R.
33. Rat anti-mouse F4/80 antigen (CI:A3-1), anti-ICAM-1.
34. Erythrocytes lysis solution (EL).
    (a) 0.83 g NH_4Cl dissolved in 100 mL distilled water.
    (b) 0.2 g Tris base dissolved in 10 mL distilled water.
    Mix 90 mL A and 10 mL B, adjust pH to 7.2 and filter with 0.22-μm filter.
35. Separation buffers (Miltenyi Biotech, Bisley, UK).
36. Thioglycollate: Thioglycollate broth is used to recruit macrophages to the peritoneal cavity of mice when injected intraperitoneally. It recruits numerous macrophages, but does not activate them.

37. Uric acid solution.
38. VarioMACS magnet (Miltenyi Biotech, Bisley, UK).

3. Methods

The descriptions of popular methods for isolation of human and mouse neutrophils according to their tissue of origin, are based on the original published techniques (8–14). In our experience, the method described in Subheading 3.1 is satisfactory for isolation of human peripheral neutrophils, and those in Subheadings 3.7 and 3.8 are best for animal experimentation (see Note 1). Elutriation (17) is also an efficient procedure, in laboratories that have access to the instrument.

3.1. One-Step Isolation of Human Neutrophils from Peripheral Blood with Polymorphprep™

1. For the best results, use whole blood treated with an anticoagulant such as EDTA (see Note 2). The blood should be used within 2 h of drawing from the donor.
2. Carefully layer (see Note 3) 5.0 mL of anticoagulated whole blood over 5.0 mL of Polymorphprep™ (see Note 4) in a 15-mL centrifuge tube. Take care to avoid mixing of the blood with the separation fluid.
3. Centrifuge the samples layered over Polymorphprep™ at $450 \times g$ for 35 min in a swing-out rotor at 20°C without brake (see Note 5).
4. After centrifugation, two leucocytes bands should be visible. The top band at the sample interface will consist of mononuclear cells and the lower band of polymorphonuclear cells; the erythrocytes are pelleted (shown in Fig. 1).
5. Harvest the lower band of PMNs into new centrifuge tube with Pasteur pipette, but avoiding mixing bands.
6. Wash cells by filling the centrifuge tube with PBS and centrifuge at $400 \times g$ for 10 min.
7. Remove supernatant and resuspend the cells in 1 mL of PBS.
8. Count the cells using a hemocytometer, and dilute for use.
9. Purity of the cell preparation can be determined by flow cytometry, viability by trypan blue exclusion, and morphology by staining of smears.

3.2. A Density Gradient Method Using Percoll for Isolation of Neutrophils from Human Peripheral Blood

1. Prepare a Percoll density gradient in a 50-mL tube as follows: 15 mL of 57% (v/v in PBS) Percoll (density 1.075 g/mL) are carefully layered over 15 mL of 67% (v/v in PBS) Percoll (density 1.088 g/mL) (see Note 6).
2. 15 mL Peripheral venous human blood is diluted with equal volume of PBS.

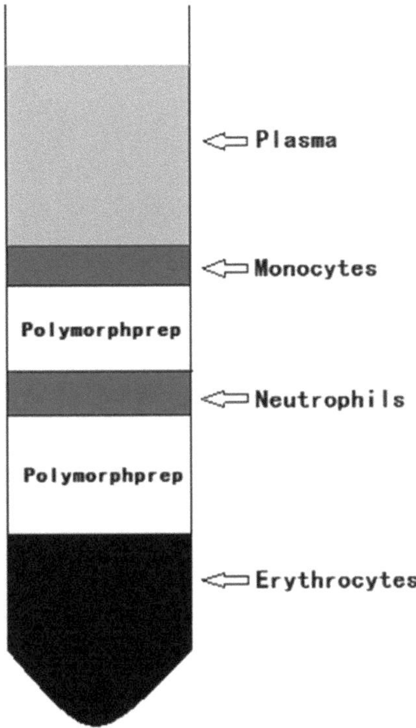

Fig. 1. One-step isolation of neutrophils from human peripheral blood.

3. 15 mL Diluted blood is layered over the gradient and centrifuged for 60 min at 400×*g* at room temperature (see Note 7).
4. Collect the cells carefully with Pasteur pipette from the interface at 57%/67% Percoll as shown in Fig. 2.
5. Cells are washed twice with PBS, and resuspended in PBS for use.

3.3. Using a Cell Preparation Tube to Isolate Neutrophils from Human Peripheral Blood (see Note 8)

1. 8 mL Peripheral blood is drawn into CPTs containing sodium heparin or sodium citrate.
2. Tubes are centrifuged for 25 min at 1,700×*g* at room temperature.
3. The cells will settle on different layers as shown in Fig. 3.
4. Discard the PBMC and plasma above the gel lock.
5. Wash the upper portion of the gel twice with ice-cold PBS.
6. Collect the erythrocyte/neutrophil mixture with a 10-mL syringe attached with an 18-gauge 1.5-in. needle by pierced through the gel lock.
7. Remove the needle and collect the cells into 50-mL tube.
8. Wash cells once with ice-cold PBS containing 2% FBS by centrifugation for 10 min at 400×*g* at 4°C.

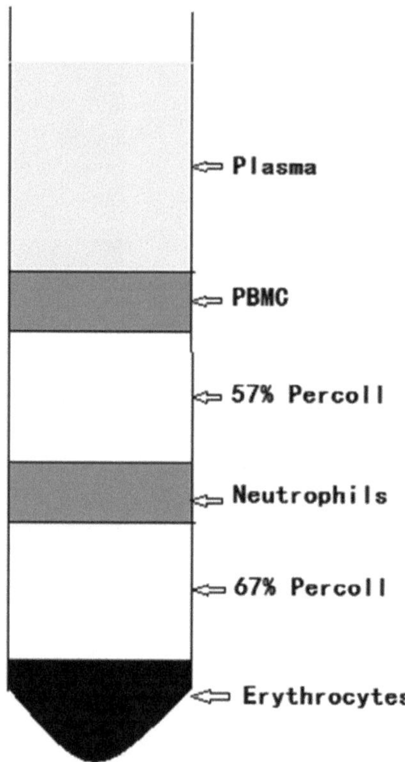

Fig. 2. Differential layering of cell populations after density gradient centrifugation for isolation of neutrophils from human peripheral blood.

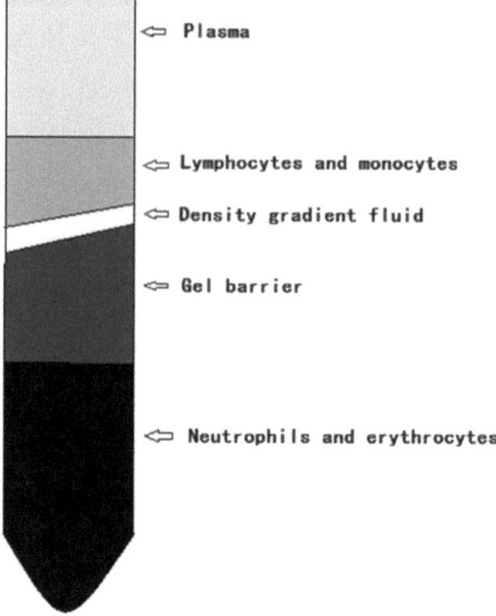

Fig. 3. Separation of cell populations from human peripheral blood using cell preparation tubes (CPT).

9. Add 10 mL EL buffer (see Note 9) for 10 min at room temperature. Mix by vortexing the tubes.
10. Wash with ice-cold PBS containing 2% FBS by centrifugation for 10 min at 400 ×*g* at 4°C.
11. Resuspend the cells in PBS for use.

3.4. Isolation of Neutrophils from Human Bone Marrow

1. Prepare Percoll gradient solutions with densities of 1.10, 1.095, and 1.085 g/mL. The bottom layer consists of 4 mL of the 1.10 g/mL Percoll solution. Two additional layers of 3 mL each with a density of 1.095 and 1.085 g/mL are then carefully applied. All solutions are allowed to come to room temperature prior to centrifugation.
2. 10 mL of bone marrow (BM) is aspirated from the posterior iliac crest into a syringe containing 1.7 mL of ACD.
3. BM is diluted to 20 mL with 0.9% NaCl.
4. The cell suspension is forced through needles with decreasing internal diameter (19, 20, and 21 gauge) to disrupt any remaining fragments.
5. 5 mL of 6% Dextran 70 in 0.9% NaCl is added to allow erythrocyte sedimentation.
6. After 60 min at 25°C, the leukocyte-rich supernatant (LRS) is removed and centrifuged at 400 ×*g* at 4°C.
7. The supernatant is discarded, and the pellet is resuspended in 8–10 mL of 0.9% NaCl.
8. 4–5 mL of the leukocyte suspension, containing 5×10^7 nucleated cells/mL, obtained from BM, is layered over the temperature-equilibrated discontinuous Percoll gradient and centrifuged at 700 ×*g* at 25°C for 30 min.
9. Three leukocyte bands will appear as shown in Fig. 4.
10. The second and third layers (L_2 and L_3) are harvested with a transfer pipette and washed in 0.9% NaCl (see Note 10).
11. Erythrocytes are lysed by adding 2–5 mL distilled water for 20 s at 25°C. Osmolarity is restored by adding an equal volume of 3.5% NaCl.
12. The cells are then washed twice with KRPG, and resuspended in PBS for use.

3.5. Isolation of Neutrophils from Human Saliva

Subjects place 15 mL of HBSS-CMF, at pH 7.4 containing 0.1% gelatin at 24°C in their mouths, swish the solution for 30 s, and expectorate into a polypropylene vessel containing 400 mL 4°C HBSS-CMF. This sequence is repeated without interruption for 20 min (about 40 cycles).

1. The collected solution is stirred (10 min) and centrifuged (at 250 ×*g* for 5 min), and the resuspended pellet is passed sequentially

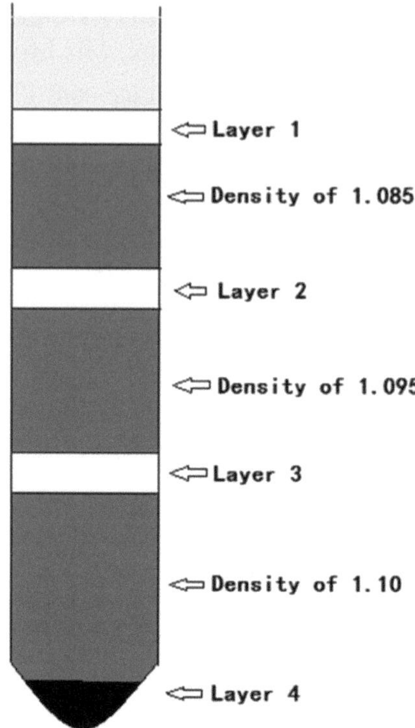

Fig. 4. Layering of cells after centrifugation on Percoll density gradients.

through a 20-μm nylon mesh and a 10-μm nylon mesh to allow separation of neutrophils from oral epithelial cells and debris.

2. The effluent containing the neutrophils is centrifuged at 250 × *g* for 5 min, and the pellet resuspended in HBSS containing Ca^{2+} and Mg^{2+} without gelatin for use.

3.6. Using Negative Immunomagnetic Separation to Isolate Murine Blood Neutrophils

1. Collect mouse blood (1 mL) by heart puncture using a heparinized syringe and transfer to dextran (3 mL, 1.25% w/v in saline).
2. Tubes are then filled to a total of 10 mL with dextran solution and inverted for 30 min at room temperature to allow erythrocytes sedimentation.
3. Transfer leukocyte-rich supernatant with Pasteur pipette to new tubes.
4. Wash cells with cold PBS without cations containing 0.5% w/v low-endotoxin BSA and resuspend in 1 mL PBS. Perform total and differential leukocyte counts.
5. Add antibodies as follows: anti-CD2 (1.5 μg/10^6 lymphocytes), anti-CD5 (2 μg/10^6 lymphocytes), anti-CD45R (10 μg/10^6 lymphocytes), anti-F4/80 antigen (2 μg/10^6 monocytes), and

anti-ICAM-1 (0.6 μg/10^6 leukocytes), and incubate at 4°C for 30 min (see Note 11).

6. Add 8 mL of buffer (provided with antibodies) and centrifuge at 300 × *g*, 4°C for 6 min to remove excess antibodies.
7. Resuspend the cells in PBS (80 μL) and incubate with goat anti-rat IgG MicroBeads (20 μL/10^7 cells) at 4°C for 15 min.
8. A chilled BS separation column is connected to a VarioMACS magnet and prepared with cold sterile water and buffer according to manufacturer's instructions.
9. The leukocyte–microbead mixture is then added to the column.
10. Neutrophils flow into a collecting tube, whereas the unwanted cells, previously labeled with magnetic beads, are retained within the metallic matrix of the column.
11. Neutrophils are then centrifuged for 6 min at 300 × *g* at 4°C.
12. The supernatant is discarded.
13. Residual erythrocytes are removed by adding 7 mL of 0.2% NaCl solution, gently inverting ten times, followed by hypertonic rescue of neutrophils with an equal volume of 1.6% NaCl solution supplemented with 0.1% glucose, and inverting once.
14. Finally, neutrophils are washed off erythrocyte debris and resuspended in PBS for use.

3.7. Isolation of Neutrophils from Mouse Bone Marrow

1. Dilute Percoll with 10× PBS at 9:1 (Percoll:PBS). Prepare gradients with diluted Percoll and 1× PBS (v/v) as 55, 65, and 80% solutions (see Note 12).
2. Prepare a Percoll gradient in a 15-mL tube by placing 4 mL 80% of Percoll at the bottom and then carefully layer 3 mL of 65% and 3 mL of 55% on top using a Pasteur pipette.
3. Euthanize the required number of mice and remove the hind limbs into a Petri dish containing RPMI.
4. After stripping off soft tissue attachments, and severing the distal tips of each extremity, bone marrow cells are flushed from femurs and tibias into RPMI in another Petri dish, using a 1-mL syringe with a 27-gauge needle.
5. Disperse cells in Petri dish by repeatedly forcing through syringe.
6. The cells are centrifuged at 200 × *g* at room temperature for 10 min, resuspended in 1–2 mL RPMI, and then layered on Percoll gradient for centrifugation at 500 × *g* and 4°C for 30 min.
7. The neutrophils (third layer) are harvested with Pasteur pipettes, washed twice with PBS, resuspended in RPMI, and counted for use.

3.8. Isolation of Neutrophils from Cell Culture

Bone marrow cells are flushed from femurs and tibias (see Note 10) using RPMI-1640 medium, centrifuged at $200 \times g$ for 10 min and resuspended in bone marrow culture medium (BMM).

1. 3×10^7 cells are inoculated into a 75-cm^2 cell culture flask in 20 mL BMM, incubated in 5% CO_2, at 37°C with 95% humidity for 3 weeks, replacing half of the medium with fresh medium every week.
2. On week 4, the culture medium is removed and an equal volume of freshly isolated bone marrow cells added into the flasks, after which they are cultured for a further week.
3. The total supernatant is collected on week 5 for isolation of neutrophils and monocytes. The cells are harvested weekly.
4. Nonadherent cells are harvested from the cultures and resuspended in 3 mL 1× HBSS with 10 mM HEPES.
5. Percoll density solutions are prepared by mixing 9 volumes Percoll with 1 volume 10× Hanks' balanced salt solution without Ca^{2+} and Mg^{2+}.
6. This solution is diluted with 1× HBSS (pH 7.2) with 10 mM HEPES to the following working concentrations: 81% (1.1002 g/mL), 65% (1.0812 g/mL), and 55% (1.0693 g/mL).
7. The cells are layered onto the gradients and centrifuged at 500 g for 40 min at 10°C.
8. The cells in the third layer (neutrophils) are harvested with Pasteur pipettes. Others are predominantly monocytes.
9. The cells are washed twice with mouse-osmolality phosphate-buffered saline (MPBS) and suspended in RPMI-1640 for use (see Note 13).

3.9. Isolation of Neutrophils from Peritoneal Exudate

1. A uric acid solution (noncrystalline form) is prepared by mixing with saline (10% wt/vol) and sonicated for 10 min. Immediately before inoculation, the milky white precipitated uric acid solution should be shaken vigorously.
2. Peritoneal exudate cells are harvested by two lavages of the peritoneal cavity with 5 mL of cold PBS, 4, 18, or 24 h after injection of 1 mL of 3% thioglycollate broth or 0.1 mL of uric acid solution into the peritoneal cavity of the mouse (see Note 14).
3. The peritoneal cells are then washed by centrifugation at $200 \times g$ for 10 min at 4°C, and hypotonic lysis performed to eliminate red blood cells, regardless the presence or absence of a visible red cell pellet (see Note 15).
4. After centrifugation and an additional wash, the cells are resuspended in KPRG, and counted for use.

4. Notes

1. Neutrophil function may differ depending on the origin of the cells. For example peritoneal exudate cells demonstrate a stronger capacity to adhere to glass slides, higher ROS production and responsiveness to chemotactic peptides than those from bone marrow (13), which may result from priming by TNF-α or complement factors induced by the eliciting agents themselves.
2. EDTA (K salt) will provide the best result; citrate is acceptable as an anticoagulant, but is inferior to EDTA, and heparin should be avoided.
3. The Polymorphprep™ may alternatively be layered underneath the blood.
4. Polymorphprep™ only works optimally with whole, undiluted blood, collected from normal healthy volunteers and used within 2 h of drawing.
5. The efficacy of Polymorphprep™ relies on the loss of water from the erythrocytes to the hyperosmotic medium. This only happens effectively at approximately 20°C. Make sure that when the refrigeration cuts in during the centrifugation, the temperature does not drop below 17°C.
6. Percoll density gradient preparation varies between laboratories. However, a simple method to prepare Percoll is as follows.

   Dilute Percoll with 10× PBS at 9:1 (Percoll:PBS) first, and then make required densities with diluted Percoll and 1× PBS:

Percoll (v/v%)	70	60	50	40	30	20
Density (g/mL)	1.090	1.077	1.067	1.056	1.043	1.031

   Always layer the gradient solutions from bottom of tube starting with the most dense solution, followed by the second and third dense, and finally carefully place the sample on top. Always make fresh gradient density Percoll within 1 week of experiment.
7. Always use a swinging bucket centrifuge, and select brake OFF when doing gradient density centrifugation, to avoid vibration, disruption of the bands of cells, and swirling of the loosely packed erythrocytes.
8. This method for isolation of neutrophils from human peripheral blood is both simple and rapid, but the cost of CPTs may make it too expensive for routine laboratory use.
9. EL buffer may alter the functions of isolated neutrophils.

10. L2 and L3 will contain 90 and 91% of mature neutrophils and accounts for 70 and 30% of total isolated PMNs, respectively.
11. Anti-CD2, anti-CD5, and anti-CD45R are chosen specifically to label lymphocytes, anti-F4/80 antigen to label monocytes, and anti-ICAM-1 as a pan-lymphocyte/monocyte antibody.
12. Percoll gradient can also be prepared in 15-mL polystyrene tubes as follows: 1.095, 1.085, and 1.070 g/mL Percoll solutions.
13. Normally, two mice are enough to set up the cell culture, which can be harvested once a week for 4–6 weeks. However, the production of neutrophils varies between mouse strains. For example, bone marrow cells from BALB/c mice generate larger numbers of neutrophils than those from CBA/CaH mice (14).
14. The maximum number of neutrophils obtainable 4 h after stimulation is approximately $1–2 \times 10^7$ cells/mouse, with a purity of 88% after induction by thioglycollate, and 95% by uric acid solution. Be aware that the elicited neutrophils will have been activated by the inducing agents.
15. This method avoids nonspecific activation that can result from the use of eliciting agents in vivo. Furthermore, red cell lysis is unnecessary, which avoids possible activation by contact with hypertonic solutions or lysis buffer, which may cause upregulation of CD11b and/or the shedding of CD62L (15, 16).

References

1. Brinkmann, V., Reichard, U., Goosmann, C., et al. (2004) Neutrophil extracellular traps kill bacteria, *Science* **303**, 1532–5.
2. Wartha, F., Henriques-Normark, B. (2008) ETosis: a novel cell death pathway, *Sci Signal* **1**, pe25.
3. Witko-Sarsat, V., Rieu, P., Descamps-Latscha, B., et al. (2000) Neutrophils: molecules, functions and pathophysiological aspects, *Lab Invest* **80**, 617–53.
4. Nathan, C. (2006) Neutrophils and immunity: challenges and opportunities, *Nat Rev Immunol* **6**, 173–82.
5. Waugh, D.J., Wilson, C. (2008) The interleukin-8 pathway in cancer, *Clin Cancer Res* **14**, 6735–41.
6. De Larco, J.E., Wuertz, B.R., Furcht, L.T. (2004) The potential role of neutrophils in promoting the metastatic phenotype of tumors releasing interleukin-8, *Clin Cancer Res* **10**, 4895–900.
7. Hickey, M.J., Kubes, P. (2009) Intravascular immunity: the host-pathogen encounter in blood vessels, *Nat Rev Immunol* **9**, 364–75.
8. Lichtenberger, C., Zakeri, S., Baier, K., et al. (1999) A novel high-purity isolation method for human peripheral blood neutrophils permitting polymerase chain reaction-based mRNA studies, *J Immunol Methods* **227**, 75–84.
9. De, A.K., Roach, S. E., De, M., et al. (2005) Development of a simple method for rapid isolation of polymorphonuclear leukocytes from human blood, *J Immunoassay Immunochem* **26**, 35–42.
10. Berkow, R.L., Dodson, R.W. (1986) Purification and functional evaluation of mature neutrophils from human bone marrow, *Blood* **68**, 853–60.
11. Ashkenazi, M., Dennison, D.K. (1989) A new method for isolation of salivary neutrophils and determination of their functional activity, *J Dent Res* **68**, 1256–61.
12. Cotter, M.J., Norman, K. E., Hellewell, P. G., et al. (2001) A novel method for isolation of neutrophils from murine blood using negative immunomagnetic separation, *Am J Pathol* **159**, 473–81.
13. Itou, T., Collins, L. V., Thoren, F. B., et al. (2006) Changes in activation states of murine

polymorphonuclear leukocytes (PMN) during inflammation: a comparison of bone marrow and peritoneal exudate PMN, *Clin Vaccine Immunol* **13**, 575–83.

14. Hu, Y., C.S. Farah, and Ashman, R.B. (2006) Effector function of leucocytes from susceptible and resistant mice against distinct isolates of *Candida albicans*, *Immunol Cell Biol* **84**, 455–60.
15. Macey, M.G., McCarthy, D. A., Vordermeier, S., et al. (1995) Effects of cell purification methods on CD11b and L-selectin expression as well as the adherence and activation of leucocytes, *J Immunol Methods* **181**, 211–9.
16. Youssef, P.P., Mantzioris, B. X., Roberts-Thomson, P. J., et al. (1995) Effects of ex vivo manipulation on the expression of cell adhesion molecules on neutrophils, *J Immunol Methods* **186**, 217–24.
17. Dodek, P.M., Ohgami, M., Minshall, D. K., et al. (1991) One-step isolation of neutrophils using an elutriator, *In Vitro Cell Dev Biol* **27A** (3 Pt 1), 211–4.

Chapter 8

Measurement of Oxidative Burst in Neutrophils

Yu Chen and Wolfgang G. Junger

Abstract

Polymorphonuclear neutrophils (PMNs) generate reactive oxygen species (ROS) during phagocytosis and in response to soluble agonists. This functional response, termed oxidative burst, contributes to host defense, but it can also result in collateral damage of host tissues. To study this important PMN response, different methods have been developed that are based on the assessment of oxidative burst by measuring intracellular ROS production or formation of ROS in the extracellular space. Among the different methods that were developed, the following two are particularly widely used because of their convenience and accuracy. The first method depends on the reduction of cytochrome c, which can be assessed by photometry, while the second method relies on changes in the fluorescence properties of dihydrorhodamine 123, which can be assessed by flow cytometry.

Key words: Oxidative burst, Polymorphonuclear neutrophils, fMLP, NADPH oxidase activity, Cytochrome c reduction, DHR oxidation

1. Introduction

Effective polymorphonuclear neutrophil (PMN) activation is important for a successful host defense. Oxidative burst, cell migration, and degranulation are some of the key functional responses that enable PMNs to accomplish their tasks in host defense. These functional responses can be triggered by receptors that recognize bacterial peptides, such as *N-formyl-Met-Leu-Phe* (*f*MLP) or inflammatory mediators, such as C5a and IL-8 (1–3). Many of these receptors, such as formyl peptide receptors (FPRs) are G protein-coupled receptors (GPCRs) that induce rapid downstream signaling responses leading to the assembly of nicotinamide adenine dinucleotide phosphate (NADPH) oxidase. This multiprotein oxidase complex is dormant in quiescent cells, where its components are segregated into the cytosolic and membrane compartments. In response to stimulation, these components rapidly assemble at cell

Robert B. Ashman (ed.), *Leucocytes: Methods and Protocols*, Methods in Molecular Biology, vol. 844,
DOI 10.1007/978-1-61779-527-5_8, © Springer Science+Business Media, LLC 2012

membranes and the enzyme becomes activated, allowing it to catalyze NADPH-dependent reduction of O_2 to form superoxide anions ($O_2.^-$) and reactive oxygen species (ROS) derived from this radical, including hydrogen peroxide (H_2O_2), hydroxyl radical (OH.), and hypochlorous acid (HClO) (4). This process termed oxidative burst is also referred to as respiratory burst, which plays an important role in innate immunity against invading microorganisms. There are several inherited disorders or deficiencies in oxidative burst. For example, chronic granulomatous disease (CGD) is caused by X-linked or autosomal-recessive inheritance and results in an inability of neutrophils to properly assemble the NADPH oxidase complex and mount appropriate oxidative burst (5). Although ROS production is critical for the killing and degradation of internalized bacteria and particles, it can also contribute to inflammatory damage of host tissue (6). In order to study the role of oxidative burst in host protection, it is important to develop efficient, simple, and highly reproducible techniques to quantify ROS generation by PMN. For instance, assessing ROS production is used to evaluate PMN function and to diagnose CGD (5). We have been using several different assays to evaluate the role of PMN and ROS in the immune response to severe trauma, shock, and sepsis (7). Below, we describe different methods we have optimized to measure oxidative burst of isolated PMN and PMN suspended in heparinized whole blood.

2. Materials

2.1. Human PMN Isolation (see Note 1)

1. Vacutainer plasma tubes with spray-coated sodium heparin.
2. Normal saline (IVNS), 0.9% sodium chloride solution, sterile, injection grade.
3. Hank's balanced salt solution (HBSS) with calcium and magnesium.
4. 20% Dextran to make 5% dextran solution, dilute 1:4 with HBSS.
5. Percoll (GE Heathcare, New York).
6. Sodium chloride (NaCl), tissue culture grade heat treated to render pyrogen free: Add 2.2 g NaCl to previously heat-treated 250-ml glass bottle; cap with two layers of aluminum foil; heat bottle with salt at 450°C overnight to render pyrogen free.
7. Percoll (100%): Add 50 ml of Percoll to bottle containing 2.2 g heat-treated NaCl; gently swirl until salt is completely dissolved, and sonicate if necessary; add remaining 200 ml of Percoll and thoroughly mix. Cap with a sterile cap and store at 4°C until use.

8. Percoll-gradient components: To obtain 73% Percoll (bottom layer), add 36.5 ml Percoll (100%) to 13.5 ml IVNS. To obtain 55% Percoll (top layer), add 27.5 ml Percoll (100%) to 22.5 ml IVNS.
9. Beckman GPR centrifuge with GH-3.8 rotor (or similar); switch off cooling function to avoid repeated temperature fluctuations during centrifugation.

2.2. Measurement of Oxidative Burst Using Cytochrome c Assay

1. Cytochrome c from bovine heart (purity ≥95%): Dissolve cytochrome c in HBSS to obtain a final concentration of 2 mM; aliquot and store at −80°C until use.
2. fMLP (Sigma, purity ≥97% by HPLC): Dissolve fMLP in sterile tissue culture-grade dimethyl sulfoxide (DMSO) to a final concentration of 10 mM; aliquot and store at −80°C until use.
3. Use aliquots of this fMLP stock solution in DMSO to prepare 100 nM fMLP working solutions in HBSS just before use. Use polypropylene rather than polystyrene tubes to minimize loss of fMLP due to adsorption.
4. DMSO (purity ≥99.7%, sterile filtered).
5. Tissue culture plates (96-well).
6. Superoxide dismutase (SOD) powder; prepare solution by dissolving in HBSS at a final concentration of 5,000 U/ml, sterile filter, aliquot, and store at −80°C until use.

2.3. Measurement of Oxidative Burst Using Dihydrorhodamine 123

1. Dissolve dihydrorhodamine 123 (DHR) to a final concentration of 30 mM in DMSO; aliquot and store at −80°C until use.
2. FACS fixing solution: Flow cytometry sheath fluid (J&S Medical Association, Inc. Framingham, MA) containing 0.5% formaldehyde (prepared with 37% formaldehyde solution, w/w).

2.4. Measurement of Oxidative Burst in Human Whole Blood

BD FACS™ lysing solution, 10× concentrated: Dilute (1:10) with distilled water.

2.5. Measurement of Oxidative Burst in Mouse Whole Blood Using DHR

1. RBC lysis buffer.
2. Heparin sodium injection, USP (1,000 USP units/ml).
3. W-peptide (WKYMVM; Phoenix Pharmaceuticals).

3. Methods

Several methods are available to measure PMN oxidative burst (8). Some conventional tests, such as chemiluminescence (9) and the reduction of cytochrome c (10), require the isolation of PMN and therefore relatively large amounts of blood. These assays allow monitoring of the production and release of extracellular superoxide anions using photometric or luminometric equipment. More recently, flow cytometry has become widely available to many researchers and therefore assays suitable for this type of equipment have been developed to assess oxidative burst. The assays based on flow cytometry have the advantage that tens of thousands of cells can be assessed in a very short period of time using small volumes of whole blood or isolated PMN (11). Here, oxidation of specific probes, such as 2′, 7′-dichlorofluorescein diacetate (DCFH) or DHR, to fluorescent derivatives is used to detect superoxide formation in individual cells (Fig. 1).

Besides the methods mentioned above, some additional assays have been described to assess oxidative burst. For example, 3′, 3′-diaminobenzidine (DAB) oxidation and *p*-nitroblue tetrazolium

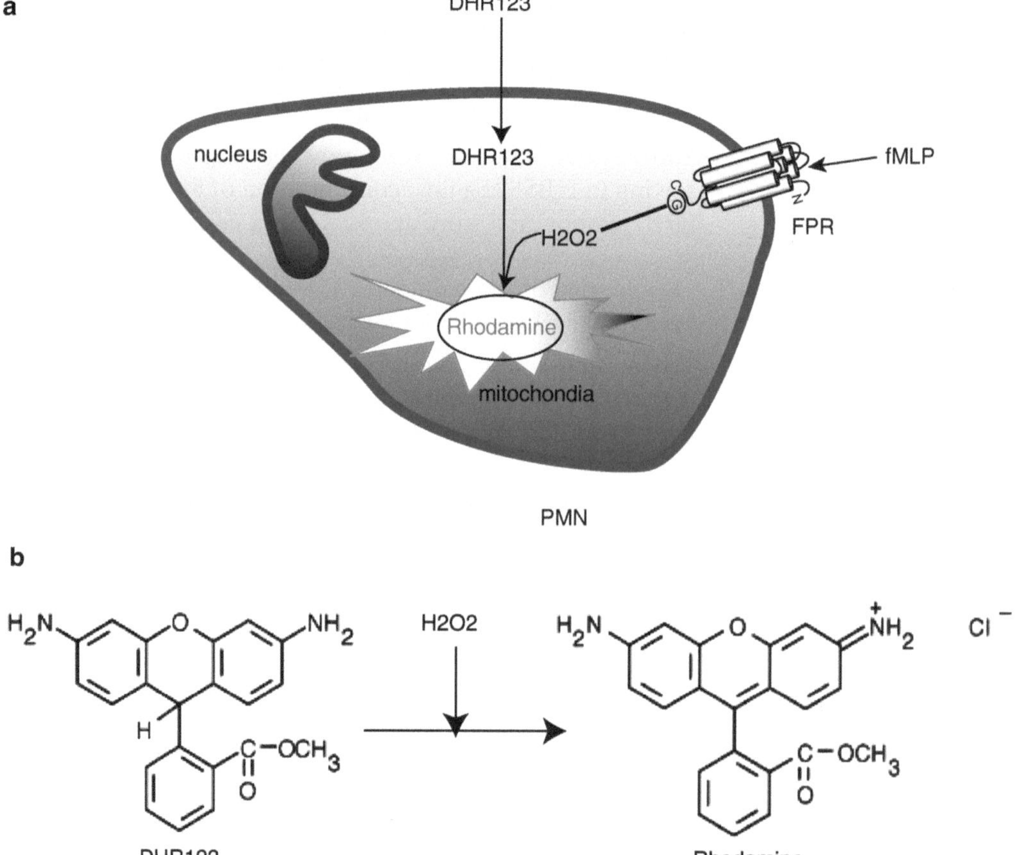

Fig. 1. Principle of assaying oxidative burst with DHR. (**a**) The freely permeable, nonfluorescent DHR 123 enters cells. Upon cell stimulation, DHR 123 is oxidized by hydrogen peroxide (H_2O_2) formed from superoxide, resulting in the formation of fluorescent dye localized in mitochondria (11). (**b**) Chemical structures of DHR 123 and the product rhodamine.

(NBT) reduction are two simple methods to measure intracellular oxygen radicals or superoxide anions through precipitation reactions (8, 12). However, these methods are comparatively cumbersome and therefore they are rarely used.

Because they are most reliable in our hands, we have extensively used and optimized the DHR method described above as well as the more traditional SOD-inhibitable reduction of cytochrome c to characterize oxidative burst activity in PMN (7, 13).

3.1. Human PMN Isolation

1. Draw blood using heparinized vacutainer plasma tubes (see Note 2).
2. Dextran sedimentation: Add 3 ml of 5% dextran per 10 ml of blood in sterile 50-ml centrifuge tube. Mix gently by inverting tubes several times. Let cells settle at room temperature for 30 min (see Note 3).
3. While the sedimentation is taking place, set up the Percoll centrifugation gradients in sterile 15-ml centrifuge tubes. You need one tube for each 10-ml aliquot of blood. Place 4 ml of 73% Percoll in a 15-ml tube and carefully layer 4 ml of 55% Percoll on top using a 5-ml serological pipette.
4. After dextran sedimentation of the blood, harvest the supernatant that contains plasma and white cells; place in fresh 50-ml centrifuge tubes, fill with HBSS, mix gently, and centrifuge at $1{,}500 \times g$ for 10 min to wash cells.
5. Remove supernatant and gently disrupt cell pellet (see Note 3). Then, add 3 ml of HBSS per 10-ml blood aliquot and carefully layer 3 ml of this cell suspension onto the Percoll gradient in the 15-ml tubes; then, spin at $400 \times g$ for 20 min at room temperature.
6. Remove top phase to just above the cell layer containing the PMN using a sterile, heat-treated Pasteur pipette and vacuum suction system. Use a serological pipette to remove the PMN layers and combine cell suspension in fresh 50-ml centrifuge tubes using no more than 10 ml of cell suspension in each 50-ml tube.
7. Fill tubes with HBSS and centrifuge at $1{,}500 \times g$ for 10 min to wash cells for the first time.
8. Remove supernatants and gently disrupt cell pellets (see Note 3). Fill 50-ml tubes with fresh HBSS and centrifuge at $220 \times g$ for 10 min for a second wash.
9. Gently disrupt pellet, add HBSS, resuspend and count cells, and then adjust cell concentration to 1×10^7/ml using fresh HBSS. Store cells at room temperature until use. Use cells as soon as possible.

3.2. Measurement of Oxidative Burst in Human PMN Using Cytochrome c in Tissue Culture Plates

1. Prepare two wells for each sample: one well for sample without and one well for sample with SOD (see Note 4).
2. PMNs (15 μl of 1.33×10^7/ml cell suspension) are added in each well of 96-well tissue culture plates prewarmed for 1 h at 37°C in a water bath.
3. Meanwhile, prepare 100 μM cytochrome c in HBSS with or without 100 nM fMLP, and prewarm to 37°C.
4. Add 5 μl SOD solution in SOD reference wells and 5 μl HBSS in sample wells. Add 80 μl prewarmed cytochrome c solution to each well and incubate at 37°C in a water bath.
5. After 10 min at 37°C, optical density changes are measured with a plate reader at a wavelength of 550 nm.
6. Calculate the relative amount of O_2^- generated using positive and negative controls.

3.3. Measurement of Oxidative Burst in Human PMN Using Cytochrome c in Cuvettes

1. Prepare two tubes (e.g., sterile 1.5-ml Eppendorf centrifuge tubes) for each sample: one for each sample and another one for each SOD reference (see Note 4).
2. PMNs (100 μl of 1×10^7/ml cell suspension) are added to each tube and HBSS is added to result in a final total volume of 1 ml per tube. The tubes are prewarmed at 37°C in a water bath for 30 min to 1 h.
3. Meanwhile, prepare 100 μM cytochrome c solution in HBSS with or without 100 nM fMLP.
4. Add 50 μl SOD solution in SOD reference tubes; add 100 μl cytochrome c solution to each tube.
5. After 10 min, stop reactions by placing tubes into an ice bath for at least 10 min. Centrifuge samples at $665\times g$ for 5 min at 4°C in an Eppendorf centrifuge and transfer supernatants to fresh Eppendorf tubes kept in an ice bath. Transfer supernatants into a cuvette with a 1-cm path length and measure optical density differences using a spectrophotometer at a wavelength of 550 nm.
6. Calculate the molar amount of O_2^- generated using the following formula (see Note 5):

$$\Delta OD_{550} = OD_{550(\text{sample})} - OD_{550(\text{SOD reference})},$$

$$\Delta OD_{550} \times 47.4 = \text{nmol } O_2^- / 10^6 \text{cells} / 10\,\text{min}.$$

3.4. Measurement of Oxidative Burst Using Dihydrorhodamine 123 and Flow Cytometry

1. Mix isolated PMN (100 μl of 1×10^7/ml cell suspension), stimuli, any other agents of interest, and HBSS in sterile 1.5-ml Eppendorf centrifuge tubes to achieve a final volume of 500 μl per tube. Prewarm tubes in water bath at 37°C for 1 h (see Note 6).

2. Open lids of tubes and add 2.8 μl DHR or HBSS as control, 25 μl fMLP or HBSS as control, onto the lid of each tube. Gently close the lids without disturbing the drops attached to the inside of the lids.
3. Using appropriate racks, convert all tubes at the same time to allow mixing of cell suspensions with DHR and stimuli added to the lids. Gently flick rack to ascertain that cell suspensions return to the bottom of each centrifuge tube, and then place tubes back into the water bath and incubate at 37°C for 20 min.
4. Place racks with tubes into an ice bath and incubate for 10 min to stop all reactions (see Note 7). Centrifuge tubes at 425 × *g* for 5 min at 4°C, remove supernatants, wash one more time with HBSS, resuspend cells in 300 μl FACS fixing solution, and keep on ice until analysis.
5. Measure samples in flow cytometer (FL1 channel with 488-nm laser) as soon as possible.

3.5. Measurement of Oxidative Burst in Human Whole Blood Using Flow Cytometry

1. Mix heparinized whole blood (100 μl), stimuli or other agents of interest, and HBSS in sterile 1.5-ml Eppendorf centrifuge tubes resulting in a final volume of 200 μl. Prewarm tubes in water bath at 37°C for 5 min (see Note 6).
2. Add 2.8 μl DHR or HBSS, 25 μl fMLP or HBSS, onto lids of tubes, mix contents as described above, and incubate at 37°C for 20 min.
3. Place tubes on ice for 10 min to stop reactions (see Note 7). Add 1 ml FACS lysing solution per tube and keep tubes on ice for 20 min to lyse completely erythrocytes.
4. Centrifuge tubes at 425 × *g* for 5 min at 4°C. Remove supernatants and wash cells twice using 1 ml HBSS. Resuspend cell pellets in 300 μl FACS fixing solution and keep on ice until analysis.
5. Measure samples using flow cytometer (FL1 channel with 488-nm laser) as soon as possible. Adjust the fluorescence gain properly using unstimulated control samples (shaded curve, Fig. 2a) and positive fMLP (100 nM) control samples (open curve, Fig. 2a). Figure 2b shows a sample scatter plot after proper adjustment of forward and side scatter detectors, allowing clear identification of the PMN population.

3.6. Measurement of PMN Oxidative Burst in Mouse Whole Blood

1. Draw mouse blood by cardiopuncture into 1-ml syringe previously rinsed with heparin.
2. Mix heparinized whole blood (50 μl), drugs of interests, and HBSS in sterile 1.5-ml Eppendorf centrifuge tubes to achieve a final volume as 600 μl. Prewarm tubes in water bath at 37°C for 5 min (see Note 6).

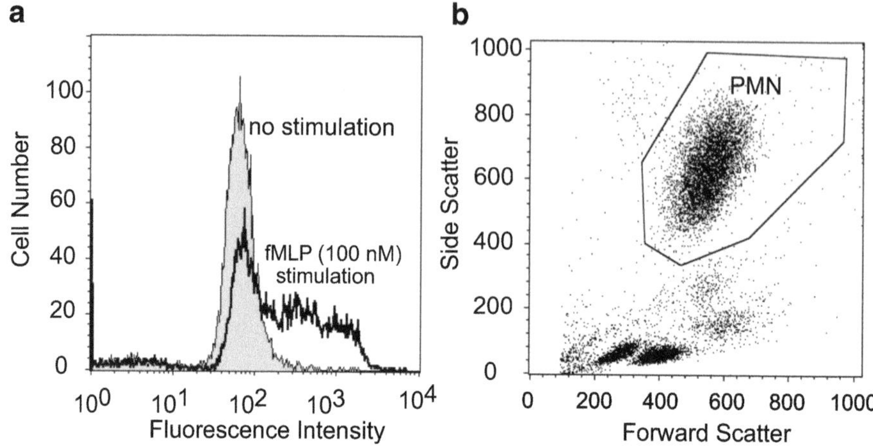

Fig. 2. Oxidative burst measurement with flow cytometry using DHR. (**a**) Fluorescence histogram profiles of the oxidative metabolic response of PMN stimulated with 100 nM fMLP (*open curve*) or of unstimulated PMN (*shaded curve*). An example of the histograms 10 min after stimulation is shown. (**b**) Gating of PMN using forward and side scatter plot of lysed human whole blood.

3. Add 3.5 μl DHR or HBSS, 6 μl w-peptide or HBSS, into tubes and mix with blood as described above. Incubate tubes in water bath for 20 min at 37°C (see Note 8).
4. Place tubes in ice bath for 10 min to stop all reactions (see Note 7). Centrifuge at 425 × *g* for 5 min at 4°C and remove supernatants. Then, add 1 ml RBC lysis buffer in each tube and keep tubes on ice for 4 min to lyse erythrocytes.
5. Centrifuge tubes at 425 × *g* for 5 min at 4°C, remove supernatants, and wash cell with 1 ml HBSS twice. Resuspend cell pellets in 300 μl FACS fixing solution and keep on ice until analysis.
6. Measure samples with flow cytometer (FL1 channel with 488-nm laser) as soon as possible. Adjust the forward and side scatter detectors in order to clearly identify PMN population and then adjust fluorescence gain using unstimulated and stimulated control samples.

4. Notes

1. Pyrogen and similar contaminants in materials that come in contact with blood or PMN result in cell aggregation during isolation, high baseline activation levels, and premature cell death. Key steps to minimize such problems are to avoid pyrogen contamination by using only tissue culture-grade reagents, pyrogen-free and sterile plasticware and other consumables,

and infusion-grade solutions and by removing endotoxin and pyrogen by baking glassware, salts, and other equipment when possible.

2. Repeated heating and cooling of cell suspensions must be avoided during cell isolation because this causes unintended cell activation. To avoid temperature fluctuations, perform all isolation steps at room temperature unless otherwise mentioned; and remember to disable cooling systems of centrifuges if necessary.
3. Mechanical stress activates cells and results in high baseline activation levels and cell death. To avoid such problems, always handle cells gently; after centrifugations, disrupt cell pellets before adding fluids to resuspend the cells. Whenever possible, avoid polystyrene tubes; the use of polypropylene plasticware is preferable to avoid cell activation and loss due to excessive cell adhesion.
4. SOD serves as a reference to exclude any signals not due to O_2^-.
5. Use the Beer–Lambert law with an extinction coefficient (ε) of 21.1 mM^{-1} cm^{-1} for reduced cytochrome c at a wave length of 550 nm to calculate molar amounts of O_2^- generated by PMN (8).
6. Controls needed in this experiment include cells only and cells with DHR but without stimulation.
7. From this step on, always keep samples in an ice water bath.
8. W-peptide is a chemotactic peptide that activates murine FPR (14).

Acknowledgments

This work was supported by NIH grants GM-51477, GM-60475, AI-072287, and AI-080582 and Congressionally Directed Medical Research Programs grant PR043034 (W.G.J.).

References

1. Rabiet, M.J., Huet, E., Boulay, F. (2007) The N-formyl peptide receptors and the anaphylatoxin C5a receptors: an overview, *Biochimie* **89**, 1089–1106.
2. Waugh, D.J., Wilson, C. (2008) The interleukin-8 pathway in cancer, *Clin Cancer Res* **14,** 6735–6741.
3. Capra, V. (2004) Molecular and functional aspects of human cysteinyl leukotriene receptors, *Pharmacol Res* **50,** 1–11.
4. Bokoch, G.M., Zhao, T. (2006) Regulation of the phagocyte NADPH oxidase by Rac GTPase, *Antioxid Redox Signal* **8**, 1533–1548.
5. Hager, M., Cowland, J.B., Borregaard, N. (2010) Neutrophil granules in health and disease, *J Intern Med* **268**, 25–34.
6. Nussler, A.K., Wittel, U.A., Nussler, N.C. et al. (1999) Leukocytes, the Janus cells in inflammatory disease, *Langenbecks Arch Surg* **384**, 222–232.

7. Naoyuki, H., Chen, Y., Rusu, C. et al. (2005) Whole-blood assay to measure oxidative burst and degranulation of neutrophils for monitoring trauma patients, *European Journal of Trauma* **31**, 379–388.
8. Dahlgren, C., Karlsson, A., Bylund, J. (2007) Measurement of respiratory burst products generated by professional phagocytes, *Methods Mol Biol* **412**, 349–363.
9. Lundqvist, H., Dahlgren, C. (1996) Isoluminol-enhanced chemiluminescence: a sensitive method to study the release of superoxide anion from human neutrophils, *Free Radic Biol Med* **20**, 785–792.
10. Cohen, H.J., Chovaniec, M.E. (1978) Superoxide production by digitonin-stimulated guinea pig granulocytes. The effects of N-ethyl maleimide, divalent cations; and glycolytic and mitochondrial inhibitors on the activation of the superoxide generating system, *J Clin Invest* **61**, 1088–1096.
11. Elbim, C., Lizard, G. (2009) Flow cytometric investigation of neutrophil oxidative burst and apoptosis in physiological and pathological situations, *Cytometry A* **5**, 475–481.
12. Schopf, R.E., Mattar, J., Meyenburg, W. et al. (1984) Measurement of the respiratory burst in human monocytes and polymorphonuclear leukocytes by nitro blue tetrazolium reduction and chemiluminescence, *J Immunol Methods* **67**, 109–117.
13. Junger, W.G., Hoyt, D.B., Davis, R.E., et al. (1998) Hypertonicity regulates the function of human neutrophils by modulating chemoattractant receptor signaling and activating mitogen-activated protein kinase p38, *J Clin Invest* **101**, 2768–2779.
14. Seo, J.K., Choi, S.Y., Kim, Y. et al. (1997) A peptide with unique receptor specificity: stimulation of phosphoinositide hydrolysis and induction of superoxide generation in human neutrophils, *J Immunol* **158**, 1895–1901.

Chapter 9

Measurement of Neutrophil Elastase, Proteinase 3, and Cathepsin G Activities using Intramolecularly Quenched Fluorogenic Substrates

Brice Korkmaz, Sylvie Attucci, Christophe Epinette, Elodie Pitois, Marie-Lise Jourdan, Luiz Juliano, and Francis Gauthier

Abstract

Neutrophil elastase, proteinase 3, and cathepsin G are three hematopoietic serine proteases, large quantities of which are stored in neutrophil cytoplasmic azurophilic granules. They act in combination with reactive oxygen species to degrade engulfed microorganisms inside phagolysosomes. Active forms of these proteases are also externalized during neutrophil activation at inflammatory sites, thus helping to regulate inflammatory and immune responses. A fraction of secreted neutrophil serine proteases (NSPs) remains bound to the external plasma membrane, where they remain enzymatically active. This protocol describes the spectrofluorometric measurement of NSP activities using sensitive ortho-aminobenzoyl-peptidyl-*N*-(2,4-dinitrophenyl) ethylenediamine fluorescence resonance energy transfer (FRET) substrates that fully discriminate between the three human NSPs. These are used to measure subnanomolar concentrations of free or membrane-bound NSPs in low-binding microplates and to quantify the activities of individual proteases in biological fluids. We describe the synthesis of FRET substrate, neutrophil purification, and kinetic experiments on activated neutrophils. The protocol for measuring NSP activity on the surface of activated neutrophils can be adapted to measure NSP activities in whole biological fluids. Such data clarify the contributions of individual NSPs to the development of inflammatory diseases. Ultimately, these proteases may be shown to be targets for therapeutic inhibitors.

Key words: Neutrophil, Serine proteases, Fluorogenic substrates, Inflammation, Therapeutic inhibitors

1. Introduction

Serine proteases are members of a large family of proteolytic enzymes that have a serine in their catalytic center which initiates the proteolytic cleavage of specific protein substrates (1). They play an essential role in processes, like inflammation, blood coagulation,

Robert B. Ashman (ed.), *Leucocytes: Methods and Protocols*, Methods in Molecular Biology, vol. 844, DOI 10.1007/978-1-61779-527-5_9, © Springer Science+Business Media, LLC 2012

and apoptosis. Immune cells produce a wide variety of serine proteases, such as neutrophil elastase (NE), proteinase 3 (PR3) and cathepsin G (CG) in neutrophils, granzymes in cytotoxic lymphocytes, and chymase and tryptase in mast cells (2, 3). NE, PR3, and CG are major components of neutrophil azurophilic granules and participate in the nonoxidative pathway of intracellular and extracellular pathogen destruction. These neutrophil serine proteases (NSPs) act intracellularly within phagolysosomes to digest phagocytized microorganisms in combination with microbicidal peptides and the membrane-associated nicotinamide adenine dinucleotide phosphate (NADPH) oxidase system, which produces reactive oxygen metabolites (4). Exposure of neutrophils to cytokines, chemoattractants, or bacterial lipopolysaccharide leads to the rapid translocation of granules to the cell surface and secretion of NE, PR3, and CG into the extracellular medium (5). A fraction of secreted NE, PR3, and CG remains at the surface of activated neutrophils (6–8). Resting neutrophils purified from peripheral blood bear variable amounts of PR3 on their surface. There is a bimodal, genetically determined distribution, with two populations of quiescent neutrophil, one of which bears the protease at its surface (9, 10). NSPs are also involved in the regulation of inflammatory processes, including chronic lung diseases (11–14). In these disorders, the accumulation and activation of neutrophils in the airways result in the secretion of excess active NSPs that destroy the lung matrix and cause inflammation. NSPs are also involved in other human disorders as a consequence of gene mutations, altered cellular trafficking, or, for PR3, autoimmune disease. Mutations in the *ELA2/ ELANE* gene encoding NE are the cause of human cyclic neutropenia and severe congenital neutropenia (15, 16). Neutrophil membrane-bound proteinase 3 is the major target antigen of antineutrophil cytoplasmic autoantibodies (ANCAs) associated with Wegener's granulomatosis (17).

Human NSPs are potential targets for anti-inflammatory treatment, but their individual contributions to the development of inflammatory diseases have not been elucidated, mainly because synthetic substrates for measuring the activity of each of them, at the cell surface and in a heterogeneous medium, have only recently become available. The partition of free and membrane-associated proteases in lung inflammatory secretions and their relative sensitivities to inhibitors appear to be important factors influencing treatment design. Measuring the activity of these proteases selectively at the membrane surface may be of great help in testing the activity of exogenous therapeutic inhibitors. We have developed and used FRET substrates bearing an ortho-aminobenzoyl (Abz) fluorescent group and an *N*-(2,4-dinitrophenyl) ethylenediamine (EDDnp) quenching group as a donor/acceptor pair, one at the N terminus and the other at the C terminus of the peptides (Fig. 1a). The method is based on measuring Abz fluorescence after the peptide

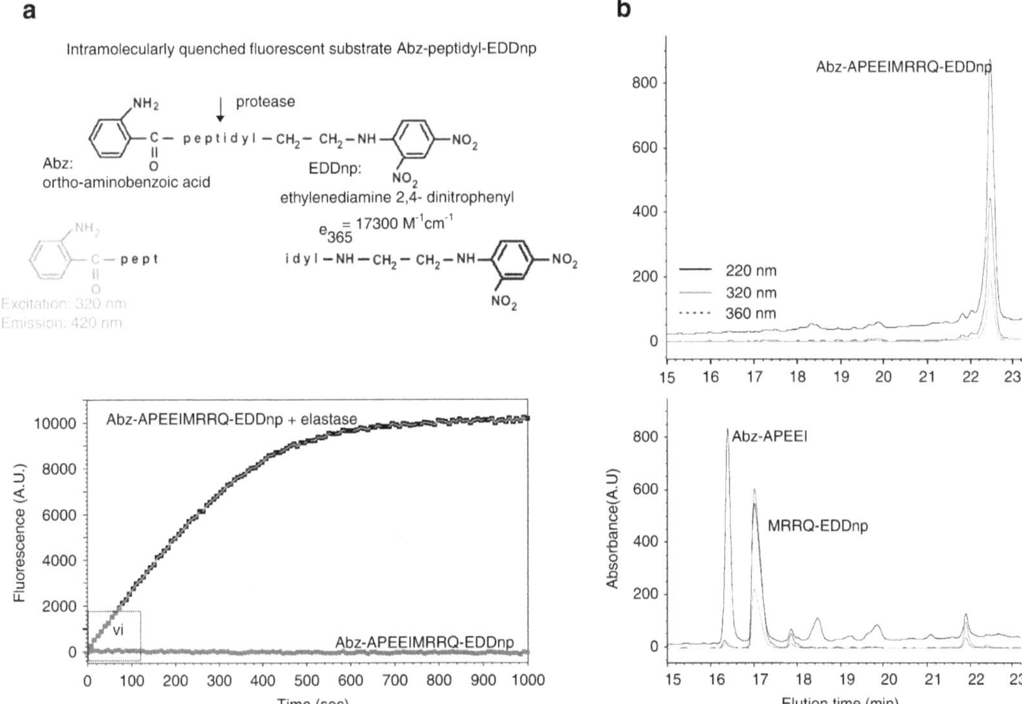

Fig. 1. (**a**) *Top*: Overall structure of fluorescence-quenched substrates that use an Abz group as fluorescence donor and an EDDnp group as quencher. Fluorescence is released upon cleavage of any peptide bond within the peptidyl sequence (*shown in light gray*). *Bottom*: Hydrolysis of Abz-APEEIMRRQ-EDDnp (20 μM) by purified elastase (10^{-9} M). Hydrolysis of the substrate by elastase is monitored for 1,000 s. (**b**) Reverse-phase HPLC (C18 cartridge) of the purified elastase substrate Abz-APEEIMRRQ-EDDnp before and after hydrolysis by elastase. Eluates are monitored at three wavelengths (220 nm, 320 nm, 360 nm). The EDDnp-containing cleavage product is selected by its absorbance at 360 nm.

moiety is cleaved at any place by the protease so that the quencher is no longer close to the Abz-bearing peptide. The amino acid sequence of the peptide segment of NSP FRET substrates is deduced from the sequences of natural substrates and suicide inhibitors of the serpin family (Table 1). The specificity constants *kcat/Km*, a kinetic parameter reflecting how efficiently an enzyme converts a substrate into product, is determined under first-order conditions. These fluorescence resonance energy transfer (FRET) substrates can be used to quantify subnanomolar concentrations of purified free NSPs in solution.

Measuring the activities of NSPs in whole biological fluids containing several proteases or in suspensions of activated neutrophils may be most important for understanding the role of each of them in inflammatory diseases. Specific measurement of NSPs allows their biodistribution to be studied and provide a clearer picture of their pathophysiological function. Thus, appropriate doses of protease-specific therapeutic inhibitors can be used to combat proteolytic tissue damage without compromising the physiological function of protease targets. This protocol describes the use of

Table 1
Specificity constants kcat/Km for the hydrolysis of specific FRET substrates by elastase, proteinase 3, and cathepsin G

		kcat/Km		
		Elastase	Proteinase 3	Cathepsin G
Substrate	Derived from	mM^{-1} s^{-1}		
Abz-APEEI MRRQ-EDDnp	PAI-1	**531**	<1	N.H.
Abz-VADnV RDRQ-EDDnp	CrmA	<1	**1,571**	<1
Abz-EPF WEDQ-EDDnp	PAR-1	<1	N.H.	**242**

sensitive FRET substrates for NSPs to measure the peptidase activities on freshly purified neutrophils that have been activated by a calcium ionophore. The protocol can also be used to measure NSP activities in whole biological fluids. In that case, FRET substrates should be totally hydrolyzed to check that no substrate has been inactivated by binding to the cells or another compound that would make it resistant to hydrolysis. It is also important to check that a single protease in whole biological fluids is involved in substrate hydrolysis by identifying the cleavage site in the FRET substrate by high-performance liquid chromatography (Fig. 1b) or by using specific inhibitors of NSPs.

2. Materials

2.1. Synthesis of Fluorogenic Substrates

1. The resin for solid-phase peptide synthesis is NovaSyn TG resin from Novabiochem. This is an amino-functionalized resin based on a PEG-polystyrene polymer that is derivatized with the cleavable carboxyfunctionalized linker *p*-((R,S)-α(1-(9-H-Fluoren-9-yl)-methoxyformamido)-2,4-dimethoxybenzyl) phenoxyacetic acid.
2. Fmoc-protected amino acids: Fmoc-Ala-OH, Fmoc-Pro-OH, Fmoc-Glu(γ-O-tert-butyl)-OH, Fmoc-Met-OH, Fmoc-Arg(Pmc)-OH, Fmoc-Val-OH, Fmoc-Asp(β-O-tert-butyl)-OH, Fmoc-norVal-OH, Fmoc-Phe-OH, Fmoc-Trp-OH, Boc-Abz, and Fmoc-Glu(α-EDDnp) are obtained from Novabiochem-Merck. Fmoc-Glu(γ-OH)-EDDnp is synthesized as described below. All other reagents for FRET peptide synthesis are from Fluka-Aldrich-Sigma Co.
3. Automated peptide synthesizer (PSSM-8, Shimadzu): The automated peptide synthesizer is programmed to recouple

each amino acid with a fivefold excess and loaded with the amounts of resin and reagents needed to obtain 100 mmol of the FRET peptides.

4. Rotary evaporator (Rotavapor R-210, Büchi).
5. F2000 spectrophotometer (Shimadzu).
6. TofSpec-E spectrometer (Micromass).
7. HPLC system including a P200 pump (Thermo Fischer Scientific) and a Spectrasystem UV3000 detector (Thermo Separation Product). The HPLC analytic pump P200 is equipped with a C18 30 mm × 2.1 mm cartridge (Merck) and coupled to a UV 6000LP detector (Thermo Fischer Scientific), allowing readings at 220 nm (peptide bond), 320 nm (Abz group), and 360 nm (EDDnp group). Peptide substrates and their hydrolysis products are eluted at a flow rate of 300 μL/min using a 20-min linear gradient of acetonitrile (0–60%) in 0.075% aqueous trifluoroacetic acid (TFA).
8. TFA.
9. *N*,*N*-Dimethylformamide (see Note 1).
10. HPLC reverse-phase C18 cartridge (2.1 × 30 mm).
11. HPLC econosil C-18 column (10 mm; 22.5 × 250 mm).
12. Ultrasphere C-18 column (5 mm, 4.6 × 250 mm).
13. Precision balances.
14. 1-mL glass spectrometric cuvettes.

2.2. Protease Activity Measurement

1. Human neutrophil elastase (Biocentrum Ltd).
2. Human PR3 (Athens Research Technology).
3. Human cathepsin G (Biocentrum Ltd).
4. Igepal CA-630 (Sigma).
5. Milli-Q PF plus water deionizing purification system.
6. Gemini XPS microplate spectrofluorometer (Molecular Devices).
7. White polypropylene-well, 96-well microplates (Hard-Shell Thin-Wall Microplates, Hardshell microplaques 96 black shell, white wells, Bio-Rad Laboratories).
8. Low-binding microtubes, 1.5 mL.
9. Disposable pipette tips.

2.3. Purification of Blood Neutrophils

1. Blood samples: 12–16 mL samples are collected into EDTA-containing tubes and used within 30 min of collection. Cells are purified from blood samples taken from healthy donors that have given their written consent.
2. Centrifuges.

3. Phosphate buffer solution (Invitrogen SARL, without Ca or Mg).
4. A23187 calcium ionophore (Sigma–Aldrich Chimie) (see Note 1).
5. Polymorphprep (AbCys).
6. Lymphoprep (AbCys).
7. EGTA (Molecular Biology grade reagent; Sigma–Aldrich Chimie (see Note 1).
8. EDTA (see Note 1).
9. $KHCO_3$.
10. NH_4Cl (see Note 1).
11. Kova® slides for cell counting.
12. 0.4% Trypan blue stain (wt/vol).
13. Coulter Epics Elite ESP flow cytometer (Beckman Coulter) equipped with a 488-nm argon laser and Expo32 software for data analyses (Beckman Coulter).
14. 15 mL TPP centrifuge tubes (ATGC Biotechnologies).
15. Monoclonal antibodies (mAbs): Mouse mAbs anti-CD63 PE, mouse mAbs anti-CD16b-FITC, and mouse IgG1-PE mouse IgG1 FITC are obtained from Beckman Coulter.

3. Methods

Abz-peptidyl-EDDnp substrates are prepared by solid-phase synthesis with the *N*-(9-fluorenyl)methoxycarbonyl (Fmoc) methodology using a multiple automated peptide synthesizer. In all FRET peptides, Fmoc-Glu(γ-OH)-EDDnp was the first amino acid to be coupled by its g-carboxyl group to the linker *p*-((R,S)-α(1-(9-HFluoren-9-yl)-methoxyformamido)-2,4-dimethoxybenzyl)phenoxyacetic acid that is attached to the NovaSyn TG resin. Once synthesis is complete, the peptides are cleaved from the resin with TFA, but in this reaction the γ-carboxyl group of Glu-EDDnp is amidated and the FRET peptides leave the resin as Abz-peptidyl-Gln-EDDnp. This synthesis strategy means that glutamine is the C-terminal residue in all FRET peptides. Substrate purity is checked by matrix-assisted laser desorption ionization time-of-flight mass spectrometry and by reverse-phase chromatography. The concentrations of Abz-peptidyl-EDDnp substrate are determined by measuring the absorbance at 365 nm using $\varepsilon_{365nm} = 17{,}300\ M^{-1}\ cm^{-1}$ for EDDnp. The hydrolysis of Abz-peptidyl-EDDnp substrates is followed by measuring the fluorescence of the Abz group at $\lambda_{ex} = 320$ nm and $\lambda_{ex} = 420$ nm.

The cleavage sites within these substrates are identified by N-terminal sequencing of the fragment conjugated to the EDDnp group, selected by its absorbance at 360 nm after reverse-phase chromatography (Fig. 1b). Freshly prepared neutrophils are suspended in detergent-free phosphate-buffered saline (PBS) in white polypropylene microplate wells selected for their low binding properties and for increasing the fluorescence signal. Measurements are performed under continuous stirring using $0.2–2 \times 10^6$ suspended activated neutrophils and an optimum substrate concentration of 20 μM to avoid intermolecular quenching that would interfere with signal detection (18, 19). The rate of hydrolysis by cell suspensions can be compared with those of soluble titrated proteases to determine the amount of each protease. One way to ensure this is to titrate the soluble or membrane-bound proteases with a specific inhibitor and compare this activity with that of a titrated free enzyme that hydrolyzes the substrate at the same rate.

3.1. Synthesis of Fluorogenic Substrates (See Notes 2 and 3)

3.1.1. Synthesis of Boc-Abz

1. Anthranilic acid (0.2 mmol) and $(Boc)_2O$ (0.24 mmol) are dissolved in dimethyl formamide (DMF) (50 mL) and triethylamine (0.38 mmol) is added.
2. The reaction mixture is stirred at room temperature (20°C) for 24 h, and then the DMF is removed by evaporation at reduced pressure in a rotary evaporator with the temperature set at 40–45°C and a diaphragm vacuum pump V-710 with the pressure set at around 5 millibars. The oily residue is obtained after evaporation for 20 min. The residue is dissolved in 10% (wt/vol) Na_2CO_3 (200 mL) and extracted with ethyl acetate (3 × 100 mL).
3. The aqueous phase is brought to pH 2 and extracted with ethyl acetate.
4. The ethyl acetate is evaporated off under vacuum using a rotary evaporator and a water vacuum pump. After complete evaporation, petroleum ether is added to crystallize the product. The crude material is crystallized by dissolving it in 100 mL ethanol and then adding 100 mL water.

3.1.2. Synthesis of EDDnp

1. Ethylene diamine (1.6 equiv.) in dioxane (360 mL) is added dropwise to 2,4-dinitro-fluorbenzene (0.8 equiv.) in dioxane (120 mL) at 4°C.
2. The reaction mixture is stirred at room temperature for 3 h, and then the dioxane is removed by evaporation.
3. Water (800 mL) is added and the resulting precipitate is collected by filtration and dissolved in ethanol (200 mL).
4. The pH is brought to 2 with 6 M HCl and the resulting precipitate collected by filtration and dried it in a glass vacuum desiccator.

3.1.3. Synthesis of Fmoc-Glu(γ-tBut)-EDDnp

1. Isobutyl chlorocarbonate (16 mmol) is added to a cooled solution (−15°C) of Fmoc-Glu(γ-tBut)-OH (16 mmol) and *N*-methyl-morpholine (16 mmol) in DMF (60 mL) and the reaction mixture stirred for 10 min.
2. This mixed anhydride solution is added to a cooled solution of EDDnp (14 mmol) in DMF (60 mL) and stirred overnight at 4°C.
3. The DMF is evaporated off under vacuum and the product precipitated with water, filtered, washed with 1 M HCl, then with 5% $NaHCO_3$, and finally with water, and then dried.

3.1.4. Synthesis of Fmoc-Glu(γ-OH)-EDDnp

1. Fmoc-Glu(g-tBut)-EDDnp (5 g) is mixed with anisole (0.5 mL) and dissolved in 20 mL TFA for 1 h at room temperature.
2. The TFA is evaporated off, the product precipitated with ethyl ether, and dried.

3.1.5. Coupling Fmoc-Glu(γ-OH)-EDDnp to Resin

1. The linker *p*-((R,S)-α(1-(9-H-Fluoren-9-yl)-methoxyformamido)-2,4 dimethoxybenzyl)phenoxyacetic acid (3 equiv.) is linked to the NovaSyn TG resin by suspending it in DMF containing *N,N'*-tetramethyl-O-benzotriazo-1-yluronium tetrafluoroborate (TBTU) (3 equiv.), 1-hydroxybenzotriazole (HOBt) (0.3 equiv.), and *N*-methyl-morpholine (9 equiv.). DMF (2 mL) and 250 mg resin are placed in a 3-mL plastic syringe with a frit column plate and the outlet connected to a membrane pump via a collecting flask.
2. The Fmoc group is removed from the linker by incubating the treated resin (twice) with 20% *N*-methyl-piperidine in DMF for 15 min.
3. Fmoc-Glu(g-OH)-EDDnp (5 equiv.) is coupled to the linker with TBTU (5 equiv.), HOBt (0.5 equiv.), and *N*-methyl-morpholine (5 equiv.) in DMF. The coupling is checked by the Kaiser test (20).
4. The Fmoc group is removed from Fmoc-Glu(γ-O-Resin)-EDDnp. The peptides are synthesized using TBTU/HOBt as coupling reagent and *N*-methyl-piperidine to remove the Fmoc group. The Boc-Abz residue is added last. See ref. 21 for details of solid-phase peptide synthesis procedures.
5. The peptide is removed from the resin and the amino acid side chains simultaneously deprotected by incubating the resin with a solution containing TFA/anisole/1,2-ethanedithiol (92:5:2) for 5 h at room temperature (20–25°C).
6. The peptides containing Arg(NGPmc) are removed by incubating with TFA/anisole/1,2-ethanedithiol/water (85:5:2:7) for 10 h. The peptide is cleaved from the linker attached to the resin to leave the γ-carboxyl group of Glu amidated. Thus, all the synthesized peptides have Gln-EDDnp as the last amino acid.

3.2. Chromatographic Analysis of Synthesized Substrates and Purification

1. The final deprotected peptides are purified by semipreparative HPLC using an Econosil C-18 column (10 mm; 22.5 × 250 mm) and a two-solvent system: (1) TFA/water (1:1,000 vol/vol) and (2) TFA/acetonitrile/water (1:900:100 by vol/vol). They are eluted at a flow rate of 5 mL/min with a 10–50% (30 min) or a 30–60% gradient of solvent B (45 min).
2. The peptides are eluted at a flow rate of 5 mL/min with a 10–50% or 30–60% gradient of solvent B over 30 or 45 min.
3. The purity of the peptides is checked by analytical HPLC using a binary HPLC system from Shimadzu with an SPD-10AV Shimadzu UV-Vis detector coupled to an Ultrasphere C-18 column (5 mm, 4.6 × 250 mm).
4. The peptides are eluted with solvent systems A1 (H_3PO_4/H_2O, 1:1,000) and B1 (ACN/H_2O/H_3PO_4, 900:100:1) at a flow rate of 0.8 mL/min and a 10–80% gradient of B1 for 15 min.
5. The HPLC column eluates are monitored by measuring their absorbance at 220 nm and their fluorescence emission at 420 nm following excitation at 320 nm.
6. The molecular masses and purity are confirmed by matrix-assisted laser desorption/ionization-time-of-flight mass spectrometry.

3.3. Determination of the Concentrations of Purified Substrates

1. A 5 mM stock solution of each FRET substrates is prepared in 30% (vol/vol) *N,N*-dimethylformamide/water using the following theoretical Mr for each substrates:

Abz-APEEIMRRQ-EDDnp: Mr = 1,473.3 Da
Abz-VADnVRDRQ-EDDnp: Mr = 1,304.04 Da
Abz-EPFWEDQ-EDDnp: Mr = 1,293.97 Da

2. These stock solutions are centrifuged for 3 min at 14,000 × *g* at 20°C and the pellets discarded.
3. The transparent yellow substrate solutions are transferred to 1.5-mL microtubes.
4. The concentration of each substrate is determined by measuring the absorbance of EDDnp at 365 nm and converting absorbance to molarity using a molar extinction coefficient e365 = 17,300 M^{-1} cm^{-1} (see Note 4). For this, 1–5 μL of stock solutions is added to PBS (final volume = 1 mL) in a spectrometer cuvette and a standard curve is constructed: it should be linear. These solutions can be prepared in advance and stored at −20°C for up to 6 months.
5. Stock solutions are diluted in PBS to obtain 1 mM working substrate solutions that are kept on ice during the enzyme essays.

3.4. Optimization of Protease Detection

1. Commercial HNE (1 mg) is dissolved in 200 μL 50 mM acetate buffer, pH 5.5, and 150 mM NaCl (can be stored at −80°C for months). An aliquot (20 μL) of stock solution is diluted in 1 mL of 50 mM HEPES, pH 7.4, 150 mM NaCl, and 0.05% (vol/vol) Igepal CA-630 to obtain a working solution (about 3×10^{-6} M). 20–30-μL single-use aliquots are prepared and stored at −80°C.
2. Commercial PR3 (100 μg) is diluted in 100 μL 50 mM MES, pH 4.5, and 700 mM NaCl to obtain a working solution (about 10^{-5} M). It can be stored at 4°C for up to 3–4 months.
3. 1 mg of commercial CG is dissolved in 400 μL of 50 mM HEPES, pH 7.4, and 50 mM NaCl to obtain a stock solution about 10^{-4} M. Store at −80°C. Working solutions are prepared by diluting 40 μL of stock solution in 400 μL of the same buffer.
4. All three proteases are titrated with α1-PI, the titer of which has been determined with p-guanidinobenzoate-titrated bovine trypsin.
5. Titrated proteases are diluted in the wells of the microplate (see Note 5) so that their final concentrations are between 10^{-11} M and 3×10^{-8} M in 150 μL of PBS (final volume). These buffer conditions are not optimal for all the proteases but are required for comparing free and membrane-bound proteolytic activities.
6. The reaction is started by adding 3 μL of the 1 mM substrate working solution for each protease (20 μM final) and the increase in fluorescence is recorded for up to 45 min, depending on the enzyme concentration. Measurements are reliable down to a final concentration of 10^{-10} M for NE and down to 10^{-9} M for CG under these conditions. The minimum concentration detected can be three to five times lower, depending on the protease, if the optimized buffer conditions indicated above are used (see Notes 6–8).

3.5. Measuring Membrane-Bound Elastase, Proteinase 3, and Cathepsin G Activity

1. The purification of the PMNs must be started within 30 min of collecting the blood samples into EDTA-containing tubes.
2. 4 mL of blood is layered over 4 mL of polymorphprep and centrifuged at 2,000 × *g* for 20 min at 20°C.
3. About 1 mL of the neutrophil-enriched band (75% granulocytes and 25% lymphocytes) is harvested and diluted with an equal volume of half-strength PBS, and the volume made up to 6 mL with PBS.
4. This 6 mL of suspension is layered over 3 mL of lymphoprep and centrifuged at 1,000 × *g* for 20 min at 20°C.

5. The neutrophil band at the bottom of the gradient is collected and diluted with 200 μL PBS. Any residual erythrocytes are lysed by mixing with 5 mL sterile water for 20 s. Osmolarity is restored by adding 5 mL 20 mM PBS + 8 mM EGTA, pH 7.4.
6. This suspension of neutrophils is centrifuged at 500 × *g* for 5 min at 20°C, the supernatant is removed, and the neutrophil pellet suspended in PBS.
7. Cell count and viability are checked by trypan blue exclusion.
8. The purity of the neutrophil preparation is checked by analyzing the forward and side scatters of the cells by flow cytometry.
9. The quality of the neutrophil preparation is checked by measuring CD63 (see Note 9), a marker of neutrophil activation, and CD16b, a constitutive marker of neutrophils, by flow cytometry. This is done as follows: (1) PMNs (5×10^5) are mixed with 20 μL monoclonal CD63-phycoerythrin antibodies and 20 μL monoclonal CD16b-FITC antibodies in 200 μL PBS–EGTA (final volume) for 20 min at room temperature in the dark. Negative controls are prepared by incubating cells with mouse IgG1-PE and IgG1-FITC instead of the specific antibodies. (2) The mixture is centrifuged at 500 × *g* for 5 min at 20°C and the pellet washed with PBS–EGTA. The pellet is centrifuged again and suspended in the same buffer.
10. Purified neutrophils are activated by incubating ~5×10^6 cells/mL in PBS containing 1 mM $CaCl_2$ and 1 mM $MgCl_2$ with the calcium ionophore A23187 (1 μM final) for 15 min at 37°C.
11. Activated neutrophils are centrifuged at 2,000 × *g* for 10 min at room temperature and then suspended in PBS–EGTA. They are kept in a cold room on a shaker.
12. The number and viability of activated neutrophils are determined by trypan blue exclusion.
13. The cell quality is checked by flow cytometry. 5×10^5 activated neutrophils are mixed with 20 μL monoclonal CD63-phycoerythrin antibodies and 20 μL monoclonal CD16b-FITC in 200 μL buffer (final volume) for 20 min at room temperature in the dark. Negative controls are cells incubated with mouse IgG1-PE and mouse IgG1-FITC instead of specific antibodies.
14. We check to ensure that all CD16b-labeled cells are labeled with an anti-CD63-PE mAb.
15. Activated human neutrophils are suspended in PBS (final concentration: 5×10^6 cells/mL) and kept at room temperature on a shaker for 2 h max (5 h for activated neutrophils). The presence of NSPs at the surface of activated neutrophil can be

checked using commercial antibodies as described in refs. 18, 19. The presence of membrane-bound PR3 on quiescent and chemically activated neutrophils is illustrated in Fig. 2.

16. The final concentrations of exogenous purified, titrated elastase, PR3, and cathepsin G in the wells are adjusted to 10^{-9} M in 150 μL PBS.

17. Activated neutrophils are added (2×10^5 to 2×10^6 cells per well) and the volume made up to 150 μL with PBS. The reaction is started by adding 3 μL of the 1 mM substrate working solution (final substrate concentration: 20 μM) (see Note 10).

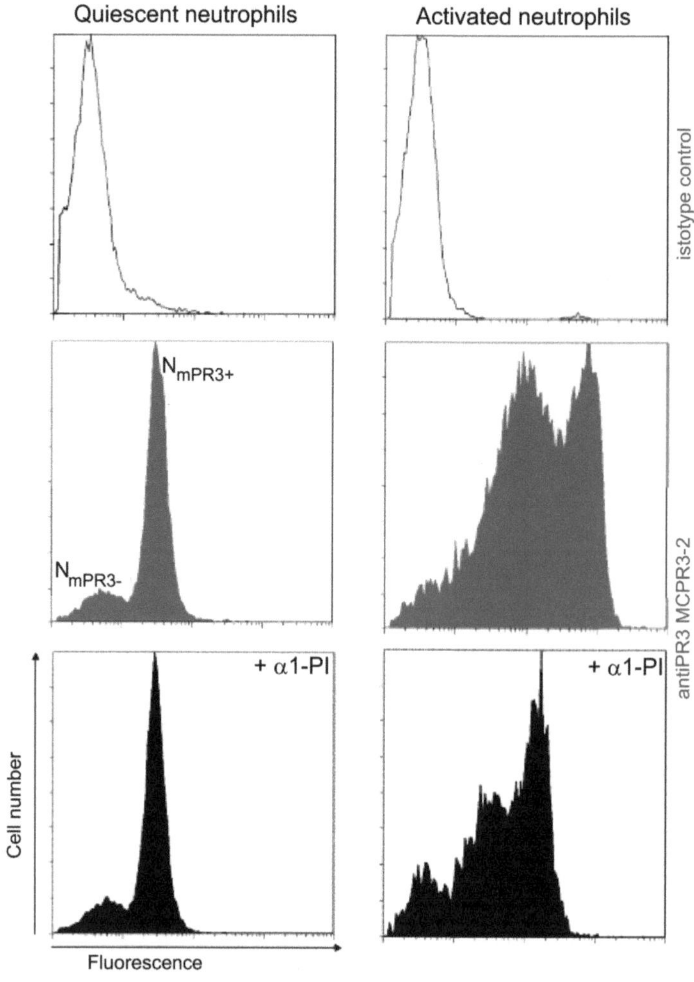

Fig. 2. Flow cytometry analysis of quiescent (*left*) and activated (*right*) purified human neutrophils before (*gray*) and after (*black*) treatment with α1-PI (1 mg/mL), as revealed with anti-proteinase 3 monoclonal antibody MCPR3-2 (1:50). Unlike elastase or cathepsin G, enzymatically inactive proteinase 3 may be present in significant amounts at the surface of quiescent cells (19). Neutrophil activation results in a surface expression of active proteases, including proteinase 3, that can be removed by α1-PI treatment to form soluble complexes. This is visualized by a decrease in fluorescence from the surface of activated cells.

4. Notes

1. *N*,*N*-Dimethylformamide and NH_4Cl are toxic. A23187, EGTA, and EDTA are irritants for the eyes, respiratory tract, and skin.
2. All FRET substrates can be stored as lyophilized powders for months at 4°C.
3. Some newly synthesized substrates are poorly soluble in DMF due to their high proportion of hydrophobic residues. Vortex the substrate suspension for 10 min, then centrifuge, and determine the substrate concentration in the supernatant by measuring its absorbance at 365 nm (-EDDnp).
4. An unstable spectrofluorometer baseline may be due to the substrate autofluorescence. The substrate should be repurified.
5. All measurements should be performed in duplicate, even though the microplate reader ensures that all measurements are carried out under the same conditions.
6. Optimal conditions for free proteases require 0.05% (vol/vol) Igepal CA-630 and higher salt concentrations, but that favors the release of proteases from their intracellular stores when using whole cells. Ensure that free proteases do not stick to the microplate wells in the absence of detergent by measuring rates of hydrolysis before and after transfer from a well to another. The proteases are first incubated in microplate wells without substrate for 30 min, and then they are transferred to another well. Activity is measured before and after the transfer.
7. Do not add any detergent that prevents sticking to glass or plastic surfaces to the reaction mixture to avoid the release of intracellular proteases.
8. An apparent decrease in the rate of hydrolysis may be due to intramolecular quenching. Lower substrate concentrations should be used.
9. The neutrophil preparation should contain 99% PMNs, no monocytes, 1% lymphocytes, and less than 20% of CD63-labeled activated cells. Care must also be taken to measure enzymatic activities using intact cells. Neutrophils are very unstable, and their surface properties may vary with time. They can be stabilized by fixing them with glutaraldehyde/formaldehyde, but we have noticed that this treatment influences the sensitivity of NSPs to inhibitors at the cell surface. Thus, experiments should be performed at least in triplicate and kinetic measurements made in duplicate for each experiment.

10. The fluorescence recording may be nonlinear if the ratio of cell number/substrate concentration is too low. This problem can be solved by increasing the number of cells rather than lowering the substrate concentration.

References

1. Hedstrom, L. (2002) Serine protease mechanism and specificity, *Chem Rev* **102**, 4501–4523.
2. Heutinck, K. M., ten Berge, I. J., Hack, C. E. et al. (2010) Serine proteases of the human immune system in health and disease, *Mol Immunol* **47**, 1943–1955.
3. Korkmaz, B., Horwitz, M. S., Jenne, D. E. et al. (2010) Neutrophil elastase, proteinase 3, and cathepsin G as therapeutic targets in human diseases, *Pharmacol Rev* **62**, 726–759.
4. Segal, A. W. (2005) How neutrophils kill microbes, *Annu Rev Immunol* **23**, 197–223.
5. Owen, C. A., Campbell, E. J. (1999) The cell biology of leukocyte-mediated proteolysis, *J Leukoc Biol* **65**, 137–150.
6. Campbell, E. J., Campbell M. A., Owen, C. A. (2000) Bioactive proteinase 3 on the cell surface of human neutrophils: quantification, catalytic activity, and susceptibility to inhibition, *J Immunol* **165**, 3366–3374.
7. Owen, C. A., Campbell, M. A., Boukedes, S. S. et al. (1995) Inducible binding of bioactive cathepsin G to the cell surface of neutrophils. A novel mechanism for mediating extracellular catalytic activity of cathepsin G, *J Immunol* **155**, 5803–5810.
8. Owen, C. A., Campbell, M. A., Boukedes, S. S. et al. (1997) Cytokines regulate membrane-bound leukocyte elastase on neutrophils: a novel mechanism for effector activity, *Am J Physiol* **272**, L385–393.
9. Halbwachs-Mecarelli, L., Bessou, G., Lesavre, P. et al. (1995) Bimodal distribution of proteinase 3 (PR3) surface expression reflects a constitutive heterogeneity in the polymorphonuclear neutrophil pool, *FEBS Lett* **374**, 29–33.
10. Schreiber, A., Busjahn, A. Luft, F. C. et al. (2003) Membrane expression of proteinase 3 is genetically determined, *J Am Soc Nephrol* **14**, 68–75.
11. Lee, W. L., and Downey, G. P. (2001) Leukocyte elastase: physiological functions and role in acute lung injury, *Am J Respir Crit Care Med* **164**, 896–904.
12. Shapiro, S. D. (2002) Proteinases in chronic obstructive pulmonary disease, *Biochem Soc Trans* **30**, 98–102.
13. Moraes, T. J., Chow, C. W., Downey, G. P. (2003) Proteases and lung injury, *Crit Care Med* **31**, S189–194.
14. Owen, C. A. (2008) Roles for proteinases in the pathogenesis of chronic obstructive pulmonary disease, *Int J Chron Obstruct Pulmon Dis* **3**, 253–268.
15. Horwitz, M., Benson, K. F., Person, R. E. et al. (1999) Mutations in ELA2, encoding neutrophil elastase, define a 21-day biological clock in cyclic haematopoiesis, *Nat Genet* **23**, 433–436.
16. Horwitz, M. S., Duan, Z. Korkmaz, B. et al. (2007) Neutrophil elastase in cyclic and severe congenital neutropenia, *Blood* **109**, 1817–1824.
17. Jenne, D. E., Tschopp, J., Ludemann J. et al. (1990) Wegener's autoantigen decoded, *Nature* **346**, 520.
18. Korkmaz, B., Attucci, S., Jourdan, M. L. et al. (2005) Inhibition of neutrophil elastase by alpha1-protease inhibitor at the surface of human polymorphonuclear neutrophils, *J Immunol* **175**, 3329–3338.
19. Korkmaz, B., Jaillet, J., Jourdan, M. L. et al. (2009) Catalytic activity and inhibition of wegener antigen proteinase 3 on the cell surface of human polymorphonuclear neutrophils. *J Biol Chem* **284**, 19896–19902.
20. Kaiser, E., Colescott, R. L., Bossinger, C. D. et al. (1970) Color test for detection of free terminal amino groups in the solid-phase synthesis of peptides, *Anal Biochem* 34, 595–598.
21. Coin, I., Beyermann, M., Bienert, M. (2007) Solid-phase peptide synthesis: from standard procedures to the synthesis of difficult sequences, *Nat Protoc* **2**, 3247–3256.

Chapter 10

The Macrophage

Chris P. Verschoor, Alicja Puchta, and Dawn M.E. Bowdish

Abstract

Macrophages are a diverse phenotype of professional phagocytic cells derived from bone-marrow precursors and parent monocytes in the peripheral blood. They are essential for the maintenance and defence of host tissues, doing so by sensing and engulfing particulate matter and, when necessary, initiating a pro-inflammatory response. Playing such a vast number of roles in both health and disease, the activation phenotype of macrophages can vary greatly and is largely dependent on the surrounding microenvironment. These phenotypes can be mimicked in experimental macrophage models derived from monocytes and in conjunction with stimulatory factors, although given the complexity of in vivo tissue spaces these model cells are inherently imperfect. Furthermore, experimental observations generated in mice are not necessarily conserved in humans, which can hamper translational research.

The following chapter aims to provide an overview of how macrophages and their parent cell-type, monocytes, are classified, their development through the myeloid lineage, and finally, the general function of macrophages.

Key words: Monocytes, Macrophages, Origin, Activation, Differentiation, Function

1. Introduction

Macrophages derived from bone-marrow precursors and parent monocytes in the peripheral blood are multi-functional cells of the innate immune system that play an important role in regulating the return of host tissues to homeostasis after tissue injury or infection. They accomplish this by engulfing and removing large particulate matter, as well as modifying the molecular and cellular makeup of their surrounding environment. In many respects, they are similar to polymorphonuclear neutrophils, the most prominent phagocytic leukocyte of the peripheral blood that specializes in the clearance of extracellular pathogens, only macrophages have a greater

Robert B. Ashman (ed.), *Leucocytes: Methods and Protocols*, Methods in Molecular Biology, vol. 844,
DOI 10.1007/978-1-61779-527-5_10, © Springer Science+Business Media, LLC 2012

capacity to modulate the inflammatory response and respond to a more varied complement of pathogens (1). Accordingly, macrophages are equipped with a broad-range of pattern-recognition receptors (PRRs), which are required for the production of an array of inflammatory and immunosuppressive cytokines, and the uptake of cellular debris and pathogenic material. To adequately perform these tasks, macrophages are highly plastic cells that can rapidly shift their phenotype based on their microenvironment.

A wealth of methodology exists to study the role of monocytes and macrophages in the many facets of disease and physiology. These include techniques to isolate and identify macrophages from biological samples, as well as assays to measure the functional capacity of these cells, in particular phagocytosis, chemotaxis, and cytokine secretion. Some researchers choose to take one step further and modify the inherent expression profile, and thereby function, of macrophages via knock-out mouse models and artificial expression technology to best answer their research questions. Essential to any study incorporating such methodologies is an understanding of macrophage biology. The following chapter aims to provide an overview of how macrophages and their parent cell-type, monocytes, are classified, the development of macrophages through the myeloid lineage, and finally, the general function of macrophages. The discussion below includes a comparison of mouse and human monocytes/macrophages, since a number of the commonly used phenotypic subset markers are not implicitly conserved across species.

2. Classification and Origins of Blood Monocytes in Mice and Men

Blood monocytes are circulating phagocytic cells with the ability to perform immune effector functions and to enter tissue spaces where they can differentiate into resident macrophages or dendritic cells. While they are precursors to macrophages, the manner in which they are classified is quite distinct.

2.1. Murine Classification of Blood Monocytes

In mice, circulating monocytes constitute 1.5–4% of the total peripheral blood leukocyte pool during the steady state (2). They have been classically defined as cells that express high levels of CD11b (Mac-1), an antigen known to be involved in chemotaxis via endothelial interaction, and CD115 (macrophage colony stimulation factor (M-CSF) receptor). Mouse monocytes may also express the F4/80 antigen at intermediate levels, particularly if they are in the process of differentiating into macrophages (3).

Murine monocytes can be subdivided based on their expression of Ly6C (4, 5), an antigen which is involved in mediating endothelial adhesion and motility in T cells (6), but whose function

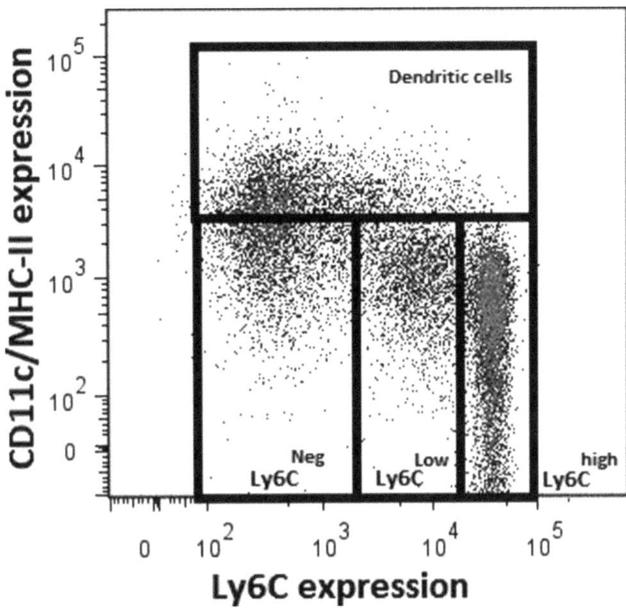

Fig. 1. Characterization of mouse peripheral blood monocytes by flow cytometry. Monocytes can be distinguished using the cell-surface markers CD11c and MHC-II, and separated into three subsets based on expression of Ly6C. Note: Cells identified as expressing CD11c/MHC-II at high levels are considered dendritic cells, and as such are not included in the monocyte subsets.

in monocytes has yet to be determined. Ly6C exhibits a broad expression pattern on monocytes, and hence, classification is not restricted to the antigen simply being referred to as positive/present ($Ly6C^{+}$) or absent/negative ($Ly6C^{-}$/$Ly6C^{neg}$). Instead, murine monocytes can be more precisely designated as expressing Ly6C at low ($Ly6C^{low}$) and high ($Ly6C^{high}$) levels (Fig. 1). This designation has been suggested to be further refined using an additional cell-surface marker CD43 to offer the following subsets: "classical" $Ly6C^{high}CD43^{low}$, "intermediate" $Ly6C^{high}CD43^{high}$, and "non-classical" $Ly6C^{high}CD43^{low}$ monocytes (7). Expression of the Ly6C antigen is typically identified via flow cytometric analysis using either antibodies directed against epitopes specific for the Ly6C molecule, or else using the Gr-1 antibody, which recognizes an epitope present on both Ly6C and a related protein, Ly6G. Since Ly6G is expressed on a number of myeloid lineage cells, especially neutrophils, reliance on the Gr-1 antibody as a unique identifying marker of monocyte subsets can be misleading (8). To complicate matters, certain rare subsets of myeloid cells [i.e. myeloid derived suppressor cells (MDSCs)) have been identified that express both Ly6C and Ly6G (9, 10). Nonetheless, the use of Ly6C has generally been proven as a useful monocyte marker. Morphological analysis can also be used to differentiate the two subsets, as the $Ly6C^{high}$ population is larger and more granulocytic than its $Ly6C^{low}$ counterpart (3, 11). Expression of

CX_3CR1 (neurotactin/fractalkine receptor) is known to correlate inversely with Ly6C expression; thus, Ly6C^{low} monocytes express the highest levels (11). Conversely, Ly6C^{low} monocytes have been identified as expressing lower levels of the monocyte chemoattractant protein-1/CCL2 (MCP-1) receptor CCR2 than their Ly6C^{high} counterparts (12). CX_3CR1 has been shown to be important in the migration of monocytes across endothelial vessels (13), while CCL2 is a well-known, potent chemoattractant of monocytes (14).

Although the characterization of mouse monocyte subsets is still in its early stages, research indicates that the murine subpopulations are functionally distinct as well. Adoptive transfer experiments have shown that Ly6C^{high} monocytes are recruited to inflamed tissue where they undergo activation and act primarily in a pro-inflammatory capacity (15). In response to acute infection or injury, Ly6C^{high} monocyte numbers in the blood compartment expand rapidly, likely in preparation for their accumulation at localized sites of injury (2). As the extent of injury/infection wanes, the circulating numbers of these monocytes decrease correspondingly (2, 15). In states of chronic inflammation such as during atherosclerosis, Ly6C^{high} monocytes have been observed to accumulate in the peripheral circulation in a progressive manner (16). Once recruited to peripheral tissues in response to bacterial infection Ly6C^{high} monocytes release pro-inflammatory mediators such as tissue necrosis factor alpha (TNF-α), inducible nitric oxide synthase (iNOS), and interleukin (IL)-12 (17–19).

Ly6C^{low} monocytes can crawl for long ranges along the endothelial layer of the vasculature and are generally believed to participate in tissue maintenance during homeostasis, homing to resting tissues where they can differentiate into resident macrophages (13, 20, 21). Whether their role is exclusively homeostatic is not clear as one recent report suggests that they are actively recruited to sites of inflammation and may even precede the arrival of their Ly6C^{high} counterparts in the early stages of the immune response (13). Consequently, they may function as effector cells in addition to being an intermediate between hematopoietic progenitors and terminally differentiated tissue macrophages.

2.2. Human Classification of Blood Monocytes

Human monocytes, which comprise approximately 10% of total peripheral blood leukocytes, are less well characterized than their murine equivalents (22). Based on experiments conducted in the 1980s, conventional classification of human blood monocyte subclasses is centred on the expression of CD14, a lipopolysaccharide (LPS) co-receptor, and CD16, an F_C gamma receptor. Using flow cytometry, these cells fall within a defined size (forward scatter) and granularity (side scatter) compartment (Fig. 2a) and commonly express markers such as human leukocyte antigen (HLA), CD115, and CD11b, although to varying degrees (7, 23).

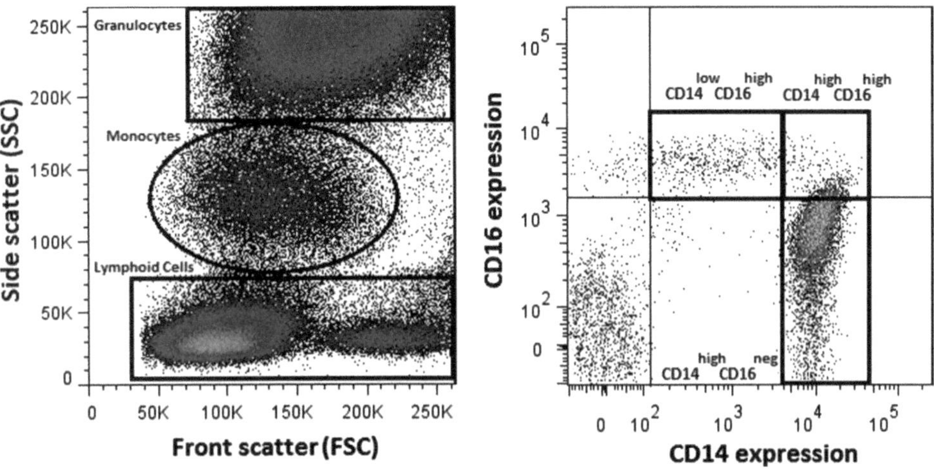

Fig. 2. Characterization of human peripheral blood monocytes by flow cytometry. Monocytes can be distinguished amongst other blood leukocytes by size (front scatter, FSC) and granularity (side scatter, SSC) (**left-hand panel**), and separated into three subsets based on the expression of the cell-surface markers CD14 and CD16 (**right-hand panel**).

Based on CD14/CD16 expression, the major subsets considered are CD14^{high}CD16^{neg} (classical), CD14^{high}CD16^{high} (intermediate), and CD14^{low}CD16^{high} (non-classical or patrolling (21)) (7) (Fig. 2b). The CD14^{high}CD16^{high} subset have also been referred to as "inflammatory" given the initial observations that CD16 expressing monocytes produce TNF-α upon stimulation with pathogen associated molecular patterns (PAMPs, i.e. LPS). However, recent evidence suggests that the distinction in inflammatory responses between the CD14^{high} monocyte subsets is not substantial enough to warrant defining only one as "inflammatory". Hence, assignment as "intermediate" monocytes may be more appropriate (2, 7).

2.3. Conservation Across Species

Human and mouse monocyte populations share many phenotypic and functional similarities. Particularly, CD14^{high} human monocytes and Ly6C(Gr-1)^{high} murine monocytes are functionally similar, as are the CD14^{low} human and Ly6C(Gr-1)^{low} murine subsets. Although no known homolog of Ly6C has been identified in human monocytes, cross-species analysis studies indicate that the expression patterns of a number of genes is conserved between the two subsets (21, 24). As with mouse monocytes, the expression of the chemokine receptors CX_3CR1 and CCR2 is also commonly used to further classify human subpopulations: monocytes lacking CD16 expression have elevated levels of CCR2, much like Ly6C^{high} monocytes, whereas CD16 expressing human monocytes, like Ly6C^{low} murine monocytes, have elevated levels of CX_3CR1 (23). Furthermore, the expression of other key monocyte markers in human and mouse subsets, including that of CD11a, CD11c, CD62L, and CD43, is similarly conserved (24).

As in mice, human monocyte subsets exhibit varied responses upon stimulation, although the nature of the stimulant plays a governing role in this respect (25). Much like the Ly6C^{high} murine subset, CD14^{high} cells have been suggested to function in a pro-inflammatory manner, and as such are highly phagocytic and produce substantial amounts of pro-inflammatory cytokines such as IL-8 and IL-6 in response to LPS (21). Within this subset, classical monocytes lacking CD16 expression produce high amounts of ROS, while intermediate CD16 expressing monocytes secrete high levels of IL-1 beta and TNF-α (21, 25, 26). Much like their murine analogues, non-classical CD14^{low} monocytes play an important role in local surveillance of tissues during the steady-state. These cells exhibit the ability to patrol endothelial vessels in a "crawling" manner, and secrete pro-inflammatory cytokines such as IL-1 beta, TNF alpha, and IL-1 receptor agonist in response to damaged/apoptotic cells and viral antigens (21, 26).

2.4. Origins and Fates of Mouse Monocytes

The common precursor of all leukocytes is the hematopoietic stem cell (HSC) pool in the bone marrow, which can differentiate into a progeny that gradually loses its self-renewal capacity and becomes restricted to a particular lineage. Traditionally, macrophage development has been described as occurring in a stepwise manner: HSC precursors in the bone marrow can develop into monocytes, which differentiate into macrophages upon recruitment to a specific tissue site. The differentiation of HSCs gives rise to two major clonogenic progenitor classes: the common lymphoid lineage, which generates T lymphocytes, B lymphocytes, and natural killer cells, and the common myeloid lineage, which generates either erythrocyte progenitors, or granulocyte/macrophage progenitors, with monocytes arising from the latter. These progenitor cell types can subsequently give rise to polymorphonuclear neutrophils, and mononuclear monocytes (Fig. 3). The mononuclear cell pool, or mononuclear phagocyte system, can further differentiate into plasmacytoid dendritic cells, classical dendritic cells, tissue-resident macrophages and recruited macrophages (27). Originally, it was thought that plasmacytoid dendritic cells were derived only from the common lymphoid lineage; however, recent literature suggests that these cells are derived from the myeloid lineage (28, 29). In the differentiation process, monocyte commitment is induced primarily by the presence of the growth factors macrophage colony stimulating factor (M-CSF) and granulocyte-macrophage stimulating factor (GM-CSF) (30).

While this paradigm of myeloid development is well supported and is still an active area of research, some investigators argue against such rigid relationships between cell-types in the myeloid lineage. Evidence for this argument includes a lack of molecular epitopes that definitively characterize a given myeloid cell-type, and striking similarities between the transcriptomes of myeloid

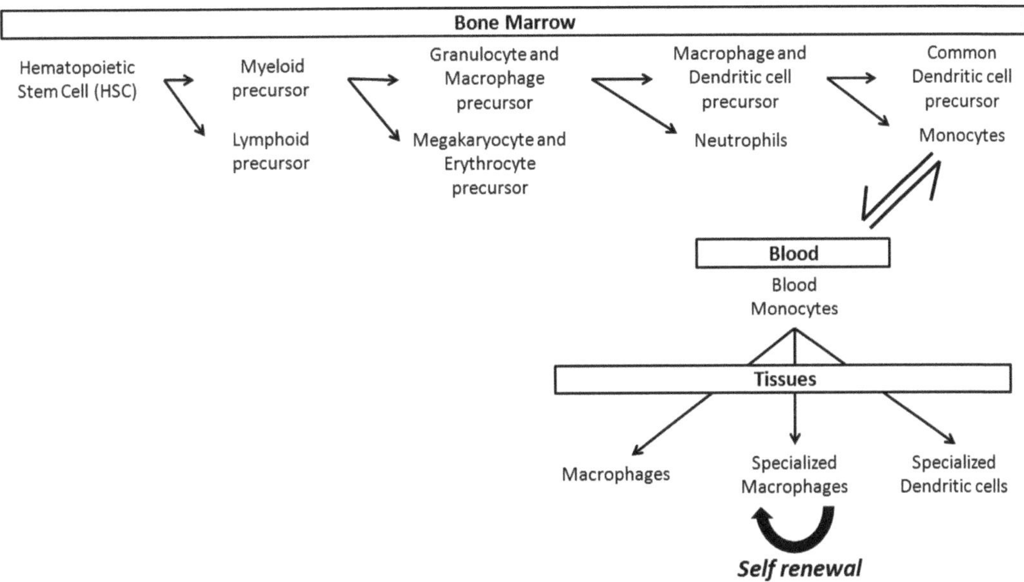

Fig. 3. An overview of the development of the myeloid lineage.

cell-types that are considered divergent by the current dogma (31). Additionally, the derivation of myeloid lineages has predominantly been proven in mice, and is not necessarily conserved in humans. Although xenogeneic in vivo transplantation models have traversed obvious ethical barriers, offering much to our understanding of the myeloid lineage in humans, there is still a great deal of hypotheses to be experimentally verified (29).

Bone marrow derived monocytes are commonly classed into two major subgroups based on their expression of the cell surface marker Ly6C and functional differences, and are believed not to proliferate (22). From the bone marrow, monocytes are mobilized into the peripheral blood via the chemokine receptors CCR2 and CX_3CR1 (32) where they circulate and await further signalling to enter tissue spaces. Mobilization is a constitutive process but can be induced or repressed in response to inflammatory signals caused by infection, for example. These signals include those that are pathogen-derived, such as cell wall components, or endogenously produced by the host, such as the pro-inflammatory cytokines IFN gamma or TNF-α (33). While it was traditionally thought that monocytes were permanently fated to blood and tissue compartments upon emigrating from the bone marrow, it has recently been shown that in the absence of external stimuli such as inflammation, $Ly6C^{high}$ monocytes can return to the bone marrow (34). It has been hypothesized that these cells may leave the circulation to provide a reservoir for the generation of $Ly6C^{low}$ monocytes, osteoclasts, or resident bone marrow dendritic cells, or to acquire antigen captured by neutrophils and B-cells also returning from the peripheral blood (34). Recent research also suggests that precursor monocytes are not limited to the bone marrow as tissue of origin. A separate reservoir of monocytes that can be found exclusively in the spleen has been identified in mice. Splenic monocytes are distinct from the resident splenic population of macrophages, and are mobilized in response to infectious cues (35). In contrast to their bone marrow-derived counterparts, their deployment to the circulation occurs independent of CCR2 (36).

3. Classification and Origins of Tissue Macrophages

Macrophages are professional phagocytes involved in the recycling and clearance of erythrocytes during the steady state, the removal of apoptotic cells and cellular debris, tissue remodelling, and host responses to infectious disease (37). They are remarkably plastic cells that can rapidly shift their physiology in response to cues generated after injury or infection (38). The surrounding microenvironment largely determines the activation phenotype of recruited macrophages, which can most simply be classified as falling within a spectrum consisting of two opposing phenotypes: classically activated, or M1, macrophages (CAMs), and alternatively activated, or M2, macrophages (AAMs) (39). Additionally, there are subsets of specialized resident macrophages whose phenotypes are uniquely adapted to their location, such as brain microglia, liver Kupffer cells, bone osteoclasts, and lung alveolar macrophages. These distinct cell types can also be skewed towards classical or alternative activation, although their differentiation in response to a given stimulus may not be analogous (40, 41).

3.1. Classically and Alternatively Activated Macrophages

Stimulation with a toll-like receptor (TLR) agonist (i.e. LPS) in the presence of interferon gamma (IFN-γ) promotes CAM differentiation (38, 42, 43). They have an enhanced capacity to present antigen, produce high amounts of nitric oxide (NO), secrete large amounts of chemokines and pro-inflammatory cytokines, and promote the expansion of T-helper 1 (Th1) lymphocytes via interactions with major histocompatibility complex (MHC)-II and stimulation by IL-12. As such, CAMs are considered vital in the defence against bacteria, but at the same time can be damaging to the host due to collateral damage brought about by the defence mechanisms they promote (39). Exposure to IL-4/IL-13 produced primarily by CD4+ T-cells promotes the differentiation of AAMs. These cells are involved in the response to parasites and fungi, and express high amounts of cytosolic arginase and extracellular matrix related proteins (44–46). The latter two characteristics provide AAMs the ability to limit inflammation and play an important role in tissue repair and hence the additional title wound-healing macrophages (38). A third subset, the regulatory or M2b and M2c macrophage has also been described. This subset is induced by immune complexes and TLR agonists, or IL-10 and glucocorticoids, functioning to dampen immune responses and inflammation (38). The in vivo relevance of these phenotypes is an active area of research.

It should be noted that, much like blood monocytes, evidence suggests that dividing activated macrophages into rigid classes is for the most part unrealistic and that these designations do not necessarily capture the nuances that exist between macrophage populations. Most in situ macrophages will lie in a more intermediate position in the activation spectrum, sharing some overlapping characteristics depending on the environmental stimuli (38). At the same type some macrophage subsets are not easily classified within this spectrum. Myeloid derived suppressor cells for example, named for their ability to suppress T-cell activation and proliferation, express NO, and arginase, hallmarks of both CAMs and AAMs (47). Macrophage foam cells generated by a dysregulated uptake of lipid compounds at sites of atherosclerotic plaques are another excellent example of a subset that exhibits an atypical phenotype with characteristics of both activation spectrums (48).

Furthermore, while it may be tempting to assume that blood monocytes that are considered "inflammatory" will have a greater propensity to differentiate into M1 macrophages, and vice-versa, this phenomenon has not been conclusively verified. There is experimental evidence to support this theory (49), but on the other hand it is known that the activation phenotype of a macrophage can be skewed quite dramatically depending on the tissue microenvironment into which it extravasates (50).

3.2. Cross-species Conservation of Macrophages Activation Spectrums

As with all cells of the myeloid lineages, characterization of the above macrophage subsets has chiefly been performed in mice. Unfortunately, there are evident discrepancies between species that can hamper macrophage classification in human studies. In mice, the hallmark of CAM induction in vitro and in vivo is NO production. In humans, controversy arises since monocyte-macrophage cell lines cannot be readily induced to express NO, although it is seen in macrophages from tissue biopsies and in blood monocyte-derived-macrophages under certain culture conditions (51, 52). Similarly in AAMs, the hallmark expression of cytosolic arginase in mice is not conserved in humans (53). Gene expression analysis across species has identified some conserved markers, such as HLA/MHC-II for the CAM phenotype and mannose receptor C, type 1 (MRC1) for the AAM phenotype, but it is yet to be determined if they are reliable markers under an array of conditions. Nonetheless, despite these divergences between species regarding classification markers, overall conservation of function appears to be retained for CAMs and AAMs (44).

3.3. Development of Tissue Macrophages from Bone-Marrow Precursors and Blood Monocytes

The migration of blood monocytes across the endothelial membrane initiates their differentiation into tissue macrophages (54). Depending on the local microenvironment, recruited monocytes have the capacity to differentiate into a macrophage subset with a tissue-specific activation state, function, and phenotype (2). While the extravasation and differentiation of blood-borne monocytic precursors was traditionally considered the sole source of tissue-resident macrophages, two additional mechanisms involved in replenishing macrophage numbers have recently been identified. These include the self-proliferation of cells in the resident compartment, and homing/proliferation of dedicated bone-marrow derived precursors to resident tissues (27).

Furthermore, the inflammatory state of the target tissue seems to influence the route of differentiation. Studies have shown that during the normal steady-state, the majority of adult tissue-resident macrophages, including alveolar macrophages (55–57), splenic macrophages (58), and liver Kupffer cells (59), are maintained through local self-renewal, independent of circulating blood monocyte populations. By contrast, during inflammation, circulating monocytic precursors, in particular the $Ly6C^{high}$ subset, travel to inflamed tissues and make substantial contributions to the macrophage population in the respective tissue compartment (60–62).

3.4. Deriving Macrophages for Immunological Studies

Given the obvious ethical barriers inherent to human immunological studies, employing laboratory mice to build our understanding of the myeloid cell lineage is and has been a necessary exercise. These barriers also cause us to rely on the in vitro manipulation of myeloid cell types that are relatively easy to acquire as a means to bridge our murine findings to our own immune system (53). At the same time, many researchers working in the murine model opt to derive or elicit certain myeloid cell types due to difficulties associated with isolation and poor yields (63). Interestingly, some of these experimental constraints have prompted researchers to consider other mammalian species, such as the domestic pig, as models for studying myeloid cell development (64).

In human studies, the most convenient source of macrophage precursors is the peripheral blood or umbilical cord blood. Monocytes can be isolated from blood by density-gradient centrifugation and differentiated into macrophages by allowing adherence to tissue culture plastic in the presence of serum, M-CSF, or GM-CSF (65). It should be noted that macrophages derived in the presence of these supplements, while similar, are not phenotypically or functionally equal. Macrophages derived in GM-CSF as compared to M-CSF differ in cell-surface marker and endogenous gene expression profiles as well as their ability to control HIV-1 viral replication (66–68), and both are inferior to those derived in human serum with regard to TNF-α secretion (69).

For murine studies, macrophages are commonly derived from ex vivo extracted bone marrow precursor cells using a combination of M-CSF and tissue culture grade plastic adherence, or harvested from the peritoneal cavities (i.e. resident peritoneal macrophages) or lungs (i.e. alveolar macrophages). To substantially increase yields, it is a common practice to induce the recruitment of macrophages to the peritoneum of naive mice using a sterile inflammatory agent (i.e. elicited macrophages) such as Bio-gel, polyacrylamide beads, or thioglycollate broth, a complex mixture of yeast components (70). While all of these macrophage subtypes are suitable models for experimental studies, they do differ in phenotype and may not respond analogously after stimulation. For example, elicited macrophages are highly phagocytic and generate large amounts of ROS, while resident peritoneal macrophages do not produce detectable levels of MHC or reactive oxygen species (ROS) (71, 72).

Some have proposed adding additional factors to culture to promote a model phenotype for a particular macrophage subset using the derived or elicited cells described above. Although it is unlikely these cells are homologous to their in vivo counterparts as it is impossible to replicate the complex cytokine milieu tissue microenvironment and the physical association between cells, they are the best experimental model that researchers have at their disposal (63).

4. An Overview of Macrophage Function

The macrophage is a fascinating cell type in that its primary role is maintaining homeostasis. This includes host defence against foreign invaders, the clearance of necrotic and apoptotic debris and tissue remodelling following injury. It performs these roles via four basic innate functions: sensing, chemotaxis, phagocytosis and repair, and adaptive stimulation. Although macrophages have the ability to promote adaptive immune responses, they are considered innate effector cells, since they do not require previous exposure to a given antigen to initiate a response.

4.1. Sensing

Macrophages use intracellular and cell-surface PRRs to sense their local environment. When bound to a given ligand, these receptors generate signals that direct the macrophage response. Unlike the antigen-specific receptor found on T and B lymphocytes for example, these innate receptors can recognize molecular patterns that may be common across a number of species. They can generally be broken down into two sensing groups: pathogen and danger signals (exogenous), and modified host proteins and lipids and necrotic/apoptotic cellular debris (endogenous). While some of the receptors falling into either of these groups are considered markers of a particular activation phenotype (for example MRC1 and dectin-1 with respect to AAMs (44, 73)), they are not mutually exclusive to either spectrum and are likely expressed on most macrophage subsets, albeit at relatively low concentrations.

The TLR family is one of the predominant pathogen sensing group of molecules and currently includes 14 members (74). Some of these include cell-surface TLR-2 and -4, which bind the PAMPs lipoteichoic acid (LTA) and LPS, respectively, and intracellular TLR-3 and -9, which bind viral and bacterial derived oligonucleotides, respectively (75). These receptors collectively promote pro-inflammatory signalling including the expression of cytokines such as IL-6, TNF-α, and IL-12 (39, 44, 75). As opposed to the TLRs that promote pro-inflammatory activities, dectin-1 and MRC1 are two commonly expressed PRRs that promote anti-inflammatory activities. Dectin-1 recognizes β-glucan polysaccharides found on fungi and some bacteria, and signals the inhibition of TNF-α and/or IL-12, and the induction of IL-10 (73, 76), whereas MRC1 binds mannose and fucose polysaccharides commonly found on fungi, bacteria and viruses, and signals the inhibition of IL-12 secretion (73, 77). Additionally, macrophages express cell-surface F_c receptors such as F_c gamma (CD16, CD32, and CD64) and F_c epsilon (CD23). These receptors bind circulating antibodies that are themselves bound to foreign antigens, leading to macrophage phagocytosis (78).

Macrophages use their ability to sense endogenous molecules to facilitate the clearance of modified host proteins and lipids and apoptotic and necrotic cell debris after a disruptive event. An example of a PRR that is integral for this function is the Tyro3, Axl, and Mer (TAM) receptor family, which bind the phosphatidylserine associated proteins Gas6 and ProS, which are associated with the recognition and uptake of apoptotic cells (79). Recognition of apoptotic or necrotic debris generally leads to phagocytosis, but can also stimulate the secretion of anti-inflammatory and immunoregulatory cytokines such as IL-10 and transforming growth factor (TGF) beta and modulate TLR signalling, all in an independent manner (80). In addition, scavenger receptors (e.g. scavenger receptor class A (SRA) and CD36) recognize both exogenous ligands (e.g. bacterial cell wall components) and endogenous ligands (e.g. oxidized low-density lipoproteins) (44, 81, 82).

4.2. Chemotaxis

Upon sensing an antigen belonging to a potentially harmful foreigner invader, macrophages stimulate the expansion of activated T cells and secrete chemokines that function to recruit appropriate effector cells to aid in their neutralization and clearance. Classically activated macrophages, who play a dominant role in anti-bacterial defence and promoting Th1-type responses, commonly secrete CCL-3 (MIP1 alpha), CCL-4 (MIP-1 beta), CCL-5 (RANTES), CXCL-9 (MIG), CXCL-10 (IP-10), and CXCL-11

(I-TAC), which are all potent chemoattractants for monocyte/macrophages, Th1 lymphocytes and natural killer cells. They also secrete CXCL-8 (IL-8), a potent chemokine for the recruitment of neutrophils, which are crucial for the resolution of many types of acute infections (39, 83). Consistent with their role in the defence against parasitic and fungal infections, AAMs commonly secrete the chemokines CCL-17 (TARC) and CCL-22 (MDC), which attract Th2 lymphocytes and natural killer cells, and CCL-24 (Eotaxin-2), which attracts eosinophils and basophils (39, 83).

4.3. Phagocytosis and Tissue Repair

To return a tissue to homeostasis after a disruptive event, the phagocytic clearance of damaged and redundant material is essential. Examples include inflammatory events triggered by infection or injury and physiological changes within the host (i.e. embryonic development and postpartum mammary gland involution (84, 85)), during which a great deal of tissue remodelling occurs. Using cell-surface receptors to identify its targets, as described above, macrophages engulf unwanted material, sequestering it within a phagosomal compartment. This compartment subsequently fuses with a lysosomal compartment, which contains a number of highly reactive and toxic molecules that facilitate the destruction of the phagosomal contents. Although ROS such as hydrogen peroxide and oxide anions play a major role in the destruction of engulfed material, NO is a major immunomodulator of this process. Nitric oxide regulates the levels and reactivity of ROS and can itself interact with ROS to produce toxic reactive nitrogen species (RNS). As a whole, NO and ROS is constitutively produced by macrophages, but upon signalling by cell-surface receptors and/or pro-inflammatory molecules the production of NO is highly induced through its parent enzyme inducible nitric oxide synthase (iNOS) (86). Returning host tissues to a homeostatic state also requires the repair and remodelling of the local environment. Returning host tissues to a homeostatic state also requires the repair and remodelling of the local environment. Alternatively activated macrophages are primarily responsible for this task, promoting extracellular matrix remodelling, cell growth, collagen production, and angiogenesis (78).

4.4. Adaptive Stimulation

The phagocytosis and subsequent destruction of foreign material by macrophages also provide a means to generate antigenic peptide sequences for presentation to T lymphocytes by way of cell-surface MHC class II receptors. Given suitable additional signalling, IL-12 or IL-4 for example, this interaction will lead to the expansion of antigen specific T lymphocytes and thus promote an adaptive immune response (87). However, unlike dendritic cells, most tissue macrophages can only present antigen and stimulate the expansion of activated T lymphocytes (88, 89).

5. Summary

To understand the innate immune response to its fullest extent, it is necessary to recognize the importance of macrophages and thus their parent cell types. Although we have made impressive strides developing a framework to understand the relationship between tissue macrophages and their progenitors, especially peripheral blood and bone marrow monocytes, it is not currently possible to conclude at which point a given phenotype is terminal. Organizing these cells into rigid classes, although appealing as a framework, is likely inaccurate, as they represent a dynamic phenotype, one that is as unique as the microenvironment in which they lie. Furthermore, classifications developed in mice are not necessarily conserved in humans. Hence, conclusions that are made regarding the particular phenotype of a monocyte or macrophage in an experimental mouse model can often lead to confusion when attempting to translate it to humans. Transcriptional profiling, systems biology, and increasingly elegant functional experiments have contributed to resolving these issues (21, 90), and will undoubtedly continue to do so in the future.

Acknowledgements

The Bowdish lab is funded by the CIHR and the MG DeGroote Institute for Infectious Disease Research.

References

1. Dale, D.C., Boxer, L. Liles, W.C,. (2008) The phagocytes: neutrophils and monocytes, *Blood*, **112,** 935–945.
2. Robbins, C.S., Swirski, F.K. (2010) The multiple roles of monocyte subsets in steady state and inflammation, *Cell Mol Life Sci*, **67**, 2685–2693.
3. Sunderkotter, C., Nikolic, T., Dillon, M.J. et al. (2004) Subpopulations of mouse blood monocytes differ in maturation stage and inflammatory response, *J Immunol* **172**, 4410–4417.
4. Hume, D.A. (2008) Differentiation and heterogeneity in the mononuclear phagocyte system, *Mucosal Immunol* **1**, 432–441.
5. Jutila, D.B., Kurk, S., Jutila, M.A. (1994) Differences in the expression of Ly-6C on neutrophils and monocytes following PI-PLC hydrolysis and cellular activation, *Immunol Lett* **41**, 49–57.
6. Hanninen, A., Maksimow, M., Alam, C., et al. (2011) Ly6C supports preferential homing of central memory CD8(+) T cells into lymph nodes, *Eur J Immunol* **41**, 634–644.
7. Ziegler-Heitbrock, L., Ancuta, P., Crowe, S., et al. (2010) Nomenclature of monocytes and dendritic cells in blood, *Blood*, **116**, e74–e80.
8. Egan, C.E., Sukhumavasi, W., Bierly, A.L. et al. (2008) Understanding the multiple functions of Gr-1(+) cell subpopulations during microbial infection, *Immunol Res* **40**, 35–48.
9. Medina-Echeverz, J., Fioravanti, J., Zabala, M., et al. (2011) Successful colon cancer eradication after chemoimmunotherapy is associated with profound phenotypic change of intratumoral myeloid cells, *J Immunol* **186**, 807–815.
10. Youn, J.I., Nagaraj, S., Collazo, M., et al. (2008) Subsets of myeloid-derived suppressor

cells in tumor-bearing mice, *J Immunol* **181**, 5791–5802.

11. Palframan, R.T., Jung, S., Cheng, G., et al. (2001) Inflammatory chemokine transport and presentation in HEV: a remote control mechanism for monocyte recruitment to lymph nodes in inflamed tissues, *J Exp Med* **194**, 1361–1373.
12. Serbina, N.V., Pamer, E.G. (2006) Monocyte emigration from bone marrow during bacterial infection requires signals mediated by chemokine receptor CCR2, *Nat Immunol* **7**, 311–317.
13. Auffray, C., Fogg, D., Garfa, M., et al. (2007) Monitoring of blood vessels and tissues by a population of monocytes with patrolling behavior, *Science* **317**, 666–670.
14. Jiang, Y., Beller, D.I., Frendl, G. et al. (1992) Monocyte chemoattractant protein-1 regulates adhesion molecule expression and cytokine production in human monocytes, *J Immunol* **148**, 2423–2428.
15. Drevets, D.A., Dillon, M.J., Schawang, J.S., et al. (2004) The Ly-6Chigh monocyte subpopulation transports Listeria monocytogenes into the brain during systemic infection of mice, *J Immunol* **172**, 4418–4424.
16. Swirski, F.K., Pittet, M.J., Kircher, M.F., et al. (2006) Monocyte accumulation in mouse atherogenesis is progressive and proportional to extent of disease, *Proc Natl Acad Sci USA* **103**, 10340–10345.
17. Serbina, N.V., Kuziel, W., Flavell, R., et al. (2003) Sequential MyD88-independent and -dependent activation of innate immune responses to intracellular bacterial infection, *Immunity*, **19**, 891–901.
18. Serbina, N.V., Jia, T., Hohl, T.M., et al. (2008) Monocyte-mediated defense against microbial pathogens, *Annu Rev Immunol* **26**, 421–452.
19. Strauss-Ayali, D., Conrad, S.M., Mosser, D.M. (2007) Monocyte subpopulations and their differentiation patterns during infection, *J Leukoc Biol* **82**, 244–252.
20. Geissmann, F., Jung, S., Littman, D.R. (2003) Blood monocytes consist of two principal subsets with distinct migratory properties, *Immunity* **19**, 71–82.
21. Cros, J., Cagnard, N., Woollard, K., et al. (2010) Human CD14dim monocytes patrol and sense nucleic acids and viruses via TLR7 and TLR8 receptors, *Immunity* **33**, 375–386.
22. Auffray, C., Sieweke, M.H., Geissmann, F. (2009) Blood monocytes: development, heterogeneity, and relationship with dendritic cells, *Annu Rev Immunol* **27**, 669–692.
23. Fung, E., Esposito, L., Todd, J.A., et al. (2010) Multiplexed immunophenotyping of human antigen-presenting cells in whole blood by polychromatic flow cytometry, *Nat Protoc* **5**, 357–370.
24. Ingersoll, M.A., Spanbroek, R., Lottaz, C., et al. (2010) Comparison of gene expression profiles between human and mouse monocyte subsets, *Blood* **115**, e10–e19.
25. Skrzeczynska-Moncznik, J., Bzowska, M., Loseke, S., et al. (2008) Peripheral blood CD14high CD16+ monocytes are main producers of IL-10, *Scand J Immunol* **67**, 152–159.
26. Mikolajczyk, T.P., Skrzeczynska-Moncznik, J.E., Zarebski, M.A., et al. (2009) Interaction of human peripheral blood monocytes with apoptotic polymorphonuclear cells, *Immunology* **128**, 103–113.
27. Geissmann, F., Manz, M.G., Jung, S., et al. (2010) Development of monocytes, macrophages, and dendritic cells, *Science* **327**, 656–661.
28. Comeau, M.R., Van der Vuurst de Vries, A.R., Maliszewski, C.R., et al. (2002) CD123bright plasmacytoid predendritic cells: progenitors undergoing cell fate conversion? *J Immunol* **169**, 75–83.
29. Giebel, B., Punzel, M. (2008) Lineage development of hematopoietic stem and progenitor cells, *Biol Chem* **389**, 813–824.
30. Lagasse, E., Weissman, I.L. (1997) Enforced expression of Bcl-2 in monocytes rescues macrophages and partially reverses osteopetrosis in op/op mice, *Cell* **89**, 1021–1031.
31. Hume, D.A. (2010) Applications of myeloid-specific promoters in transgenic mice support in vivo imaging and functional genomics but do not support the concept of distinct macrophage and dendritic cell lineages or roles in immunity, *J Leukoc Biol* doi: 10.1189/jlb.0810472.
32. Combadiere, C., Potteaux, S., Rodero, M., et al. (2008) Combined inhibition of CCL2, CX3CR1, and CCR5 abrogates Ly6C(hi) and Ly6C(lo) monocytosis and almost abolishes atherosclerosis in hypercholesterolemic mice, *Circulation* **117**, 1649–1657.
33. Baldridge, M.T., King, K.Y., Goodell, M.A. (2011) Inflammatory signals regulate hematopoietic stem cells, *Trends Immunol* doi: 10.1016/j.it.2010.12.003.
34. Varol, C., Landsman, L., Fogg, D.K., et al. (2007) Monocytes give rise to mucosal, but not splenic, conventional dendritic cells, *J Exp Med* **204**, 171–180.
35. Swirski, F.K., Nahrendorf, M., Etzrodt, M., et al. (2009) Identification of splenic reservoir monocytes and their deployment to inflammatory sites, *Science* **325**, 612–616.

36. Tsou, C.L., Peters, W., Si, Y., et al. (2007) Critical roles for CCR2 and MCP-3 in monocyte mobilization from bone marrow and recruitment to inflammatory sites, *J Clin Invest* **117**, 902–909.
37. Erwig, L.P., Henson, P.M. (2007) Immunological consequences of apoptotic cell phagocytosis, *Am J Pathol* **171**, 2–8.
38. Mosser, D.M., Edwards, J.P. (2008) Exploring the full spectrum of macrophage activation, *Nat Rev Immunol* **8**, 958–969.
39. Benoit, M., Desnues, B., Mege, J.L. (2008) Macrophage polarization in bacterial infections, *J Immunol* **181**, 3733–3739.
40. Dorger, M., Munzing, S., Allmeling, A.M., et al. (2001) Phenotypic and functional differences between rat alveolar, pleural, and peritoneal macrophages, *Exp Lung Res* **27**, 65–76.
41. Itoh, K., Udagawa, N., Kobayashi, K., et al. (2003) Lipopolysaccharide promotes the survival of osteoclasts via Toll-like receptor 4, but cytokine production of osteoclasts in response to lipopolysaccharide is different from that of macrophages, *J Immunol* **170**, 3688–3695.
42. Meltzer, M.S. (1981) Macrophage activation for tumor cytotoxicity: characterization of priming and trigger signals during lymphokine activation, *J Immunol* **127**, 179–183.
43. Pace, J.L., Russell, S.W., Torres, B.A., et al. (1983) Recombinant mouse gamma interferon induces the priming step in macrophage activation for tumor cell killing, *J Immunol* **130**, 2011–2013.
44. Martinez, F.O., Helming, L. and Gordon, S. (2009) Alternative activation of macrophages: an immunologic functional perspective, *Annu Rev Immunol* **27**, 451–483.
45. Stein, M., Keshav, S., Harris, N., et al. (1992) Interleukin 4 potently enhances murine macrophage mannose receptor activity: a marker of alternative immunologic macrophage activation, *J Exp Med* **176**, 287–292.
46. Edwards, J.P., Zhang, X., Frauwirth, K.A., et al. (2006) Biochemical and functional characterization of three activated macrophage populations, *J Leukoc Biol* **80**, 1298–1307.
47. Youn, J.I., Gabrilovich, D.I. (2010) The biology of myeloid-derived suppressor cells: the blessing and the curse of morphological and functional heterogeneity, *Eur J Immunol* **40**, 2969–2975.
48. Newby, A.C., George, S.J., Ismail, Y., et al. (2009) Vulnerable atherosclerotic plaque metalloproteinases and foam cell phenotypes, *Thromb Haemost* **101**, 1006–1011.
49. Swirski, F.K., Libby, P., Aikawa, E., et al. (2007) Ly-6Chi monocytes dominate hypercholesterolemia-associated monocytosis and give rise to macrophages in atheromata, *J Clin Invest* **117**, 195–205.
50. Khallou-Laschet, J., Varthaman, A., Fornasa, G., et al. (2010) Macrophage plasticity in experimental atherosclerosis, *PLoS One* **5**, e8852.
51. Daigneault, M., Preston, J.A., Marriott, H.M., et al. (2010) The identification of markers of macrophage differentiation in PMA-stimulated THP-1 cells and monocyte-derived macrophages, *PLoS One* **5**, e8668.
52. MacMicking, J., Xie, Q.W. and Nathan, C. (1997) Nitric oxide and macrophage function. *Annu Rev Immunol* **15**, 323–350.
53. Murray, P.J. and Wynn, T.A. (2011) Obstacles and opportunities for understanding macrophage polarization, *J Leukoc Biol* doi: 10.1189/jlb.0710409.
54. Randolph, G.J., Beaulieu, S., Lebecque, S., et al. (1998) Differentiation of monocytes into dendritic cells in a model of transendothelial trafficking, *Science* **282**, 480–483.
55. Landsman, L., Varol, C., Jung, S. (2007) Distinct differentiation potential of blood monocyte subsets in the lung, *J Immunol* **178**, 2000–2007.
56. Sawyer, R.T., Strausbauch, P.H., Volkman, A. (1982) Resident macrophage proliferation in mice depleted of blood monocytes by strontium-89, *Lab Invest* **46**, 165–170.
57. Tarling, J.D., Lin, H.S., Hsu, S. (1987) Self-renewal of pulmonary alveolar macrophages: evidence from radiation chimera studies, *J Leukoc Biol* **42**, 443–446.
58. Wijffels, J.F., de Rover, Z., Beelen, R.H., et al. (1994) Macrophage subpopulations in the mouse spleen renewed by local proliferation, *Immunobiology* **191**, 52–64.
59. Crofton, R.W., Diesselhoff-den Dulk, M.M., van Furth R. (1978) The origin, kinetics, and characteristics of the Kupffer cells in the normal steady state, *J Exp Med* **148**, 1–17.
60. Arnold, L., Henry, A., Poron, F., et al. (2007) Inflammatory monocytes recruited after skeletal muscle injury switch into antiinflammatory macrophages to support myogenesis, *J Exp Med* **204**, 1057–1069.
61. Matute-Bello, G., Lee, J.S., Frevert, C.W., et al. (2004) Optimal timing to repopulation of resident alveolar macrophages with donor cells following total body irradiation and bone marrow transplantation in mice, *J Immunol Methods* **292**, 25–34.
62. Nahrendorf, M., Swirski, F.K., Aikawa, E., et al. (2007) The healing myocardium sequentially mobilizes two monocyte subsets with divergent and complementary functions, *J Exp Med* **204**, 3037–3047.

63. Geissmann, F., Gordon, S., Hume, D.A., et al. (2010) Unravelling mononuclear phagocyte heterogeneity, *Nat Rev Immunol* **10**, 453–460.
64. Fairbairn, L., Kapetanovic, R., Sester, D.P., et al. (2011) The mononuclear phagocyte system of the pig as a model for understanding human innate immunity and disease, *J Leukoc Biol* doi: 10.1189/jlb.1110607.
65. Davies, J.Q., Gordon, S. (2005) Isolation and culture of human macrophages. In *Basic cell culture protocols*. 3rd ed edition. Edited by Helgason, C.D., Miller, C.L. and Totowa, N.J. Humana Press 105–116.
66. Hashimoto, S., Suzuki, T., Dong, H.Y., et al. (1999) Serial analysis of gene expression in human monocytes and macrophages, *Blood*, **94**, 837–844.
67. Lee, B., Sharron, M., Montaner, L.J., et al. (1999) Quantification of CD4, CCR5, and CXCR4 levels on lymphocyte subsets, dendritic cells, and differentially conditioned monocyte-derived macrophages, *Proc Natl Acad Sci USA* **96**, 5215–5220.
68. Matsuda, S., Akagawa, K., Honda, M., et al. (1995) Suppression of HIV replication in human monocyte-derived macrophages induced by granulocyte/macrophage colony-stimulating factor, *AIDS Res Hum Retroviruses* **11**, 1031–1038.
69. Kreutz, M., Krause, S.W., Hennemann, B., et al. (1992) Macrophage heterogeneity and differentiation: defined serum-free culture conditions induce different types of macrophages in vitro, *Res Immunol* **143**, 107–115.
70. Davies, J.Q., Gordon, S. (2005) Isolation and culture of murine macrophages, In *Basic cell culture protocols*. 3rd ed edition. Edited by Helgason, C.D., Miller, C.L. and Totowa, N.J., Humana Press 91–104.
71. Kondo, M., Weissman, I.L., Akashi, K. (1997) Identification of clonogenic common lymphoid progenitors in mouse bone marrow, *Cell* **91**, 661–672.
72. Zhang, X., Goncalves, R., Mosser, D.M. (2008) The isolation and characterization of murine macrophages, *Curr Protoc Immunol* Unit 14.1. doi: 10.1002/0471142735.im1401s83.
73. Taylor, P.R., Martinez-Pomares, L., Stacey, M., et al. (2005) Macrophage receptors and immune recognition, *Annu Rev Immunol* **23**, 901–944.
74. Kumar, H., Kawai, T., Akira, S. (2011) Pathogen recognition by the innate immune system, *Int Rev Immunol* **30**, 16–34.
75. Trinchieri, G., Sher, A. (2007) Cooperation of Toll-like receptor signals in innate immune defence, *Nat Rev Immunol* 7, 179–190.
76. Brown, G.D., Gordon, S. (2001) Immune recognition. A new receptor for beta-glucans, *Nature* **413**, 36–37.
77. Knutson, K.L., Hmama, Z., Herrera-Velit, P., et al. (1998) Lipoarabinomannan of *Mycobacterium tuberculosis* promotes protein tyrosine dephosphorylation and inhibition of mitogen-activated protein kinase in human mononuclear phagocytes. Role of the Src homology 2 containing tyrosine phosphatase 1, *J Biol Chem* **273**, 645–652.
78. Varin, A., Gordo, S. (2009) Alternative activation of macrophages: immune function and cellular biology, *Immunobiology* **214**, 630–641.
79. Lemke, G., Burstyn-Cohen, T. (2010) TAM receptors and the clearance of apoptotic cells, *Ann N Y Acad Sci* **1209**, 23–29.
80. Savill, J., Dransfield, I., Gregory, C., et al. (2002) A blast from the past: clearance of apoptotic cells regulates immune responses, *Nat Rev Immunol* **2**, 965–975.
81. Bottcher, A., Gaipl, U.S., Furnrohr, B.G., et al. (2006) Involvement of phosphatidylserine, alphavbeta3, CD14, CD36, and complement C1q in the phagocytosis of primary necrotic lymphocytes by macrophages, *Arthritis Rheum* **54**, 927–938.
82. Chen, X.W., Shen, Y., Sun, C.Y., et al. (2011) Anti-class A scavenger receptor autoantibodies from systemic lupus erythematosus patients impair phagocytic clearance of apoptotic cells by macrophages in vitro, *Arthritis Res Ther* **13**, R9.
83. Mantovani, A., Sica, A., Sozzani, S., et al. (2004) The chemokine system in diverse forms of macrophage activation and polarization, *Trends Immunol* **25**, 677–686.
84. O'Brien, J., Lyons, T., Monks, J., et al. (2010) Alternatively activated macrophages and collagen remodeling characterize the postpartum involuting mammary gland across species, *Am J Pathol* **176**, 1241–1255.
85. Rae, F., Woods, K., Sasmono, T., et al. (2007) Characterisation and trophic functions of murine embryonic macrophages based upon the use of a Csf1r-EGFP transgene reporter, *Dev Biol* **308**, 232–246.
86. Wink, D.A., Hines, H.B., Cheng, R.Y., et al. (2011)Nitric oxide and redox mechanisms in the immune response, J Leukoc Biol doi: 10.1189/jlb.1010550.

87. Jensen, P.E. (2007) Recent advances in antigen processing and presentation, *Nat Immunol* **8**, 1041–1048.
88. Antoine, J.C., Prina, E., Courret, N., et al. (2004) Leishmania spp.: on the interactions they establish with antigen-presenting cells of their mammalian hosts, *Adv Parasitol* **58**, 1–68.
89. Kamada, N., Hisamatsu, T., Honda, H., et al. (2009) Human CD14+ macrophages in intestinal lamina propria exhibit potent antigen-presenting ability, *J Immunol* **183**, 1724–1731.
90. Martinez, F.O., Gordon, S., Locati, M., et al. (2006) Transcriptional profiling of the human monocyte-to-macrophage differentiation and polarization: new molecules and patterns of gene expression, *J Immunol* **177**, 7303–7311.

Chapter 11

Generation and Characterization of *MacGreen* Mice, the *Cfs1r*-EGFP Transgenic Mice

R. Tedjo Sasmono and Elizabeth Williams

Abstract

Macrophage colony-stimulating factor (CSF-1) regulates the differentiation, proliferation, and survival of cells of the mononuclear phagocyte system. The activity of CSF-1 is mediated by the CSF-1 receptor (CSFIR, CD115) that is encoded by c-*fms* (*Csf1r*) protooncogene. The c-*fms* gene is expressed in macrophage, trophoblast cell lineages, and to some extent granulocytes. A reporter gene construct containing 3.5-kb of 5′ flanking sequence and the downstream intron 2 of the c-*fms* gene directed expression of enhanced green fluorescent protein (EGFP) to cells expressing the c-*fms* gene including the macrophages and trophoblasts. EGFP was detected in trophoblasts from the earliest stage of implantation. During embryonic development, EGFP expression highlighted the large numbers of c-*fms* positive macrophages in most organs. These embryonic macrophages contribute to organogenesis and tissue remodeling. In adult c-*fms* EGFP transgenic mice, which have been called the *MacGreen* mice, EGFP expressed in all tissue macrophage populations and permitted convenient detection of tissue macrophages as well as facilitates their isolation from various tissues.

Key words: Transgenic mice, Green fluorescent protein, Macrophages

1. Introduction

Many aspects of the macrophage biology can be studied using animal model. The transgenic mice have been used in various studies. In this chapter, a line of transgenic mice in which the CSF-1R (encoded by c-*fms* protooncogene) promoter directs expression of the enhanced green fluorescent protein (EGFP) to cells of the macrophage lineage (1, 2) is described. The expression of the EGFP transgene closely resembles that of the F4/80 antigen, but extends to F4/80 negative populations of macrophages (1). All known tissue macrophages in the transgenic mice highly express the EGFP. Figure 2 shows a selection of examples of the expression of the EGFP transgene in organs of the mouse, highlighting the numbers,

Robert B. Ashman (ed.), *Leucocytes: Methods and Protocols*, Methods in Molecular Biology, vol. 844,
DOI 10.1007/978-1-61779-527-5_11, © Springer Science+Business Media, LLC 2012

locations, and characteristics morphology of the cells of the mononuclear phagocyte system. Here, we demonstrate that the *fms*-EGFP reporter gene provides a definitive marker for cells of the mononuclear phagocyte lineage throughout embryonic development and in all adult tissues. The *MacGreen* mice represent a unique and invaluable resource for the study of macrophage biology. The transgene marker can be applied conveniently for detection as well as purification of macrophage lineage cells from all tissues in the body for phenotypic analysis.

2. Materials

2.1. Transgenic Mice Generation

1. p7.2*fms*-EGFP transgene construct (Fig. 1).
2. Restriction endonucleases, these are for removal of plasmid backbone from the construct before microinjection.
3. Agarose gel and 1× TAE buffer (20× TAE stock: 96.8 g/L Tris-HCl, 14.9 g/L EDTA, 22.8 mL/L glacial acetic acid), to be used in transgene purification.
4. QIAquick gel extraction kit (QIAGEN), to be used in transgene purification.
5. Isopropanol for gel extraction.

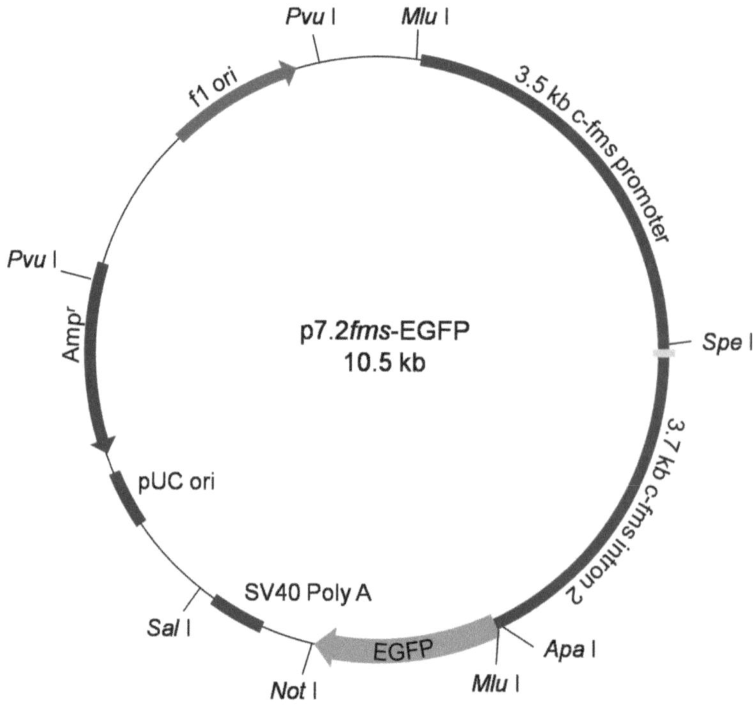

Fig. 1. Map of the p7.2*fms*-EGFP plasmid construct.

6. 3 M Sodium acetate, pH 5 for gel extraction.
7. Egg donor mouse strain: F1 hybrid (C57BL/6 female×CBA male). C57BL/6 mice are used for transgenic mice propagation.
8. Pregnant mare serum gonadotropin (PMSG; Folligon, Lyppard) for mice superovulation.
9. Human chorionic gonadotropin (hCG; Chorulon, Lyppard) for mice superovulation.
10. 70% Ethanol, general material for sterilization during mouse procedures.
11. M2 and M16 media (Sigma-Aldrich) for microinjection.
12. Hyaluronidase (Sigma-Aldrich) for cumulus cell digestion during microinjection preparation.
13. Microinjection buffer (5–10 mM Tris–HCl, pH 7.4, 0.1–0.25 mM EDTA).
14. 2.5% (v/v) Avertin or other suitable mouse anesthetic agents.
15. Proteinase K (Sigma-Aldrich) for tail tips digestion.
16. Tail digestion buffer (100 mM NaCl, 10 mM Tris–Cl, pH 8, 25 mM EDTA, pH 8, and 0.5% SDS).
17. Absolute ethanol for DNA extraction.
18. Tris–EDTA (TE) buffer (10 mM Tris–HCl, 1 mM EDTA) for DNA dilution.
19. TE-buffered phenol (Sigma-Aldrich) for genomic DNA extraction (genotyping).
20. 24:1 Chloroform–isoamyl alcohol (Sigma-Aldrich) for genomic DNA extraction (genotyping).
21. PCR primers for EGFP detection (genotyping): EGFP-F: 5′-CTGGTCGAGCTGGACGGCGACG-3′ and EGFP-R: 5′-CACGAACTCCAGCAGGACCATG-3′.
22. PCR Buffer (10 mM Tris–HCl, pH 8.8, 1.5 mM $MgCl_2$, 50 mM KCl, and 0.1% Triton X-100) (Invitrogen).
23. 10 mM dNTPs mixture (Invitrogen).
24. *Taq* DNA Polymerase (Invitrogen).
25. Gel loading dye: 10× (0.125% bromophenol blue, 0.25% xylene cyanol, 30% glycerol).
26. Ethidium bromide (Sigma-Aldrich).

2.2. EGFP Expression Characterization in Embryo and Adult Tissues

1. Phosphate-buffered saline (PBS) (137 mM NaCl, 2.7 mM KCl, 10 mM Na_2PO_4, and 1.8 mM KH_2PO_4).
2. Complete TC media for macrophage culture: RPMI 1640 (Invitrogen) supplemented with 10% heat-inactivated fetal calf serum (FCS) (Invitrogen), 30 U/mL penicillin (Invitrogen), and 100 μg/mL streptomycin (Invitrogen), store at 4°C.

3. Human recombinant colony-stimulating factor-1 (CSF-1) (Chiron Corp).
4. Ca^{2+}/Mg^{2+}-free Hanks' balanced salt solution (HBSS) (Invitrogen).
5. Osteoclast medium: MEM α-modification medium (Invitrogen) supplemented with 10% heat-inactivated FCS, 30 U/mL penicillin, and 100 μg/mL streptomycin (Invitrogen).
6. Enzyme cocktail for tissue disaggregation: 0.1 U/mL Collagenase and 0.8 U/mL Dispase (Roche has premixed enzyme) and 5 U/mL DNase type II (Sigma-Aldrich) in PBS/10% FCS or HBSS/10%FCS or RPMI 1640 (or any other suitable tissue culture medium).
7. Nycoprep 1.068 or Nycoprep 1.070 (Nycodenz) gradient centrifugation.
8. 4% w/v Paraformaldehyde (PFA) (Sigma-Aldrich) in PBS for tissue fixation. To make, add 4 g of PFA to 100 mL of PBS. Heat to dissolve. Store at 4°C.
9. 18% Sucrose in PBS for tissue cryopreservation.
10. Tissue-Tek optimal cutting temperature (OCT) compound (Sakura Finetek, Japan) for tissue cryosection.
11. Tissue-Tek plastic cryomolds (Sakura Finetek).
12. DAKO fluorescent mounting medium (DAKO Corporation), store at 4°C.
13. Liquid nitrogen or ethanol–dry ice bath.
14. 20 mM EDTA/PBS for epidermal sheet preparation.

3. Methods

3.1. Generation of Transgenic Mice

3.1.1. Transgene Construction and Preparation for Pronuclear Injection

Plasmid p7.2*fms*-EGFP minigene (Fig. 1) contains the 3.5-kb mouse c-*fms* exon 2 promoter, exon 2, the whole (3.7 kb) intron 2, and 12-bp exon 3 driving the EGFP gene. The backbone of this construct is the pGL2-Basic vector (Promega). The luciferase version of this construct and its detailed construction has been described elsewhere (3); however, it has been renamed from p6.7*fms*-GL2 to obtain accurate DNA fragment size as published in GenBank [AF290879]. Construction is done by replacing the luciferase gene of the p7.2*fms*-GL2 with the EGFP gene derived from pEGFP-N1 vector (Clontech). This is achieved by cloning the *Kpn* I-*Sal* I EGFP fragment into *Apa* I-*Sal* I fragment of p7.2*fms*-GL2, utilizing *Kpn* I-*Apa* I oligonucleotide linker (Fig. 1). The ATG start codon in the first coding exon (exon 2) in this construct has been disrupted and engineered to become a *Spe* I site to prevent possible transcription interference with the EGFP start codon.

For pronuclear injection, the DNA transgene must be clean and of high purity. The plasmid backbone in the transgene construct must be removed as the vector sequences are toxic for mouse embryos. The removal is done by restriction enzyme digestion with suitable enzymes, and the vector-free transgene is repurified following agarose gel electrophoresis. The QIAquick gel extraction kit (QIAGEN) can be used for this purpose.

The following is the QIAquick gel extraction method as described by the manufacturer:

1. Digest the construct with appropriate restriction enzymes; separate the DNA on low percentage agarose gel (e.g., 0.8%).
2. Excise the DNA fragment of interest, i.e., the transgene without the backbone, from the agarose gel with a clean, sharp scalpel.
3. Weigh the gel slice in a colorless tube. Add 3 volumes Buffer QG to 1 volume gel (100 mg–100 μL). For >2% agarose gels, add 6 volumes Buffer QG.
4. Incubate at 50°C for 10 min (or until the gel slice has completely dissolved).
5. Vortex the tube every 2–3 min to help dissolve gel.
6. After the gel slice has dissolved completely, check that the color of the mixture is yellow (similar to Buffer QG without dissolved agarose). If the color of the mixture is orange or violet, add 10 μL 3 M sodium acetate, pH 5, and mix. The color of the mixture will turn yellow.
7. Add 1 gel volume of isopropanol to the sample and mix.
8. Place a QIAquick spin column in a provided 2-mL collection tube.
9. To bind DNA, apply the sample to the QIAquick column and centrifuge for 1 min until all the samples have passed through the column. Discard flow-through and place the QIAquick column back into the same tube. For sample volumes of >800 μL, load and spin again.
10. Add 0.5 mL Buffer QG to the QIAquick column and centrifuge for 1 min. Discard flow-through and place the QIAquick column back into the same tube.
11. To wash, add 0.75 mL Buffer PE to QIAquick column and centrifuge for 1 min. Discard flow-through and place the QIAquick column back into the same tube.
12. Centrifuge the QIAquick column once more in the provided 2-mL collection tube for 1 min at 17,900 × *g* to remove residual wash buffer.
13. Place QIAquick column into a clean 1.5-mL microcentrifuge tube.

14. To elute DNA, add 50 μL microinjection buffer to the center of the QIAquick membrane and centrifuge the column for 4 min. For increased DNA concentration, add 30 μL microinjection buffer (0.2 mM EDTA, pH 7.5, 5 mM Tris–Cl, pH 4) to the center of the QIAquick membrane, let the column stand for 1 min, and then centrifuge for 1 min. After the addition of microinjection buffer to the QIAquick membrane, increasing the incubation time to up to 4 min can increase the yield of purified DNA.
15. Measure the concentration of the DNA by using spectrophotometer. Adjust the DNA concentration to 2 ng/μL for microinjection (see Note 1).

3.1.2. Harvesting Fertilized Eggs

F1 hybrid (C57BL/6 female × CBA male) mice are used as embryo donor mice. Animal House light cycle is 0500–1900. Three to four-week-old F1 female mice are superovulated with intraperitoneal injection of 5 IU PMSG (Folligon, Lyppard) on day 1 at 2 pm; followed on day 3 with 5 IU hCG (Chorulon, Lyppard) at 12 pm. Females are then mated with F1 males and left overnight. Females are checked for the presence of vaginal plugs the following morning. The presence of the positive plug is indicative of those females which have mated with the males.

1. Females are euthanized by cervical dislocation and then placed on their backs on absorbent toweling. The abdomen is sprayed with 70% ethanol (this will reduce the likelihood of mouse fur entering the abdominal cavity).
2. Pick up the skin with blunt forceps, cutting through the lower abdominal wall, internal fascia, and lateral wall. Retract the skin to expose the gut.
3. To expose the ovary and oviducts, displace the gut contents to over the thorax. Using blunt forceps pick up the oviduct just distal to the uterotubal junction. Applying traction laterally, stretch out the uterus and oviduct so that you can identify the junction between the ovary and the oviduct. Cut at this junction and at the uterotubal junction to remove the oviduct. Repeat on the other side.
4. Place all oviducts in a 35-mm diameter tissue culture dish half-filled with M2 medium.
5. Locate swollen ampulla and tear open with fine forceps to release the cumulus mass.
6. Add 300 μg/mL hyaluronidase to the dish and swirl to disperse hyaluronidase evenly throughout. Leave for a few minutes, observing the cumulus cells as they detach from the eggs. Once the eggs are stripped of the cumulus cells, remove them from the Petri dish to another dish with 5× 20 μL M2 wash drops.

7. Wash embryos three times in M2 medium, followed by several washes in M16 medium droplets.
8. Place washed eggs in a M16 culture dish (prepared prior to culling the mice so as to let the M16 media temperature and gas equilibrate) and leave in incubator (37°C, 5% CO_2) until they are ready to inject.

3.1.3. Pronuclei Injection

Pronuclei microinjection is carried out according to a method described by Gordon et al. (4). The following is the brief method of pronuclear injection to be used as a guide only. Detailed method of transgenic mice generation can also be obtained from for example refs. 5, 6.

1. Prepare the holding pipette and microinjection needles. Detailed method is as described in (5, 6).
2. Dilute DNA to 2 ng/μL prior to injection with microinjection buffer and place 20 μL into a 1.5-mL tube. Place the injection needle into the tube—capillary action will draw the DNA up through the internal filament to the tip of the needle—or by the use of a long, thin pipette tip inserting DNA up through the injection needle to the tip.
3. Take the M16 culture dish containing the eggs from the incubator and under the dissecting microscope, separate the eggs which have visible 2 pronuclei (2PN). The presence of the 2PN indicates that the egg has been fertilized. These are the eggs which will be used for the microinjection. Load into the injection chamber approximately 30–40 eggs.
4. Draw the egg onto the holding pipette, and using the tip of the injection needle, spin/move the egg so that the larger male pronucleus is in the mid plane of the egg, close to the zona. Align the injection needle so that it is in the same focal plane as the 2PN. With a swift forward action, push the injection needle through the zona, oolemma, and nuclear membrane, into the pronucleus. Expel DNA into the pronucleus until you see it swell up to twice its size. Quickly and swiftly remove the injection needle form the egg.
5. Repeat until all eggs in the chamber have been injected. Return these injected eggs to a new drop within the culture dish and load chamber with more 2PN. Perform subsequent rounds of injection until all pronuclei have been injected.
6. Assess eggs for lysis and remove those eggs to a separate drop. Viable injected eggs can then be embryo transferred into 0.5-day post-coitus (dpc) pseudopregnant mother straightaway or cultured overnight (M16, 37°C, 5% CO_2) where the two cell embryos are transferred.

3.1.4. Embryo Transfer/ Implantation

Viable embryos are transferred into the oviduct of 0.5 dpc pseudo-pregnant foster mothers.

1. All viable eggs/embryos are loaded into a Petri dish with several drops of M2 media.
2. Anesthetize pseudopregnant mouse by intraperitoneal injection of 2.5% (v/v) Avertin. Place a drop of eye gel/ointment on each eye to prevent the eyes from drying out during surgery.
3. Load the 10–15 eggs/embryos into a fine drawn glass pipette and put aside well out of the way from being knocked.
4. Lay the mouse on its abdomen and make a 1–2 cm incision along the midline of the mouse, just below the last rib.
5. Locate the ovarian fat pad, ovary, and oviduct under the fascia and make another small incision. Then, delicately pull out the fat pad, ovary, oviduct, and first section of the uterus. Clamp fat pad with a Serafine clip and lay across the back of the mouse.
6. Place the mouse under the dissection microscope, orientating the mouse so that the oviduct is in view. Locate and tear open the bursa overlying the infundibulum and retract bursa to expose the opening. Change the orientation of the mouse if need be to ensure that the end of the infundibulum and opening is lying horizontally. Insert the pipette tip into the infundibulum opening and gently expel embryos into the lumen.
7. Slowly withdraw the pipette and watch for any media flowing out of the opening. Blow out the transfer pipette into a media drop to make sure that all embryos were deposited in the oviduct. Gently return the oviduct to the abdominal cavity. Reposition the skin incision over the oviduct on the other side and repeat the embryo transfer.
8. Close the incision with wound clips, wrap the mouse loosely in a tissue and place in a fresh cage on a heating pad. Monitor at intervals until the mouse recovers from the anesthetic.
9. Progeny are born 19–20 days post embryo transfer.

3.1.5. Genomic DNA Extraction from Tail Tip Biopsies

Pups delivered by foster mothers are weaned at the age of 2 weeks. The genotype of the mice is determined by PCR detection of EGFP gene integrated in the genome (see Note 2). The genomic DNA template for PCR is extracted from mice tail tips. The following is the phenol–chloroform genomic DNA extraction method.

1. A small amount of permanent ink is placed under the skin of the paw. Using a combination of dots on each paw you can identify genotype sampled to each mouse. Using a scalpel blade cut 0.2 cm in length from the tip of the tail and put into a

labeled 1.5-mL tube. Use safe-lock tubes to prevent leakage during subsequent phenol–chloroform extraction.

2. Digest freshly cut tail tips with Proteinase K (600 μg/mL) in 400 μL digestion buffer overnight at 55°C heating block/water bath.
3. On the following day, spin down the tubes briefly and add an equal volume of TE-buffered phenol and then vortex the mixture until the solution becomes white.
4. Centrifuge for 5 min at maximum speed.
5. Carefully transfer the aqueous phases into fresh tubes and add 400 μL of 24:1 chloroform–isoamyl alcohol, vortex and spin for 3 min at maximum speed. Again, transfer aqueous solutions into fresh tubes.
6. Precipitate DNA by addition of 2× volumes of absolute ethanol followed by mixing the mixture slowly. Typically, DNA strands are visible during this procedure.
7. Centrifuge for 15 min at maximum speed to pellet the DNA. The DNA pellet may or may not be visible after centrifugation. Remove the supernatant carefully using 1-mL pipette tip.
8. Wash DNA pellets once with 500 μL of 70% ethanol and recentrifuge for 5 min at maximum speed.
9. Remove the supernatant carefully and place tube upside down on paper towel to air-dry the pellet.
10. Resuspend DNA in the desired volume (approximately 50–150 μL) of TE buffer and the DNA is ready for subsequent experiments.

3.1.6. PCR Screening of Positive Founders/TG Litter

The genotype of the mice is determined by PCR amplification of EGFP gene integrated in the genome of transgenic mice (see Note 2). The EGFP PCR detection will generate a 600-bp PCR fragment. The following is the PCR detection protocol using primers amplifying the EGFP gene, performed according to standard method.

1. Prepare a mixture of PCR reagents in 0.2- or 0.5-mL tubes containing a pair of oligonucleotide primers (10–25 μM each), 1× PCR Buffer, 10 mM dNTPs mixture, 1 U of *Taq* DNA Polymerase, and sterile ddH$_2$O in total volume reaction of 25 μL.
2. Add 0.2–0.5 μL of genomic DNA extracted from tail tips as PCR template.
3. Perform amplification in thermocycler with the following PCR condition: 95°C for 5 min, followed by 95°C from 1 min, 60°C for 1 min, 72°C for 1 min and cycled for 30 times, and ended by extra extension for 5 min at 72°C.

4. To detect the amplified DNA, separate the PCR products on 1–1.5% agarose gel in 1× TAE buffer. Mix the PCR product with gel loading dye and electrophoresed through the gel.
5. Visualize the PCR fragments by ethidium bromide staining (0.5 pg/mL) and exposure to ultraviolet. Successful PCR amplification will generate a 600-bp PCR products.
6. Record the gel photographs on UV transilluminator.

3.2. Characterization of c-fms-EGFP Transgenic Mice

3.2.1. Trophoblast Primary Cell Culture

Beside the mononuclear phagocyte family, the c-*fms* gene is also expressed in trophoblasts (7, 8). In the *MacGreen* mice, the trophoblasts express the EGFP. Examination of EGFP-expressing trophoblasts can be performed in by culturing the ectoplacental cones of 6.5 dpc embryo in TC and visualized under stereomicroscope. The following method describes the isolation and culture of trophoblasts according to method described by Albieri and Bevilacqua (9).

1. Cull the mice by cervical dislocation.
2. Dissect the ectoplacental cones (EPC) from the uteri of the pregnant mice.
3. Dissect free the conceptuses from the decidua.
4. Subsequently, culture the EPC on each well of 24 wells TC plates, in MEM α-modification medium supplemented with 10% FCS, 20 U/mL penicillin, and 20 μg/mL streptomycin, in the presence or absence of 10^4 U/mL of CSF-1.
5. Examine the trophoblast under the Olympus IX70 fluorescent inverted microscope with standard fluorescein isothiocyanate (FITC) filter after 48 h when the trophoblast cells attach to the TC plate and begin their outgrowth.

In later development of the placenta, e.g., in 12.5 dpc, expression of EGFP in trophoblasts can be easily detected by whole mount placenta examination under fluorescence microscope, and this can be used as a marker to differentiate between TG or non-TG embryos before more detailed examination be performed. The other method of examination of placental trophoblasts is by placenta tissue section (performed as in Subheading 3.2.9). The placental trophoblasts EGFP expression is shown in Fig. 2a.

3.2.2. Whole Embryo Examinations

The expression of EGFP in embryonic macrophages can be examined in whole embryo or in embryo sections. For whole mount embryo examination, freshly dissected embryos can be observed directly without tissue fixation or fixed with 4% PFA in PBS and visualized in PBS or phenol red-free HBSS under an inverted Olympus IX70 fluorescent microscope. An example of EGFP expression in embryonic macrophages is shown in Fig. 2b. For examination of embryonic macrophages in tissue sections, embryo is fixed in 4% PFA in PBS and tissue cryosection is performed as in Subheading 3.2.9.

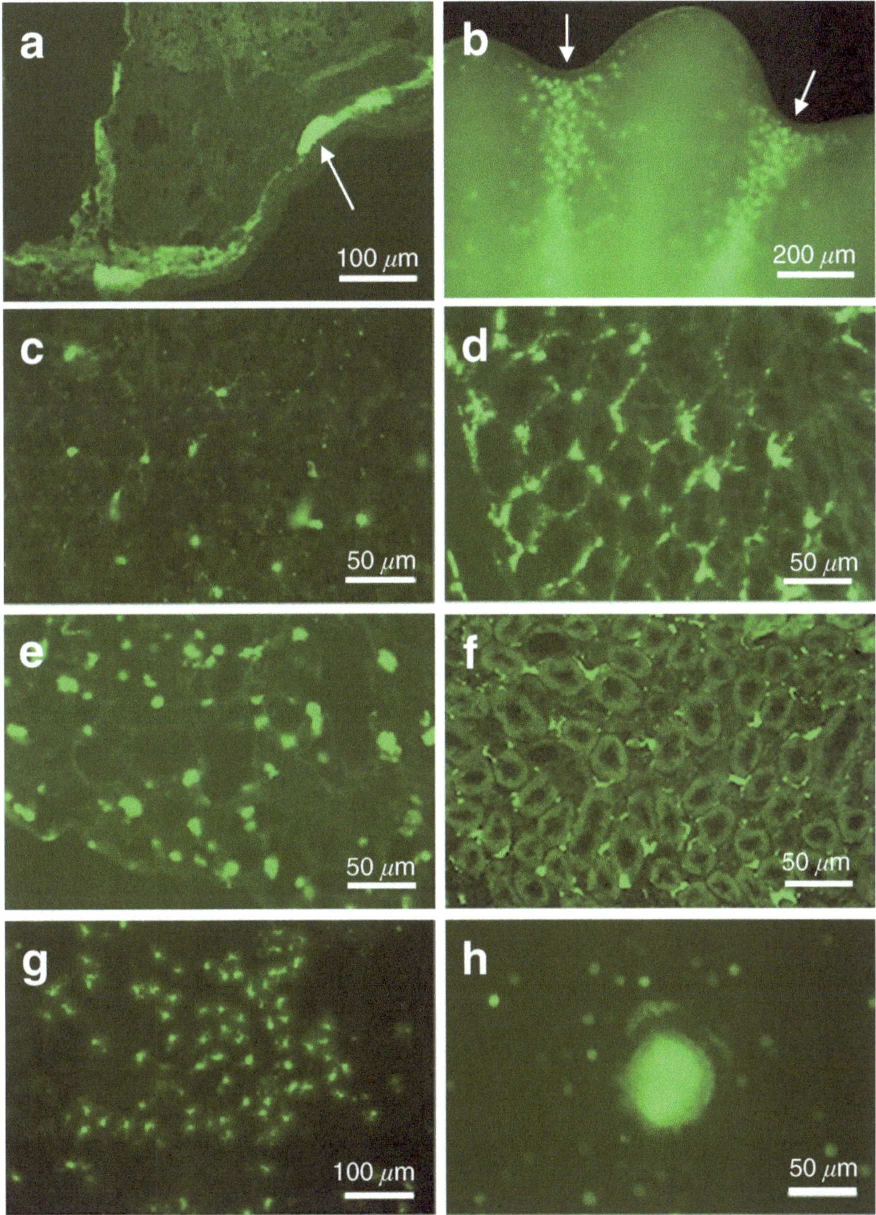

Fig. 2. Expression the EGFP in the *MacGreen* mice. (**a**) Shows the cross section of 12.5 dpc placenta showing the expression of EGFP in trophoblast giant cells (*arrow*). (**b**) Depicts the whole embryo examination of 13.5 dpc embryo, showing the expression of EGFP in embryonic macrophages in the interdigital areas (*arrows*) of the developing footplate. (**c**) Shows the remarkably ramified morphology of microglia, the macrophages of the brain expressing EGFP. (**d**) Shows the EGFP-expressing intestinal macrophages, as shown in cross section of colon crypts. (**e**) Shows the lung section with pulmonary interstitial macrophages expressing the EGFP. (**f**) Is the cross section of kidney, showing the EGFP-expressing macrophages reside in the cortex area of the kidney. (**g**) Shows the expression of EGFP in Langerhans cells examined in the epidermal sheet of the ear. (**h**) Shows the EGFP expression in osteoclast grown in TC dish.

3.2.3. Bone Marrow-Derived Macrophages

Bone marrow cells are obtained by flushing the marrow from femurs. The cells are then differentiated into macrophages in tissue culture media in the presence of CSF-1. Differentiation usually takes about 7 days.

1. Cull adult mice by cervical dislocation or CO_2 asphyxiation and make an incision along the inner thigh of the hind limbs.
2. Expose each femur and dissect femurs from mouse. Do this by teasing away muscle tissue from the femur. The initial goal is to remove the muscles which hold the femur in place, so that it can be easily dislocated.
3. Dislocate the femur. Do this by inserting forceps in the pelvic region and giving a slight twist. The femur should readily dislocate if muscle tissue has been adequately removed.
4. Cut through the tibia (i.e., below the knee joint) to extract the femur. By cutting below the knee joint, sterility of bone marrow is maintained, as the joint provides a barrier to contamination.
5. Sterilize the bone in 70% ethanol in a tissue culture dish inside a TC hood. A few minutes sterilization in Ethanol will help in removing the remaining muscle from the bone (see Note 3).
6. Open the dissected femurs in both ends using a pair of scissors. Hold the bone using a forceps over an open 50-mL tube, and flush out marrow cells with complete RPMI media by the use of 27-G needles fitted in 10-mL syringe. To maximize the number of cells obtained, gently work the needle up and down along the marrow cavity while flushing with complete RPMI media. When the femur turning white, this is an indication that the majority of marrow cells have been removed from the bone.
7. Disaggregate cell clumps by pipetting it up and down several times, and grow the cell suspensions in four 10-cm bacteriological plastic dishes using TC media in the presence of CSF-1.
8. On day 3, change the media and continue to culture up to 7 days, when typically more than 95% of cells are macrophages (10, 11).
9. Bone marrow-derived macrophages (BMM) can be harvested using cell scraper, or alternatively, by squirting media on the cell monolayer using 10-mL syringe fitted with18-G needle.
10. Perform analysis for example by FACS. Typically, more than 95% of BMM will express EGFP.

3.2.4. Resident Peritoneal Macrophages

The following protocol is used for isolating resident peritoneal macrophages, or alternatively for obtaining macrophages infiltrating the peritoneal cavity in response to inducing agents such as Brewer thioglycollate broth (see Note 4):

1. Cull mouse by CO_2 inhalation. This method is preferred than the cervical dislocation, since the latter method can cause peritoneal contamination with blood.

2. Sterilize mouse by spraying the abdomen with 70% ethanol.
3. Create a small incision on the abdomen skin, and then open up the skin with both hands to expose the intraperitoneal cavity. Spray the peritoneum with 70% ethanol.
4. Perform lavage by injecting 10 mL PBS into peritoneal cavity using 18-G needle fitted in 10-mL syringe. Avoid puncturing internal organs/intestines by inserting the needle gently and not too deeply. Rupture of internal organ/s or blood vessels will contaminate the peritoneal sample. Squirt the PBS into the peritoneal cavity.
5. Take out needle slowly. The peritoneal fat can clog the peritoneum hole and prevent the PBS leaking out of the hole.
6. Wash the mouse peritoneal cavity using the injected PBS by holding the mouse and shaking/rocking it around gently. Massaging the abdomen gently may also help dislodging the peritoneal macrophages.
7. Harvest the macrophages by inserting the needle into the upper part of the abdomen while avoiding puncturing the internal organs by inserting the needle horizontally.
8. Aspirate the PBS slowly, aiming to get at least 7 mL of PBS. Collect the lavage fluid in 50-mL tube.
9. The procedure can be repeated with another 10 mL PBS. Squirting the PBS around the peritoneal cavity may help to dislodge all the remaining macrophages.
10. Macrophages can be pelleted by 5 min centrifugation of the fluid at $400 \times g$, resuspended in 0.5–1 mL PBS and ready for subsequent experiments, e.g., by FACS, in which typically about 82% of cells are expressing EGFP.

3.2.5. Pulmonary Macrophages

The lung contains two main populations of macrophages, i.e., the alveolar macrophages and the stromal/interstitial macrophages. The alveolar macrophages can be isolated by washing the bronchoalveolar spaces with saline. The method of alveolar macrophages isolation according to Holt (12) is described here.

1. Cull mouse by CO_2 inhalation. Cervical dislocation is not recommended since it may damage the trachea. Lay the mouse on its back and pin on a cork or Styrofoam board, and dampen the animal fur with 70% ethanol.
2. Using scissors, make a small incision in the skin at the abdomen, tear skin around entire body, and peel skin upward to expose thoracic cage and neck.
3. Make small incision in the diaphragm to expose the thoracic cavity.
4. Dissect tissue from neck to expose trachea. Avoid large blood vessels around the shoulder and thyroid gland.

5. Make an incision below the trachea to allow insertion of surgical string/cotton thread.
6. Place a length of cotton thread beneath the trachea and make a small incision using a sharp scissors on the trachea in between of cartilaginous rings to allow passage of cannula into trachea. Do not cut trachea all the way through.
7. Carefully cannulate the trachea in situ using intravenous catheter tubing (of about 5 cm in length) attached into 1-mL syringe. Alternatively, a blunt 21-G needle can be used. The blunt needle can be bent in the middle of the shaft to facilitate the cannulation.
8. Tie firmly the cannula with cotton thread, and instill 1 mL of sterile PBS or HBSS supplemented with 60 mM EDTA followed by fluid collection by gentle aspiration. Repeat the procedure 3–4 washes per animal. Pool experimental group animal lavages on ice.
9. The total BAL fluid recovered will be around 2.3–2.8 mL. Centrifuge the fluid at 400×*g* for 10 min and resuspend the cell pellet in 0.5 mL of PBS.
10. Use immediately for subsequent experiment, e.g., flow cytometry analysis. Typically, 85–88% of alveolar macrophages isolated by this method will be expressing EGFP.

3.2.6. Pleural Macrophages

These cells present within the pleural cavity between the visceral and parietal pleural that covers the surface of the lung and thoracic cage. The following is the brief method of isolation of pleural macrophages:

1. Open the mouse abdomen to expose the diaphragm.
2. Carefully puncture the diaphragm and insert the lavage apparatus as described above into the cavity.
3. Wash the cavity for at least three times with 1 mL PBS or HBSS supplemented with 60 mM EDTA and collect the fluid.
4. Centrifuge the fluid at 400×*g* for 5 min, collect the cell pellet and resuspend in 0.5 mL PBS and use immediately for FACS. Typically, 40% of pleural cells expressing EGFP.

3.2.7. Culture of Osteoclasts

Bone resorptive osteoclasts share progenitors with macrophages and their production is CSF-1 dependent. The c-*fms* gene is expressed in mature osteoclasts, and in *MacGreen* mice, the osteoclasts expressed EGFP. For EGFP examination in culture, osteoclasts are obtained from femurs of 1 to 2-day-old neonatal mice.

1. Cull neonates by decapitation.
2. Dissect the femoral bones and remove all the muscle tissues.
3. Finely mince bones using scalpel blades on TC dish, add 10 mL of TC media and homogenize the minced bones.

4. Remove bone particulates by straining through 100-μm mesh cell strainers (BD-Falcon).
5. Place suspended cells on either bone slices or tissue culture plastic and allow the cells to attach for one hour prior to the removal of the nonadherent cell fraction, leaving a cell population enriched in primary osteoclasts.
6. Grow the cells in MEM-α-modification medium supplemented with 10% FCS, 30 U/mL penicillin, and 100 μg/mL streptomycin, in the presence or absence of 10^4U/mL of CSF-1.
7. Examine the EGFP expression in osteoclasts directly on TC dish under fluorescent inverted microscope. Figure 2h is an example of EGFP-expressing osteoclasts.

3.2.8. Enzymatic Disaggregation of Mouse Organs for Macrophage Isolation

Resident tissue macrophages, e.g., from spleen, liver, lung, thymus, intestines, or other organs, can be isolated by enzymatic digestion and disaggregation of tissues. Organs can be perfused with warm (37°C) enzyme cocktail by injecting the solution into the tissues following organs dissection (see Note 5).

1. Perform perfusion using 1-mL syringe fitted with 27-G needles.
2. Place the dissected organs in TC dishes and inject the enzyme cocktail.
3. Incubate the perfused organs in 37°C for about 30 min.
4. Following incubation, mince the organs finely using sterile scalpel blades.
5. Incubate the minced tissues in the enzymes cocktail for about 1–1.5 h at 37°C.
6. After incubation, disperse the cells by pipetting the solution thoroughly using 5-mL pipette, and strain the cell suspension through 100 μm or 40-μm mesh cell strainers (BD-Falcon).
7. Wash the cells in PBS and ready for subsequent experiments such as FACS (see Note 6).

3.2.9. Isolation of Macrophages from Intestinal Lamina Propria

Intestinal macrophages are isolated from small intestines according to the method described by Pavli et al. (13).

1. Cull mouse by cervical dislocation and open the abdomen.
2. Dissect-free small intestines from the mouse by pulling the intestines using forceps.
3. Expel the contents of the intestine and split intestines lengthwise using blunt point scissor.
4. Wash intestine in cold PBS and cut in 1-cm segments.
5. Remove epithelial cells by stirring the tissues in Ca- and Mg-free Hanks' balanced salt solution for 1.5 h in baffled Erlenmeyer flask at 37°C, with media changes in every 5 min

in the first three washes, followed by media changes every 10–15 min until the solution becomes clear.

6. Mince tissues in Petri dish using scalpel blades as fine as possible.
7. Transfer to 20 mL complete RPMI 1640 media per <500 mg tissues in 10-cm glass Petri dishes.
8. Perform enzymatic digestion using 0.1 U/mL collagenase, 0.8 U/mL Dispase (Roche), and 5 U/mL DNase type II for 1.5 h with occasional gentle agitation.
9. Mechanically disaggregate digested tissues by pipetting thoroughly and straining through 100-μm mesh cell strainers (BD-Falcon) to remove particulate matter.
10. Pellet cells by centrifugation, wash in PBS, and resuspend in RPMI 1640 media.
11. Perform gradient centrifugation to separate the high- and low-density cells using Nycoprep 1.068 or Nycoprep 1.070. Perform this by underlaying the resuspended cells with 2–3 mL of Nycoprep solution and centrifuging the gradient at 600 × *g*, RT, for 20 min.
12. Collect cells in the interphase between RPMI 1640 media and Nycoprep solution using glass pipette, transfer to fresh tubes, resuspend in PBS and ready for FACS analysis.

3.2.10. Examination of EGFP Expression in Tissue Sections

Macrophages take up residence in virtually every tissue of the body. Macrophages from different tissues exhibit a wide range of phenotypes with regard to their morphology, antigenic expression, and function (14). The expression of EGFP in macrophages from various tissues of the mice can be visualized using fluorescence microscope examination of tissue section. Figure 2 is an example of the EGFP expression in brain microglia (panel c), macrophages in the intestinal lamina propria (panel d), in the lung (panel e), and kidney (panel f) of the *MacGreen* mice visualized in tissue sections. The following protocol is the cryosection method for EGFP examination in various tissues.

1. Dissect tissue from mouse and rinse briefly in saline solution (PBS or HBSS) in a Petri dish.
2. Transfer tissue to 4% PFA at room temperature and allow to “fix” for at least 2 h. Keep in mind that for fixation to occur the PFA solution must penetrate the tissue. May have to allow longer fixing time for more dense tissue or for tissues with low surface-area/volume ratios.
3. Remove PFA solution, and immerse tissue in 18% sucrose in PBS for cryopreservation. Store overnight at 4°C in the dark to prevent photobleaching of the EGFP.
4. Transfer tissue from sucrose solution to a dry Petri dish.

5. Using a scalpel blade, cut a representative slice (from the middle of the tissue, i.e., requires two cuts). Ensure that the slice has two flat surfaces.
6. Place the tissue slice with cross-section face-up, into a labeled (with pencil, not pen) plastic sample holder. Cover the tissue slice with OCT medium, ensuring no bubbles formed.
7. Transfer the sample holder containing the tissue slice to dry ice–absolute ethanol slurry. The embedding medium will become opaque when it has set (see Note 7).
8. When the embedding medium is completely set (all white), transfer the sample holder to −80°C for storage. Embedded tissue should be stored at least overnight, but can be stored at −80°C for a long period (weeks).
9. Frozen sections (8- to 16-μm thickness, depend on the tissue types) are cut at −12°C with a LEICA cryostat model CM305O (Leica Instrument, Germany) following the instrument manual.
10. Wash the glass slides with the tissue sections on it for 5 min in a slide jar of PBS on the mechanical shaker. In the case of EGFP samples, ensure that the jar is wrapped in aluminum foil to minimize exposure to light. The purpose of the PBS wash is to melt/dissolve the mounting medium in the sections.
11. Dry out the slide. To remove PBS, flick slide and then wipe with a lint-free tissue underneath, and then wipe down each side of the top of the slide, being careful not to disturb the sections.
12. Drop of fluorescent mounting media (DAKO) in a streak along the slide. Place coverslip gently over the top ensuring no bubbles formed.
13. Wrap in a lint-free tissue and apply pressure using a tip-box, to remove excess mounting media.
14. Store completed slides in slide box. The slides must be analyzed on the same day they are prepared, as they cannot be stored.
15. Visualize the fluorescence of the EGFP under an inverted fluorescent microscope (e.g., Olympus AX70 or IX70 microscope) with standard FITC filter set.

3.2.11. Examination of Langerhans Cell in the Epidermal Sheet of the Skin

The Langerhans cells in the skin require CSF-1R for their development and express EGFP in *MacGreen* mice. Besides using tissue section, the EGFP expression can be examined in the epidermal sheet of the ear of the mice. The examination is done according to a method described by Price et al. (15). An example of EGFP examination in Langerhans cells in the epidermal sheet of the ear is shown in Fig. 2g.

1. Cull mice by cervical dislocation and the cut a portion of the ear.

2. Prepare the epidermal sheet by separating the ear into dorsal and ventral halves with forceps under dissecting microscope.
3. Incubate the dorsal ear halves in 20 mM EDTA/PBS for 1 h.
4. Separate the epidermal sheet and mount on microscope slides.
5. Examine the EGFP fluorescence in Langerhans cells under fluorescence microscope with or without coverslips.

4. Notes

1. Do not freeze diluted DNA. It is better to dilute the DNA just before microinjection. It is absolutely crucial that DNA be very pure for injection—i.e., free of phenol, alcohol, agarose, enzymes, etc.—to help keep the injected embryos viable. It is also important that DNA be free of particulate matter that could clog injection pipettes, so it is recommended to centrifuge the DNA for extended period (e.g., 30 min) at maximum speed just before injection.
2. The described extraction method is to provide high-quality/pure DNA that can be also be used in Southern blot analysis, as this method can also be performed to confirm the genotype of founders by using the standard method, e.g., as described by Sambrook et al. (16). EGFP PCR product is used as a probe. DNA obtained from simpler DNA extraction method, e.g., by isopropanol precipitation, is sufficient to be used in PCR detection. For quick screening of the transgenic mice litter, green-fluorescent monocytes can be observed by visualizing blood smear from tail tip prepared on microscope slides examined under fluorescence microscope.
3. Avoid prolonged exposure to ethanol as it may cause bone to become brittle, causing it to shatter when the bone is cut.
4. Thioglycollate-elicited peritoneal macrophages (TEPM) can be isolated using the above method, following intraperitoneal injections of 1 mL of 10% thioglycollate broth (Sigma-Aldrich). Peritoneal lavage can be performed after 3–4 days stimulation.
5. Alternatively, enzyme cocktail could be intracardially perfused into the whole mouse using standard cardiac perfusion method (17). For lung, enzyme perfusion can be done by inserting blunt 21-G needles in the trachea as in the process of collecting bronchoalveolar lavage macrophages.
6. Gradient centrifugation to separate the high- and low-density cells can be performed to enrich the macrophage number, e.g., using Nycoprep 1.068 or Nycoprep 1.070 (Nycodenz). This

can be done by underlaying resuspended cells with 2–3 mL of Nycoprep solution and centrifuging the gradient at 600 × *g* at room temperature for 20 min. Cells in the interphase between RPMI media and Nycoprep solution are collected using glass pipette, transferred into fresh tubes, resuspended in PBS and ready for subsequent analysis.

7. Ensure that no ethanol washes over the top of the tissue slice. Ethanol will alter the composition of the embedding medium, making it fragile and prone to disintegration during cutting with the cryostat.

Acknowledgments

We would like to thank Professor David A. Hume for his supervision during the course of this study and the Institute for Molecular Bioscience, the University of Queensland, Brisbane, Australia for providing facilities for the generation and characterization of the *MacGreen* mice.

References

1. Sasmono, R. T., Oceandy, D., Pollard, J. W. et al. (2003) A macrophage colony-stimulating factor receptor-green fluorescent protein transgene is expressed throughout the mononuclear phagocyte system of the mouse, *Blood* **101**, 1155–1163.
2. Sasmono, R. T., Ehrnsperger, A., Cronau, S. L. et al. (2007) Mouse neutrophilic granulocytes express mRNA encoding the macrophage colony-stimulating factor receptor (CSF-1R) as well as many other macrophage-specific transcripts and can transdifferentiate into macrophages in vitro in response to CSF-1, *J Leukoc Biol* 82, 111–123.
3. Yue, X., Favot, P., Dunn, T. L. et al. (1993) Expression of mRNA encoding the macrophage colony-stimulating factor receptor (c-fms) is controlled by a constitutive promoter and tissue-specific transcription elongation, *Mol Cell Biol* **13**, 3191–3201.
4. Gordon, J. W., Scangos, G. A., Plotkin, D. J. et al. (1980) Genetic transformation of mouse embryos by microinjection of purified DNA, *Proc Natl Acad Sci USA* 77, 7380–7384.
5. Awasthi, A., and Kuchroo, V. K. (2009) Th17 cells: from precursors to players in inflammation and infection, *Int Immunol* **21**, 489–498.
6. Nagy, A., Gertsenstein, M., Vintersten, K.et al. (2003) *Manipulating the Mouse Embryo: A Laboratory Manual* 3ed., Cold Spring Harbor Laboratory Press.
7. Hunter, C. A., Roberts, C. W., Alexander, J. (1992) Kinetics of cytokine mRNA production in the brains of mice with progressive toxoplasmic encephalitis, *European Journal of Immunology* **22**, 2317–2322.
8. Sherr, C. J. (1990) Colony-stimulating factor-1 receptor, *Blood* **75**, 1–12.
9. Albieri, A., Bevilacqua, E. (1996) Induction of erythrophagocytic activity in cultured mouse trophoblast cells by phorbol myristate acetate and all-trans-retinal, *Placenta* **17**, 507–512.
10. Hume, D. A., Gordon, S. (1983) Optimal conditions for proliferation of bone marrow-derived mouse macrophages in culture: the roles of CSF-1, serum, Ca2+, and adherence, *J Cell Physiol* **117**, 189–194.
11. Tushinski, R. J., Oliver, I. T., Guilbert, L. J. et al. (1982) Survival of mononuclear phagocytes depends on a lineage-specific growth factor that the differentiated cells selectively destroy, *Cell* **28**, 71–81.
12. Holt, P. G. (1979) Alveolar macrophages. I. A simple technique for the preparation of high numbers of viable alveolar macrophages from small laboratory animals, *J Immunol Methods* **27**, 189–198.
13. Pavli, P., Woodhams, C. E., Doe, W. F. et al. (1990) Isolation and characterization of antigen-presenting dendritic cells from the

mouse intestinal lamina propria, *Immunology* **70**, 40–47.

14. Sasmono, R. T., Hume, D. A. (2004) The Biology of Macrophages, in *The Innate Immunity Response to Infection* (Kaufmann, S. E., Medzhitov, R., and Gordon, S., Eds.), The American Society of Microbiology press.
15. Price, A. A., Cumberbatch, M., Kimber, I. et al. (1997) Alpha 6 integrins are required for Langerhans cell migration from the epidermis, *J Exp Med* **186**, 1725–1735.
16. Sambrook, J., and Russel, D. W. (2001) *Molecular Cloning: A Laboratory Manual*, Cold Spring Harbour Laboratory Press,, Cold Spring Harbour, New York.
17. Zeller, R. (2001) Fixation, embedding, and sectioning of tissues, embryos, and single cells, *Curr Protoc Pharmacol* doi: 10.1002/0471141755.pha03ds07.

Chapter 12

Generation of Mouse Bone Marrow-Derived Macrophages

Silvia Manzanero

Abstract

Isolation of resident macrophages from mouse tissues involves complex procedures for a small yield. This is inconvenient for many functional macrophage assays, which require large numbers of relatively homogeneous cells. An alternative method is the culture of bone marrow cells in vitro with appropriate growth factors, to allow the differentiation of precursor cells into large numbers of macrophages. This procedure is easy and inexpensive except for the use of M-CSF, the macrophage colony stimulating factor, and it is characterised by high yield and reproducibility. Once obtained, bone marrow-derived macrophages (BMMs) can be used for a considerable number of functional and structural assays and are commonly regarded as a model for the role of resident macrophages in the innate immune system.

Key words: Bone marrow, Macrophage, Mouse, Macrophage colony stimulating factor, CSF-1, Innate immune system

1. Introduction

Tissue-resident macrophages are an essential component of the innate immune system. They derive from blood monocytes, which migrate to the tissues and differentiate, and establish throughout the tissue to accomplish functions of defence against pathogens as well as homeostasis (1). These functions are varied and for this reason their study is important, but their sparse presence in the tissue makes isolation hard and yields insufficient for common assays. An alternative strategy to study macrophages is in vitro differentiation, a convenient way to obtain a highly pure and abundant population of cells. In vitro differentiation can be achieved from blood monocytes, but in the mouse, a much more efficient technique involves the use of bone marrow. In it, there is an abundance of precursor cells that respond to growth factors by rapidly dividing and producing millions of differentiated macrophages from a single mouse in just a few days.

Robert B. Ashman (ed.), *Leucocytes: Methods and Protocols*, Methods in Molecular Biology, vol. 844,
DOI 10.1007/978-1-61779-527-5_12,

The growth factor most extensively used in this method is M-CSF, also known as colony stimulating factor 1 (CSF-1). It specifically affects the proliferation and differentiation of committed precursor cells into macrophages and it is essential for macrophage survival and function (2). The resulting population of macrophages is quite homogeneous and the procedure reproducible. For this reason, this method is widely used to test these innate immune cells on functional assays such as phagocytosis, pathogen killing, antigen presentation and cytokine production in the presence of different stimuli. Gene function can be assessed by comparing BMMs from *knock-out* mice against those from isogenic controls. Protein, DNA or RNA can be isolated and studied, and numerous other assays performed.

It is important to note that although BMMs are a widely used, invaluable model for resident macrophages, they are not like resident macrophages. An indication of this is that resident macrophages themselves are not homogeneous, for they differentiate to suit the environment of each tissue and the pathogens each tissue is likely to encounter (1).

2. Materials

1. 6–9 week mice (see Note 1).
2. RPMI 1640 tissue culture medium, supplemented with 1× GlutaMAX (Gibco) and 10% heat-inactivated Foetal Bovine Serum (FBS, see Note 2).
3. Saline solution: 0.9% sodium chloride solution for irrigation, sterile (Baxter). Keep sterile and store at 4°C.
4. Purified recombinant M-CSF (CSF-1), available from protein specialised companies. Dilute upon arrival, aliquot and store at –80°C. Before use, thaw and add to tissue culture medium. Store this supplemented medium at 4°C for up to 2 months.
5. Surgical scissors and forceps: one clean set and one sterile set.
6. 70% ethanol solution.
7. 27-gauge needles.
8. 20-ml syringe.
9. Low-lint tissues (e.g. KimWipes, Kimberly-Clark).
10. Standard bacterial Petri dishes, sterile.
11. Three 50-ml sterile tubes.

3. Methods

1. Warm up tissue culture medium and tissue culture medium supplemented with rM-CSF in water bath, at 37°C.
2. Sacrifice mouse by approved method and spray lower body thoroughly with 70% ethanol solution. Pull both back legs apart until cracks are heard, indicative of femurs being disjoined from the hips.
3. With clean scissors and forceps, cut skin around one of the back legs. Pull skin down towards paw and remove.
4. Pierce leg with scissors and tear muscle alongside bone by opening scissors. Do not cut through muscle as there is danger of cutting through bone. Repeat this on both sides of femur and tibia until both bones are roughly clean.
5. Cut ligaments between femur and hip. Cut bone below the ankle joint. Place femur and tibia in a tube of ice-cold, sterile saline solution.
6. Repeat the process with the other back leg.
7. From this stage all steps must be done in sterile conditions and with sterile equipment. Wipe femur and tibia by rubbing with low-lint tissues (these need not be sterile) to remove attached tissue. Place the bones in a dish containing a small amount of 70% ethanol.
8. Prepare a syringe full of 37°C medium (RPMI 1640, supplemented with GlutaMAX and 10% FBS), with a 27-gauge needle and a sterile 50-ml tube.
9. Separate tibia from femur by bending slightly at the knee joint. Discard the knee. Hold femur with sterile forceps and cut the top end with sterile scissors (see Note 3). Insert the needle through the cut end and flush bone marrow with medium into a sterile tube. While flushing, move the needle up and down while scraping the inside of the bone. Do this until the bone appears clear. This should use approximately 5 ml per bone. Discard bone.
10. Hold tibia with sterile forceps and cut the top and bottom ends with sterile scissors. Insert the needle through the knee end and repeat the process. Repeat with remaining bones.
11. Centrifuge the cell suspension at $150 \times g$, 5 min. Discard the supernatant and replace with RPMI, supplemented with GlutaMAX, 10% FBS and recombinant M-CSF (see Notes 4 and 5). Pipette up and down several times to disaggregate bone marrow (see Note 6).
12. Label Petri dishes at a ratio of two per bone. Pipette cell suspension into dish and complete volume to 10 ml per plate

with supplemented medium with M-CSF. Incubate cells at 37°C, 5% CO_2 for 5 days (see Note 7).

13. On day 5, wash cells twice with 5 ml of saline solution at room temperature (see Note 8). Scrape the cells off the dish with a cell scraper, transfer to a sterile tube and count. There should be about 5×10^6 per dish (10^7 per bone). Centrifuge at $150 \times g$ for 5 min and resuspend in the desired volume of medium with M-CSF. The cells are ready to be assayed (see Note 9). Cells can be washed and fresh medium added; however, it is recommended that they are used within 2 days.

4. Notes

1. Differences have been noted between BMMs from male and female mice (3). To prevent variation, using sex-matched, as well as age-matched, mice is recommended.
2. It is important for the FBS to be low in endotoxin, and less than 2 EU/ml is recommended. Endotoxin is a strong macrophage activator and its presence could alter the results. Other components of FBS could also activate or inhibit activation of BMMs (4), so it is convenient to test a few batches for activation of unstimulated macrophages.
3. The femur does not need to be cut at the knee end. Removing the knee is sufficient to open the bone at that end.
4. To assess the optimal concentration of rM-CSF, titration and testing should be done prior to experiment. A concentration range from 1 ng/ml to 1 μg/ml is recommended. After 5 days, the concentration that produces yields close to 10^7 BMMs per bone should be chosen.
5. A more affordable option to rM-CSF is the use of L929 cell-conditioned medium. However, the concentration of M-CSF will vary between batches and L cells will also produce varying amounts of GM-CSF, affecting the consistency of results (5).
6. Some fractions of bone marrow might not easily disaggregate and clots might appear. These large fractions should not interfere with the growth of BMMs and will easily wash on day 5.
7. Bone marrow-derived macrophages show a foamy appearance in culture. This is due to the fact that they do not only use M-CSF, but they also internalise it and degrade it. M-CSF internalisation causes cytoplasmic vacuolation, which is evident under the microscope (6).
8. The main criterion for separation of macrophages from other cells in this model is that macrophages adhere to regular dishes

not treated for tissue culture, while other cells do not. However, different types of plastic result in different rates of adhesion. It is important to make sure that macrophages are not lifting off the dish during washes.

9. Assays must be done in medium with M-CSF, because macrophages need M-CSF not only for growth and differentiation but also for survival (2).

References

1. Handel-Fernandez, M.E., and Lopez, D. M. (2000) Macrophages in tissues, fluids and immune response sites, In: Paulnock, D.M. (ed) Macrophages. Oxford University Press, Oxford, pp 1–30.
2. Stanley, E.R., Guilbert, L.J., Tushinski, R.J. et al. (1983) CSF-1--A mononuclear phagocyte lineage-specific hemopoietic growth factor, *J Cell Biochem* **21**, 151–159.
3. Bhasin, J.M., Chakrabarti, E., Peng, D-Q. et al. (2008) Sex specific gene regulation and expression QTLs in mouse macrophages from a strain intercross, *PLoS ONE* **3**(1), e1435. doi:10.1371/journal.pone.0001435.
4. Mills, C.D., Kincaid, K., Alt, J.M et al. (2000) M-1/M-2 Macrophages and the Th1/Th2 Paradigm, *J Immunol* **164**, 6166–6173.
5. Pang, Z.J., Chen, Y. and Xing, F.Q. (2001) Effect of L929 cell-conditioned medium on antioxidant capacity in RAW264.7 cells, *Br J Biomed Sci* **58**, 212–216.
6. Tushinski, R.J., Oliver, I.T., Guilbert, L.J. et al. 1982) Survival of mononuclear phagocytes depends on a lineage-specific growth factor that the differentiated cells selectively destroy, *Cell* **28**, 71–81.

Chapter 13

Isolation and Differentiation of Monocytes–Macrophages from Human Blood

Dipti Vijayan

Abstract

The prevalence of fungal infections remains high, and it is associated with significant mortality and morbidity. Macrophages are heterogeneous population of effectors enriched in regions of *Candida* colonization. These cells sense *Candida*, and are critical in the resolution of these infections. Here, we describe how macrophages are generated in the presence of colony-stimulating factor-1 (CSF-1); an important cytokine required for the survival, proliferation and ex-vivo differentiation of monocytes to macrophages.

Key words: Macrophages, CSF-1, *Candida* infections, Monocyte differentiation, MACs separation

1. Introduction

Candida infections present a serious clinical burden with overwhelming episodes diagnosed each year (1, 2). The resistance of the fungus to antifungal drugs have prompted researchers to investigate the host–pathogen dynamics for developing strategies that counterattack the invading pathogen (3).

Macrophages are key immune regulators that rapidly infiltrate the infected area for fungal killing and secretion of effective mediators that prime adaptive responses (4, 5). Fungicidal activity by macrophages occurs via both oxygen-dependent and independent mechanisms. While the former mechanism occurs by the release of toxic oxygen metabolites; O_2^- and $H_2O_2^-$, in the latter, the macrophages are adhered onto a collagen matrix that facilitates significant death of *Candida* (6). IFN-γ, a potent cytokine secreted by T cells, augments the fungicidal activity of the macrophages (7). Activated macrophages release a pool of cytokines that mediate local and systemic effects. Most importantly, TNF-α and IL-12 direct helper $CD4^+$ T cells towards protective Th1 immunity. Thus,

Robert B. Ashman (ed.), *Leucocytes: Methods and Protocols*, Methods in Molecular Biology, vol. 844, DOI 10.1007/978-1-61779-527-5_13, © Springer Science+Business Media, LLC 2012

C. albicans colonization triggers the recruitment and activation of macrophages to the infected area, and any inactivation of these cells enhances the susceptibility to *Candida* infections (8, 9).

Monocytes are continuously flushed out of the bloodstream to enter body tissue, where they undergo differentiation to macrophages, thereby acting as precursors to phagocytes like macrophages (10). Differentiation of macrophages occurs in the presence of several haematopoietic cytokines, colony-stimulating factor-1 (CSF-1) being the master switch regulating this differentiation. The CSF-1 receptor is a family of protein kinases implicated in the regulation of monocyte–macrophage transition (11). This chapter provides details on the human monocytic differentiation from peripheral blood, triggered by an external signal, CSF-1.

2. Materials

1. Ficoll Hypaque (GE Healthcare).
2. 0.9% saline.
3. 0.4% trypan blue.
4. Magnetic-activated cell sorting (MACs) buffer containing 0.9% saline supplemented with 0.5% heat inactivated foetal bovine serum (FBS) and 2 mM EDTA.
5. Anti-human CD14 beads (Miltenyi Biotec).
6. MACs apparatus containing separation columns and magnets (Miltenyi Biotec).
7. Anti-human CD14-FITC antibody (Miltenyi Biotec) fluorescence-activated cell sorting (FACs) buffer containing 0.9% saline supplemented with 5% heat inactivated serum/bovine serum albumin and 0.1% sodium azide.
8. RPMI medium supplemented with 10% heat inactivated FBS 1% glutamine and 100 ng/ml CSF-1 (Peprotech).
9. Flow cytometer.

3. Methods

3.1. Ficoll-Hypaque Density Centrifugation to Isolate PBMC (from RBC and PMN)

1. Before beginning, bring Ficoll-Hypaque to room temperature.
2. Dilute whole blood with an equal volume of saline.
3. Aliquot 10 ml Ficoll-Hypaque into 50-ml tubes.
4. Gently overlay 20 ml diluted whole blood onto 10-ml Ficoll in the tube using a 25-ml pipette (see Note 1).

5. Centrifuge for 25 min, 180–200 × *g*, room temperature, no brake.
6. After centrifugation, set temperature in centrifuge to 10°C for next spin.
7. Carefully aspirate the plasma layer using 10-ml pipette, ensuring that the tip does not touch the monolayer.
8. Collect the monolayer into 50-ml tubes using a 10-ml pipette. Collect as far down as possible without disturbing RBC pellet.
9. Wash with cold saline (up to the 50 ml mark) and centrifuge, 10 min, 400 × *g*, 10°C.
10. Carefully discard supernatant by tipping out gently, leaving approx 3–5 ml in the tube, and flick pellet to resuspend. Then, wash again with cold saline (up to 50 ml), centrifuge, 6 min, 400 × *g*, 10°C, brake on.
11. Discard supernatant. Pool into fewer tubes. Wash cells with cold saline and centrifuge, 6 min, 400 × *g*, 10°C, brake on.
12. Discard supernatant and resuspend pellet in 5 ml saline.
13. Pool samples and finally resuspend in 10 ml to count cells using 0.4% trypan blue (two parts sample: one part dye).

3.2. MACS Separation for Positive Selection of CD14⁺ Monocytes

1. Centrifuge cells at 400 × *g* for 10 min and completely remove the supernatant.
2. Resuspend pellet in 40 μl buffer/10^7 total cells (example: Donor 1 had 10.5×10^7 cells. Therefore, cells should be suspended in 10.5×40 μl MACS buffer = 420 μl.) (see Note 2).
3. Add 10 μl CD14 microbeads/10^7 cells. (Example: If donor had 10.5×10^7 PBMCs, add 10.5×10 μl = 1.25 ml of beads).
4. Mix and incubate for 15 min at 4°C. Set centrifuge temperature to 4°C (see Note 3).
5. Make up sample to 50 ml with MACS buffer, spin 400 × *g* for 5 min at 4°C.
6. Resuspend pellet in 250 μl MACS buffer, up to 10^8 cells (i.e. if donor has 10.5×10^7 cells, then total volume of the buffer will be- $250 \times 1.05 \times 10^8 = 262.5$ μl buffer).
7. Pre-rinse LS column with 3 ml or MS column with 500 μl of pre-cooled MACS buffer (see Note 4).
8. Add 500 μl cell suspension to LS column and run through, wash with 3 × 3 ml MACS buffer OR.

   Add 500 μl cell suspension to MS column and run through, wash with 3 × 500 μl MACS buffer.
9. Elute positively selected cells in 5 ml by removing column (for LS columns) or 1 ml of MACs buffer (for MS columns) from magnetic field and applying plunger force.

10. Wash eluted cells in cold MACS buffer. Take an aliquot during last wash to stain for FACS.
11. This aliquot is washed twice in FACs buffer, then labelled with CD14-FITC antibody for 15 min at 4°C, in the dark. Centrifuge at 400 × *g* for 3 min, wash twice in FACs buffer, and analyse by flow cytometry.
12. Resuspend the remaining fraction in RPMI supplemented with 10% heat inactivated FBS, 1% glutamine, and 100 ng/ml CSF-1.

3.3. Monocyte Differentiation into Macrophages

1. Culture monocytes in medium for approximately 6 days.
2. Monitor cells, but change/replenish medium at day 5, retaining adherent cells (see Note 5).
3. Harvest on day 6 and re-plate for experiments—retain adherent cells.
4. Yield of MDM from $CD14^+$ monocytes: anywhere from 30 to 70%.

4. Notes

1. Overlaying of diluted blood samples over the Ficoll should be done slowly by placing the Ficoll tube at an angle of 45° and gently adding blood so it flows on top of the Ficoll.
2. All solutions for MACs separation should be pre-cooled and kept on ice. The magnets used for separation should also be cooled before using.
3. Incubation of the beads must be carried out as stated to avoid any non-specific binding.
4. LS or MS columns are cell separators designed by Miltenyi Biotec for the isolation of cells. While the LS column permits separation of 1×10^8–2×10^9 labelled cells, the MS columns have a capacity between 10^7 and 2×10^8 cells.
5. MDM differentiation is donor dependent. While some donors demonstrate "macrophage-like" sometimes you will see many non-adherent, "monocyte-like" cells. The non-adherent cells can be obtained and replated using CSF-1.

References

1. Richardson, M.D. (2005) Changing patterns and trends in systemic fungal infections, *J Antimicrob Chemother* **56** (Suppl 1), i5–i11.
2. Lehrnbecher, T., Frank, C., Engels, K.et al. (2010) Trends in the postmortem epidemiology of invasive fungal infections at a university hospital, *J Infect* **61**, 259–65.
3. Chaffin, W.L., Lopez-Ribot, J.L., Casanova, M. et al. (1998) Cell wall and secreted proteins of *Candida albicans*: identification, function, and expression, *Microbiol Mol Biol Rev* **62**, 130–80.
4. Mansour, M.K., Levitz S.M. (2002) Interactions of fungi with phagocytes, *Curr Opin Microbiol* **5**, 359–65.

5. Villar, C.C., Dongari-Bagtzoglou, A. (2008) Immune defence mechanisms and immunoenhancement strategies in oropharyngeal candidiasis, *Expert Rev Mol Med* **10**, e29.
6. Newman, S.L., Bhugra, B., Holly, A. et al. (2005) Enhanced killing of *Candida albicans* by human macrophages adherent to type 1 collagen matrices via induction of phagolysosomal fusion, *Infect Immun* **73**, 770–7.
7. Baltch, A.L., Bopp, L.H., Smith, R.P. (2005) Effects of voriconazole, granulocyte-macrophage colony-stimulating factor, and interferon gamma on intracellular fluconazole-resistant *Candida glabrata* and *Candida krusei* in human monocyte-derived macrophages, *Diagn Microbiol Infect Dis* **52**, 299–304.
8. Newman, S.L. Holly, A. (2001) *Candida albicans* is phagocytosed, killed, and processed for antigen presentation by human dendritic cells, *Infect Immun* **69**, 6813–22.
9. Farah, C.S., Elahi, S., Pang, G. et al. (2001) T cells augment monocyte and neutrophil function in host resistance against oropharyngeal candidiasis, *Infect Immun* **69**, 6110–8.
10. Torosantucci, A., Romagnoli, G., Chiani, P. et al. (2004) *Candida albicans* yeast and germ tube forms interfere differently with human monocyte differentiation into dendritic cells: a novel dimorphism-dependent mechanism to escape the host's immune response, *Infect Immun* **72**, 833–43.
11. Imamura, K., Dianoux, A., Nakamura, T. et al. (1990) Colony-stimulating factor 1 activates protein kinase C in human monocytes, *EMBO J* **9**, 2423–8, 2389.

Chapter 14

In Vitro Measurement of Phagocytosis and Killing of *Cryptococcus neoformans* by Macrophages

André Moraes Nicola and Arturo Casadevall

Abstract

Macrophages are pivotal cells in immunity against a wide range of pathogens. Their most important property, as suggested by their name, is to ingest pathogens, leading to their killing, the release of inflammatory mediators and antigen processing. On the other hand, macrophages can also be exploited by microbes as a niche for survival in the host, as exemplified by *Cryptococcus neoformans*. This encapsulated yeast is an important cause of meningoencephalitis in immunocompromised people, particularly those with AIDS. Using culture and microscopy techniques, we present here methods that can be used to quantify phagocytosis of *C. neoformans* and its killing by macrophages, as well as the viability of the phagocyte after interaction.

Key words: Macrophage, J774 cells, Phagocytosis, Microscopy, *Cryptococcus neoformans*, Giemsa, Trypan blue, Colony forming units

1. Introduction

Macrophages are cells of the immune system that specialize in clearing self and non-self material via ingestion (phagocytosis) and degradation. In addition to this innate immunity effector role, macrophages also have two additional important functions. They can secrete many cytokines and chemokines that modulate the immune response and present antigens via MHC II, thus making this cell a crucial link between innate and adaptive immunity.

Macrophages play a central role in immunity against several microbial pathogens. In this chapter, we focus on one of them, *Cryptococcus neoformans*, an opportunistic pathogen that causes about 600,000 deaths/year (1). *Cryptococcus neoformans* cells are surrounded by a polysaccharide capsule, which is their most important virulence attribute (2). The presence of the capsule hinders

Robert B. Ashman (ed.), *Leucocytes: Methods and Protocols*, Methods in Molecular Biology, vol. 844,
DOI 10.1007/978-1-61779-527-5_14, © Springer Science+Business Media, LLC 2012

access to structures on the surface of the yeast cell that are recognized by macrophage phagocytic receptors, blocking its ingestion by phagocytic cells unless opsonins are present (3). This fact permits detailed studies about phagocytosis mediated by different means, such as antibodies or complement.

The interaction between macrophages and *C. neoformans* can be studied with a wide range of different experiments, each addressing a different aspect (4). All of them start with infection of macrophages with *C. neoformans* in tissue culture plates, detailed in Subheading 3.1. The kinetics of phagocytosis is quantified by the phagocytosis assay (Subheading 3.2), in which the number of macrophages with internalized *C. neoformans* cells is counted on a microscope after staining. The fate of the phagocytosed fungal cells is measured by plating them and counting colony forming units (CFUs) using the *C. neoformans* killing assay (Subheading 3.3), whereas the effects of such interaction on the macrophage itself are evaluated by trypan blue staining followed by microscopy on the macrophage viability assay (Subheading 3.4).

2. Materials

1. Tissue culture medium: Dulbecco's modified Eagle's medium (DMEM) with 4.5 g/L glucose, L-glutamine and sodium pyruvate supplemented with 10% fetal calf serum, 10% NCTC-109 (Gibco/Invitrogen, Grand Island, NY), 1× MEM nonessential amino acids, 100 IU/mL penicillin and 100 μg/mL streptomycin.
2. IFN-γ: Recombinant murine interferon-γ, produced in *Escherichia coli* (Roche, Mannheim, Germany). Prepare small aliquots and keep frozen at −80°C for prolonged storage. Individual aliquots may be kept at −20°C for immediate use.
3. Lipopolysaccharide (LPS) diluted in PBS to a concentration of 1 mg/mL. Aliquot and store at −20°C.
4. Cold methanol: Store an aliquot of methanol in an explosion-proof −20°C freezer until needed.
5. Phosphate-buffered saline (PBS): 137 mM NaCl, 2.7 mM KCl, 1.5 mM KH_2PO_4, 8.5 mM Na_2HPO_4.
6. Giemsa stain: Purchased as concentrated stock (Ricca Chemical Company, Arlington, TX). Dilute 20 times in water and, if necessary, filter the diluted dye to remove precipitates before addition to the plate.
7. Sabouraud dextrose broth and agar.
8. Trypan blue 0.4% solution.

9. Mouse serum complement (Pel-Freez Biological, Rogers, AR): aliquot in microcentrifuge tubes and stock at −80°C. Remove aliquots immediately before use and discard leftovers, as complement proteins are unstable.
10. Cellstripper solution (Mediatech Inc, Manassas, VA).

3. Methods

3.1. Infection of Macrophages with Cryptococcus neoformans

This protocol describes the preparation of 96-well plates containing macrophages and their infection with *C. neoformans*. These plates will be used in all three experiments described later.

1. Grow J774 cells (see Note 1) in tissue culture medium; 1–2 days prior to the experiment, split the cells so that they will be confluent at the day of the experiment. Use plastic Petri dishes that have not been tissue culture treated so that the cells can be more easily detached.
2. Maintain a stock of *C. neoformans* isolate H99 (see Note 2) in glycerol at −80°C. In the week prior to starting the experiment, inoculate a Sabouraud dextrose agar plate with *C. neoformans* from the stock. Grow for 2 days at 30°C and maintain at 4°C until used.
3. Day 1: Macrophages are plated to achieve a final density of 50,000 cells per well at the time of infection and activated if necessary. To achieve this, one volume of a macrophage suspension with 25,000 cells is mixed with one volume of medium that is supplemented with two times the final concentration of IFN-γ and LPS.
4. Remove the medium from the Petri dish containing J774 cells. Add 5 mL of fresh tissue culture medium into the dish and pipette up and down to dislodge the cells.
5. Count a 1:10 dilution of the cell suspension on a hemocytometer. One full Petri dish usually yields about 10^7 cells. Resuspend them at 5×10^5 cells/mL in feeding medium. You will use 50 μL of this cell suspension per well.
6. In a separate tube, prepare tissue culture medium that has been warmed to 37°C and supplemented with 200 U/mL recombinant IFN-γ and 1 μg/mL LPS. You will use 50 μL of this supplemented medium per well.
7. In a 96-well tissue culture treated plate, add 50 μL of the supplemented medium to each well that will be used (see Note 3). Add 50 μL of the cell suspension and mix by tapping gently against the side of the plate. This will result in wells with 2.5×10^4 J774 cells in 100 μL of medium containing 100 U/mL IFN-γ and 500 ng/ml LPS (see Note 4).

8. Prepare one additional row in the 96-well plate with 50 μL of the supplemented medium and 50 μL of the cell suspension. These wells will be used as controls (see Note 5).
9. Incubate at 37°C with 10% CO_2 for 24 h.
10. Prepare the *C. neoformans* culture by inoculating a single colony in 10 mL of Sabouraud dextrose broth. Grow at 37°C with 150 rpm shaking.
11. Day 2: Prepare a tube with feeding medium supplemented with 200 U/mL IFN-γ, 1 μg/mL LPS and an opsonin at four times the final concentration (see Note 6). You will use 50 μL of this medium per well.
12. Collect 1 mL of the overnight *C. neoformans* culture into a microcentrifuge tube. Wash twice by centrifuging for 2 min at 400 × *g*, removing the supernatant and resuspending in tissue culture medium.
13. Count a 1:100 dilution of the cells and dilute to 10^6 cells/mL. You will use 50 μL of this cell suspension per well.
14. Add 50 μL of the supplemented medium to each well containing J774 cells on the 96-well plate. Add 50 μL of the *C. neoformans* suspension to each well. This will result in infection of the J774 cells at 1:1 effector-to-target ratio (see Note 7).
15. Return to the tissue culture incubator for phagocytosis to happen. The incubation time will depend on which experiment is to be performed (see Subheadings 3.2–3.4 below).

3.2. Phagocytosis Assay

1. Day 1: Prepare a 96-well plate with macrophages and *C. neoformans* cells as described above.
2. Day 2: Incubate the *C. neoformans* and J774 cells prepared using the protocol in Subheading 3.1 for 2 h (see Note 8).
3. To fix the cells, remove the medium and add 200 μL of ice-cold methanol into each well. Incubate at room temperature for 30 min. In the meantime, prepare the staining solution by diluting the concentrated Giemsa stain 20 times in distilled water.
4. Remove the methanol and wash the wells twice with 200 μL of PBS. The washes should be gentle to avoid detaching the macrophages from the plate.
5. Add 100 μL of the diluted Giemsa stain to each well. Incubate at 4°C for at least 2 h, but preferentially overnight.
6. Remove the stain and wash the wells again twice with PBS. Use an inverted microscope to observe the cells (Fig. 1) and count the proportion of macrophages that have internalized *C. neoformans* or the phagocytic index (see Notes 9 and 10).

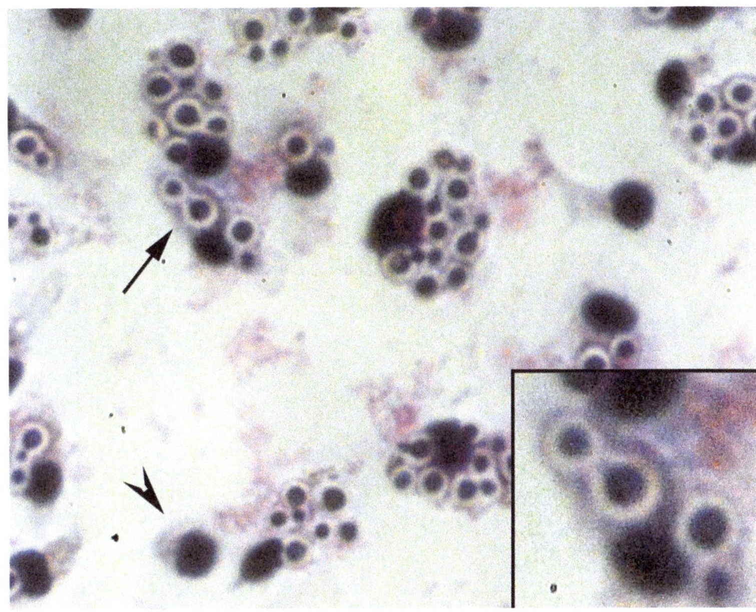

Fig. 1. Photomicrograph of a phagocytosis assay done with primary murine bone marrow-derived macrophages and *Cryptococcus neoformans* isolate 24067. The Giemsa stain results in *light staining* of the macrophage cytoplasm and *dark purple* staining of the macrophage nucleus and *C. neoformans* cells. The capsule is not stained but can be observed as a *white halo* surrounding the ingested fungi, as in the cell pointed with an *arrow* and magnified in the *inset*. The *arrowhead* indicates a macrophage that has not phagocytosed any yeast cell.

3.3. *Cryptococcus neoformans* Killing Assay

Prepare at least one plate of Sabouraud dextrose agar and one microcentrifuge tube with 400 μL of sterile PBS for each well on the experiment. Also, prepare Sabouraud dextrose agar plates to count the number of CFUs on the inoculum (see Note 11).

1. Day 1: Prepare a 96-well plate with macrophages and a *C. neoformans* cells as described in Subheading 3.1.
2. Day 2: Incubate the plates containing *C. neoformans* and J774 cells between 2 and 24 h (see Note 12).
3. In the meantime, prepare the agar plates to count CFUs on the inoculum. Into three microcentrifuge tubes, pipette 950 μL of sterile PBS. Add 50 μL of the fungal suspension used to infect the macrophages to each one. Plate 5 μL of each dilution on a separate Sabouraud dextrose agar plate. Incubate the plates at 30°C for 2 days.
4. After the desired incubation time, collect the 200 μL of supernatant from each well of the 96-well plate into a labeled microcentrifuge tube that already has 400 μL of sterile PBS. Add 200 μL of sterile distilled water into each well and lyse the macrophages for 30–40 min at 37°C.

5. Vigorously pipette up and down to collect every *C. neoformans cell* on the well and add the liquid to the respective microcentrifuge tube. Rinse the well with 200 μL sterile PBS and add to the respective microcentrifuge tube. The tubes should now have a final volume of 1 mL and contain all of the *C. neoformans* cells from each well.
6. Plate 5 μL of the suspension in each microcentrifuge tube in a Sabouraud dextrose agar plate and incubate at 30°C for 2 days to count CFUs (see Note 13).

3.4. Macrophage Viability Assay

1. Day 1: Prepare a 96-well plate with macrophages and a *C. neoformans* cells as described in Subheading 3.1.
2. Day 2: Incubate the plates containing *C. neoformans* and J774 cells between 2 and 24 h.
3. After the desired incubation time, remove and discard the medium and add 100 μL of warm Cellstripper solution to each well. Return to the tissue culture incubator for 5 min.
4. Pipette up and down to detach the macrophages and transfer to a microcentrifuge tube. Add 100 μL of the trypan blue solution and count cells on a microscope using a hemocytometer in up to 5 min (see Note 14). Dead macrophages will be stained blue, whereas live ones will be colorless.

4. Notes

1. In this chapter, we describe experiments using the J774A.1 macrophage-like cell line (5) (ATCC number TIB-67). This line, derived from a murine reticular sarcoma, has been used extensively with *C. neoformans* and other pathogens and is very easy to handle and maintain. Other types of macrophages such as RAW 264.7 murine and THP-1 human monocytic cell lines can be used, as can primary macrophages. The only important alteration to keep in mind is that the number of macrophage cells to be plated on day 1 (Subheading 3.1) has to be adjusted according to the replication rates of the specific macrophage, such that on day 2 there are 50,000 macrophages per well.
2. We use isolate H99 (ATCC number 208821). Other *C. neoformans* isolates can be used with no additional alterations. The methods described here can also be used with other microorganisms, albeit adaptations might be necessary.
3. The number of wells with macrophages is dependent on the experimental design and objectives. It is desirable to have at least three individual wells for each condition to be tested plus the control wells. Because cells have to be counted rapidly in

trypan blue, it is advisable to keep the number of conditions tested in a macrophage viability assay as low as possible.

4. To activate the macrophages, we use the Th1 cytokine IFN-γ and the toll-like receptor agonist LPS. Whether or not to activate the cells should ideally be decided after pilot experiments are done in both conditions. Moreover, the cytokine, concentrations and time stated in the protocol should be used as a guide only. For example, primary murine peritoneal macrophages do not need activation, whereas primary bone marrow-derived macrophages do. Activation can even have deleterious effects: in our hands, the fungistatic activity of J774 cells as measured in the *C. neoformans* killing assay was often reduced.
5. Several controls are essential in order to properly interpret the results from these experiments:
   - Phagocytosis assay: (a) wells having macrophages only and (b) wells having macrophages and *C. neoformans*, but no opsonin. In (b), less than 2% of the macrophages should have ingested *C. neoformans* cells.
   - *Cryptococcus neoformans* killing assay: (a) contamination control with wells having macrophages only, which should have no colonies at all after plating for CFUs and (b) wells with *C. neoformans* only, which show how much the fungal cells grow at the particular conditions.
   - Macrophage viability assay—wells with macrophages only. The viability of these cells should be higher than 90%.
6. Commonly used opsonin are IgG antibodies that bind to the capsule, used at a final concentration of 10 μg/mL, and mouse serum, at a final concentration of 20%.
7. The effector-to-target ratio (proportion of macrophage to *C. neoformans* cells) has to be carefully planned according to the experimental objectives and can range between 1:10 and 10:1. Higher ratios tend to result in more efficient phagocytosis and *C. neoformans* killing, whereas lower ratios tend to increase the impact on macrophage viability.
8. In presence of opsonins, macrophages phagocytose *C. neoformans* within minutes. The proportion of macrophages with internalized fungal cells increases during the first 2 h of interaction, when a plateau is reached (6).
9. The addition of a film with square divisions to the eyepiece can help counting the macrophages with internalized fungi. Alternatively, photographing the wells with a digital camera attached to the microscope and then counting on a computer screen can be easier. In any case, at least 100 macrophages from two or more fields should be analyzed.

10. There are two ways to quantify phagocytosis; the choice of which one to use will depend on the experimental objectives. The most straightforward is by scoring macrophages as either having or not internalized fungi, which gives the percent phagocytosis. Alternatively, the number of ingested fungal cells can be counted as well to determine the phagocytic index, which is defined by the number of internalized fungal cells divided by the total number of macrophages.
11. Having the plates and tubes labeled and organized prior to starting the experiment saves time and greatly decreases the chance of mistakes, especially in large experiments with many different conditions being tested simultaneously.
12. The incubation time for *C. neoformans* killing assays depends on the properties you want to test. Phagosomal acidification and fusion with lysosomes happen as soon as 15 min after phagocytosis and *C. neoformans* replicates approximately every 3 h. Thus, 2 h incubations are better to evaluate the macrophage's fungicidal activity, whereas 24 h incubations evaluate fungistatic activity.
13. The volume of the fungal suspension to be plated varies with the conditions being tested. In experiments done in the presence of fungicidal drugs, for instance, it is necessary to plate a larger volume; on the other hand, the control wells with *C. neoformans* only in 24 h experiments usually have about ten times more *C. neoformans* cells than the inoculum, so it is necessary to dilute the cell suspension prior to plating.
14. Trypan blue is a vital stain, but it is also toxic to cells. Prolonged exposure to trypan blue leads to artificially low viability measurements, so it is very important to only add the dye immediately before analyzing on the microscope and count the cells within at most 5 min.

Acknowledgments

This work was supported by funds from NIH grants AI033142, AI033774, and HL059842.

References

1. Park, B.J., Wannemuehler, K.A., Marston, B.J. et al. (2009) Estimation of the current global burden of cryptococcal meningitis among persons living with HIV/AIDS, *Aids* **23**, 525–530.
2. Zaragoza, O., Rodrigues, M.L., De Jesus, M. et al. (2009) The capsule of the fungal pathogen *Cryptococcus neoformans*, *Adv Appl Microbiol* **68**, 133–216.
3. Del Poeta, M. (2004) Role of phagocytosis in the virulence of *Cryptococcus neoformans*, *Eukaryot Cell* **3**, 1067–1075.
4. Mukherjee, S., Lee, S.C., Casadevall, A. (1995) Antibodies to *Cryptococcus neoformans* glucuronoxylomannan enhance antifungal activity of murine macrophages, *Infect Immun* **63**, 573–579.
5. Ralph, P., Prichard, J., Cohn, M. (1975) Reticulum cell sarcoma: an effector cell in antiody-dependent cell-mediated immunity, *J Immunol* **114**, 898–905.
6. Macura, N., Zhang, T., Casadevall, A. (2007) Dependence of macrophage phagocytic efficacy on antibody concentration, *Infect Immun* **75**, 1904–1915.

Chapter 15

Measuring the Inflammasome

Olaf Groß

Abstract

Inflammasomes are multiprotein complexes whose activity has been implicated in physiological and pathological inflammation. The hallmarks of inflammasome activation are the secretion of the mature forms of Caspase-1 and IL-1β from cells of the innate immune system. This protocol covers the methods required to study inflammasome activation using mouse bone marrow-derived dendritic cells (BMDCs) as a model system. The protocol includes the generation and handling of BMDCs, the stimulation of BMDCs with established Nlrp3 inflammasome activators, and the measurement of activation by both ELISA and western blot. These methods can be useful for the study of potential inflammasome activators, and of the signaling pathways involved in inflammasome activation. General considerations are provided that may help in the design and optimization of modified methods for the study of other types of inflammasomes and in other cell types.

Key words:, Inflammasome, Nlrp3, Caspase-1, Interleukin-1, Bone marrow-derived dendritic cells

1. Introduction

Inflammasome activity has been causally linked to the induction of numerous inflammatory responses, which can be either beneficial or harmful for the organism (1). Among the harmful inflammatory responses are particle-induced sterile inflammation, caused by host-derived particles such as monosodium urate (MSU) crystals (2), which are involved in the pathogenesis of gout, as well as environmental and industrial particles such as asbestos, silica (3), and metallic nanoparticles (4), which induce lung inflammation upon inhalation. In addition, dominant gain-of-function mutations in inflammasome components are associated with certain hereditary periodic fever syndromes (5). Accumulating evidence also implicates inflammasome activity in numerous other disease conditions involving chronic inflammation, including cancer (6, 7) and the development of metabolic diseases such as type 2 diabetes (8, 9), atherosclerosis (10), and inflammatory bowel diseases (11).

Robert B. Ashman (ed.), *Leucocytes: Methods and Protocols*, Methods in Molecular Biology, vol. 844,
DOI 10.1007/978-1-61779-527-5_15, © Springer Science+Business Media, LLC 2012

Conversely, inflammasome activation after pathogen recognition can have beneficial effects for the host (12). Likewise, inflammasome activation may enhance vaccine efficacy (13). Though the inflammasome can mediate cleavage of other cellular targets, most of the effects mentioned above have been attributed to the ability of the inflammasome to cause the maturation and secretion of the proinflammatory cytokine IL-1β (14).

The inflammasome has largely been studied in innate immune cells of the myeloid lineage, such as macrophages and dendritic cells. Inflammasomes typically consists of three components: (1) a cytoplasmic sensor molecule, (2) Caspase-1, and (3) in most cases, the adaptor protein ASC (1). Several distinct inflammasomes, containing different sensor molecules and activated by distinct stimuli, have been described: the Aim2-inflammasome is activated by cytoplasmic DNA (15–18), while a Rig-I inflammasome has been suggested to be active in response to viral RNA (19). Nlrp1 responds to anthrax lethal toxin (20), whereas Nlrc4 is activated by *Salmonella* species and other Gram-negative bacteria with type III or IV secretion systems (21). These activators, either directly or indirectly, are thought to induce changes in their respective sensor molecules (1). The activated sensor can then recruit the other components of the inflammasome, leading to the formation of a multiprotein complex containing clustered Caspase-1. This leads to the autoproteolytic activation of Caspase-1, which can then cleave its various targets, including IL-1β (1). The best studied and also the most remarkable inflammasome complex contains the sensor molecule Nlrp3 (Cryopyrin/Nalp3/Cias1/Pypaf1) (22). It is unusual in that it is activated in response to an ever-expanding list of pathogen- and damage-associated molecular patterns (PAMPs and DAMPs). This includes not only live pathogens as well as individual pathogen-derived toxins, but also the aforementioned environmental and endogenous particles, and extracellular ATP (23). As the actual mechanisms of Nlrp3 inflammasome activation remain largely obscure, it is unclear how Nlrp3 is able to integrate signals derived from so many diverse activators.

As pro-IL-1β is absent in resting cells, inflammasome activation represents the second step in the generation of the cleaved, bioactive IL-1β (14). The first step is the upregulation of the expression of the IL-1β gene, a process that in this context is referred to as priming. Depending on the cell type, priming can also augment the expression of Nlrp3 and potentially other factors involved in inflammasome activation (24). This occurs, for example, in response to the activation of membrane-bound pattern recognition receptors (PRRs) and involves the activation of transcription factors, including NF-κB (14, 25). Figure 1 shows an example of the upregulation of intracellular IL-1 (but not the constitutively expressed Caspase-1) by bone marrow-derived dendritic cells (BMDCs) after a 3-h priming period with various amounts of LPS.

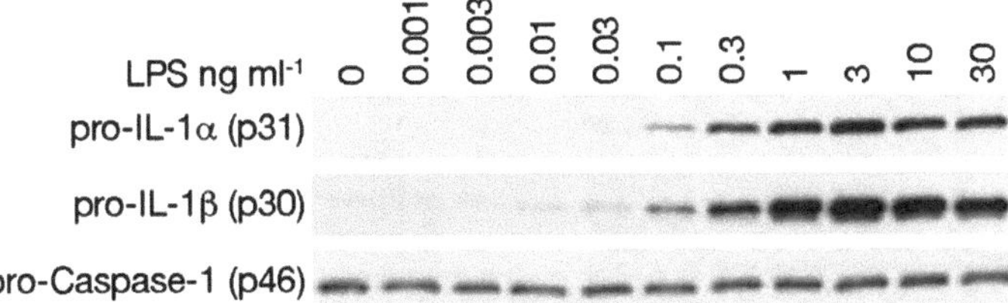

Fig. 1. **Priming effect of LPS treatment on pro-IL-1α and pro-IL-1β production.** Bone marrow-derived dendritic cells (BMDCs) were prepared as described in this protocol and treated for 3 h with increasing doses of LPS, as indicated. The cells were washed with PBS, lysed in SDS sample buffer, and subjected to western blot analysis for the presence of pro-IL-1α, proIL-1β, and pro-Caspase-1.

In the absence of Caspase-1 activity, the inactive pro-form of IL-1β is retained in the cytoplasm. Following cleavage, both mature IL-1β and mature Caspase-1 are secreted from the cells by a mechanism that is poorly understood (26). The release of these two proteins represents the hallmark of inflammasome activation. This protocol describes a standardized method to induce and measure this process.

As it is not feasible to foresee the objectives of all readers, this protocol is written as an outline. Modifications may have to be made in order to suit the goals of each experiment, and modified protocols may require further optimization. These may include not only the choice of stimuli and mouse genotypes and inhibitors, but might also involve the cell type used or even the selection of another method of measurement. Should the latter be required, a paragraph at the end of Subheading 4 summarizes some currently available alternative read-out methods.

The rationale for the key parameters of the method presented here are summarized below.

The choice of cell type: This protocol utilizes primary, murine BMDCs for several reasons. First, BMDCs are highly sensitive to inflammasome activators, respond quickly, and secrete large amounts of IL-1 and Caspase-1. In direct comparison, bone marrow-derived macrophages (BMDMs) secrete less than 10% of the amount of IL-1β produced by BMDCs (Fig. 2a). Another benefit of using primary murine cells is that gene-targeted mouse lines are available for inflammasome components, as well as many candidate inflammasome regulators. Finally, primary cells are preferred as they are more physiologically relevant. Immortalized cell lines can behave quite different from primary cells as they harbor alterations in many signal transduction pathways. Notably, inflammasome- and apoptosis-activating pathways both lead to the activation of caspases, and there is a form of inflammatory cell death termed pyroptosis, that is driven by Caspase-1 (27). The caspase-activating cell death pathways of many cell lines are reprogrammed, and this may influence

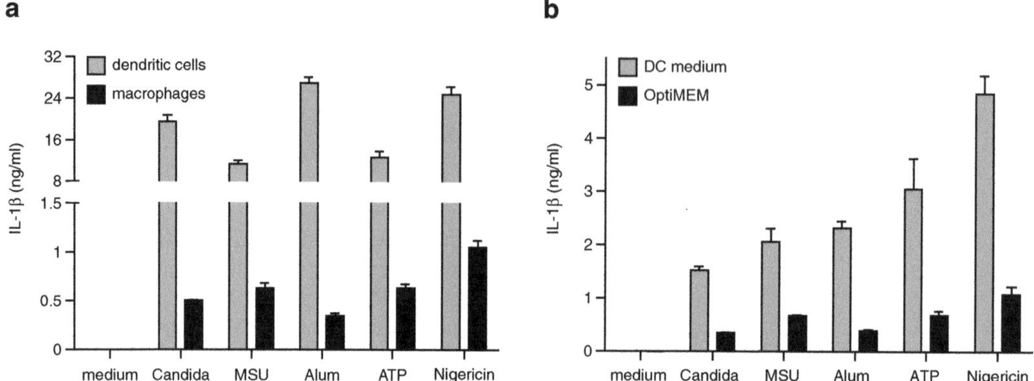

Fig. 2. **Comparison of cell types and conditions for inflammasome activation.** (**a**) BMDCs were prepared as described here and bone marrow-derived macrophages (BMDMs) were generated in the same manner from the bone marrow of the same mouse by using 10 ng/ml of recombinant murine M-CSF instead of GM-CSF. Cells were differentiated for 8 days and, for BMDMs only adherent cells were used while for BMDCs, both floating and adherent cells were included in the assay. Cells were stimulated with 10^6 cells/ml of live *Candida albicans*, 300 μg/ml of MSU or alum for 5 h or with 5 mM ATP or 5 μM nigericin for 1 h as described in this protocol. IL-1β was measured from cell-free undiluted and 1:5 diluted supernatants by ELISA. (**b**) BMDCs were collected at day 5 of differentiation, then primed and stimulated either in DC medium or in OptiMEM (without additional supplements) as in (**a**).

inflammasome activation, and could give misleading results. For example, the macrophage cell line RAW264.7 does not express ASC, while the neutrophil-like HL-60 cells do not express Nlrp3 or Caspase-1. Not surprisingly, these cell lines are irresponsive to many inflammasome activators. In contrast, human monocyte-like THP-1 cells do express inflammasome components and are very sensitive to some inflammasome activators, but rather unresponsive to others such as ATP.

The importance of detecting cleavage and secretion: Both cleavage AND secretion of Caspase-1 and IL-1β are important events in inflammasome activation, especially since cleavage is currently thought to be a prerequisite for secretion (28). In principle, the commercially available ELISA kits should detect both pro- and cleaved IL-1β. Therefore, they may not distinguish between specific, inflammasome-dependent release of the cleaved form and the accidental release of the pro-form (e.g., due to cell death). In order to specifically demonstrate inflammasome activation, one should perform a western blot to show that both cleaved IL-1β and cleaved Caspase-1 are present in the cellular supernatants. However, if multiple experiments with similar conditions and activators are performed, it may be sufficient to show both western blot and ELISA data for just one key experiment. This would demonstrate that under your experimental conditions, the ELISA data for IL-1β correlate with the secretion of cleaved IL-1β and Caspase-1. Herein, however, I suggest a two-step analysis, in which all supernatant samples are first measured by ELISA, then both supernatants and lysates of selected samples (based on ELISA data) are analyzed by western blot. This approach combines the advantages of both methods. While ELISA allows precise quantification of IL-1β, western blots provide information about the inflammasome-dependent cleavage of IL-1β.

Measurability: A major challenge in establishing inflammasome assays can be to reach a point of reliable, reproducible measurability. Two approaches can be applied: increasing the amount of IL-1β and Caspase-1 secreted, and/or increasing the sensitivity of the assay. An important consideration is the cell type used, as the amount of IL-1β and Caspase-1 secreted is highly variable between different cell types. Furthermore, optimizing the stimulus concentration and duration of stimulation may be necessary when characterizing new inflammasome activators. Up to a certain point, increasing the cell density will also increase the concentrations of Caspase-1 and IL-1β in the supernatant. Another approach used to increase the amount of IL-1β and Caspase-1 per sample is to precipitate the supernatants before western blotting. However, this requires the use of serum-reduced medium, which can decrease the amount IL-1β and Caspase-1 secreted and thereby counteract the gain achieved by this approach (Fig. 2b). A summary of this method is provided at the end of Subheading 4.

The sensitivity of an ELISA can be increased by shifting the IL-1β standard curve samples to lower concentrations than recommended by the manufacturer. This method requires incubation of the samples on the ELISA plate overnight at 4°C, a more thorough final wash step before adding the substrate, and a longer development time. Background problems also affect the sensitivity of the western blot. Here, longer washing, or the use of more stringent wash buffers or different secondary antibodies can help. The type and protein loading capacity of the polyacrylamide gel used, as well as the blotting conditions and the substrate and development systems used can also influence the signal strength. Some of the options for increasing sensitivity at this point in the procedure are mentioned in Subheading 4. However, the most important factor for western blot sensitivity is probably the choice of a primary antibody. Unfortunately, some of the published antibodies for mouse IL-1β and Caspase-1 are polyclonal, and not all are commercially available. Figure 3 shows a direct comparison of some antibodies commonly used.

Useful controls: Any experiment is only as good as its controls. The choice of appropriate controls is also crucial when performing inflammasome experiments. A simple way to ensure equal cell numbers and equal priming is to perform control western blots from cell lysates and probe for pro-caspase-1 (constitutively expressed) and for pro-IL-1 or Nlrp3 (induced upon priming). A caveat here is that some inflammasome activators can induce cell death, or are otherwise stressful to the cells. Also, secretion of IL-1

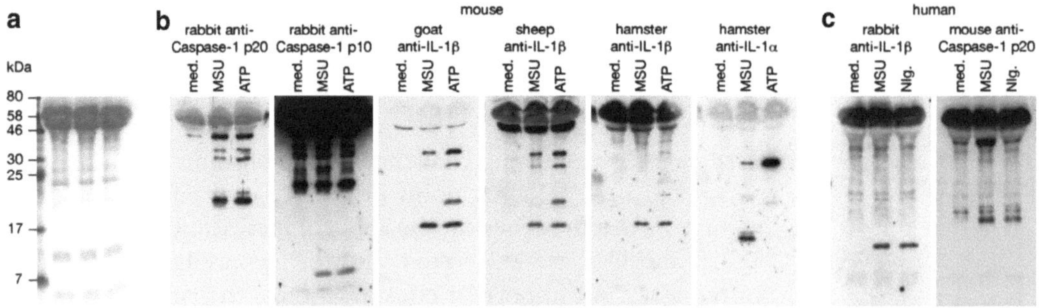

Fig. 3. **Example western blots from inflammasome assays, comparing some currently available antibodies.** (**a**) An example of a Ponceau stain of a western blot performed with samples containing 10% FCS. (**b**) Western blot analysis of murine BMDC supernatants in DC medium from cells either left unstimulated (med.) or stimulated with 300 μg/ml of MSU for 5 h or with 5 mM ATP for 1 h. Antibodies used were (from *left* to *right*): rabbit polyclonal anti-mouse Caspase-1 p20 (generated by Peter Vandanabeele, Ghent); rabbit polyclonal anti-mouse Caspase-1 p10 (Santa Cruz, SC514); goat polyclonal anti-mouse IL-1β (R&D Systems, AF-401-NA); sheep polyclonal anti-mouse IL-1β (generated by GSK); hamster monoclonal anti-mouse IL-1β (eBioscience, clone B122); hamster monoclonal anti-mouse IL-1α (eBioscience, clone ALF-161). (**c**) Human primary blood monocytes were rested after purification in the presence of 100 ng/ml recombinant human M-CSF (Immunotools) in complete RPMI. The cells were stimulated in the same medium with 300 μg/ml of MSU for 5 h or with 5 μM nigericin for 1 h or left unstimulated (med.). Cell-free supernatants were subjected to western blot analysis using the following antibodies (from *left* to *right*): rabbit polyclonal anti-human IL-1β (Cell Signaling; #2022); mouse monoclonal anti-human Caspase-1 p20 (Adipogen). Note: In addition to the antibodies used here, monoclonal antibodies directed against the p20 and p10 fragments of murine caspase-1 (clones Casper-1 and Casper-2, respectively, from Adipogen) recently became available, and also work well in these assays.

and Caspase-1 after inflammasome activation can significantly reduce the amount of the pro-forms of these proteins left intracellularly. These factors can lead a reduction of the signal from stimulated as compared to primed but unstimulated cells.

Another informative control can be to perform, *in a parallel experiment*, stimulation and measurement of an inflammasome-independent pathway such as LPS-induced TNF production. This is especially useful when working with cells derived from uncharacterized mouse model systems, or with inhibitors that display toxicity or other side effects.

Finally, it might be useful to monitor the secretion of IL-1α. IL-1α and IL-1β activate a common receptor and, under certain conditions, IL-1α is cosecreted along with IL-1β and can contribute to an inflammatory response (4). Figures 1 and 3, therefore, also show blots probed with an antibody against IL-1α. These measurements can help to connect in vivo and in vitro findings, and may give insight toward the possible contribution of IL-1α to the observed phenomena. In addition, monitoring inflammasome-independent IL-1α secretion in response to inflammasome activating stimuli can be a suitable control to evaluate side effects or toxicity of potential inflammasome inhibitors. As the antibodies for Caspase-1, IL-1β, and IL-1α were generated in different host species, the panel used here offers the attractive option to detect all of these secreted factors on the same membrane.

As it is the nature of a protocol, the following instructions are specific. However, it is most likely not critical if, for example, the wash buffer contains 0.5% or 0.2% Tween-20. It is therefore advisable that laboratories, in which some of the methods presented here are already in use, stay with their established methods, reagents, and equipment. The recipes for the standard buffers used in this protocol are nonetheless provided to serve as a reference for troubleshooting. Any aspect that has appeared to be critical for the performance of the method is discussed in Subheading 4.

2. Materials

2.1. Differentiation of Murine BMDCs

1. Mice.
2. Ice.
3. 12-Well cell culture plates.
4. Flushing medium: RPMI1640, containing 100 units/ml penicillin/streptomycin (Invitrogen).
5. Scissors and forceps.
6. Paper towels.
7. 70% Ethanol.
8. $n+2$ sterile 50-ml tubes (n being the number of mice).
9. n 100-μm cells strainers.
10. n 20-ml syringes.
11. n 2-ml syringes.
12. n 22-G needles.
13. n ml of red blood cell (RBC) lysis buffer (eBioscience).
14. Trypan blue solution (Invitrogen).
15. Hemocytometer (Neubauer improved).
16. $n\times 50$ ml of DC medium (see Note 9) RPMI1640 with Glutamax (Invitrogen), containing 10% FCS, 100 units/ml penicillin/streptomycin, 50 μM β-mercaptoethanol, 10 mM HEPES, and 20 ng/ml of recombinant murine GM-CSF (Immunotools).
17. $n\times 5$ sterile, non-tissue culture treated 10-cm Petri dishes.
18. Cell culture facility and equipment including a 37°C, 5% CO_2, water-jacketed cell culture incubator and a laminar flow hood.

2.2. Stimulation

1. 12-Tip multichannel pipette with a volume of at least 50–200 μl per tip.
2. 96-Well flat-bottom cell culture plates.
3. 100 μg/ml *E. coli* K14 ultrapure LPS (Invivogen, in water, store at –20°C).

4. Stimuli of choice, e.g., 1 M ATP (Sigma; 1 g ATP in 1.8 ml water; store at −20°C), 10 mM nigericin (Sigma; 5 mg in 670 μl Et-OH; store at 4°C), 40 mg/ml alum (Pierce; store at 4°C or −20°C), 40 mg/ml silica (Alfa Aesar; sterilize with EtOH and resuspend in water; store at −20°C).
5. Inhibitors of choice, e.g., 1 M APDC (Enzo or Sigma; 10 mg in 60 μl water; make fresh), 2.5 M KCl (Sigma; 1 g in 5.3 ml RPMI; store at −20°C), 200 mM Glibenclamide (Enzo or Sigma; 10 mg in 100 μl DMSO; make fresh), 50 mM Z-VAD-fmk (Merck; 10 mg in 100 μl DMSO).
6. DC medium (see Subheading 2.1).
7. Cell culture centrifuge with swing-out buckets to hold 96-well plates.
8. 3× and 1× SDS sample buffer (see below, dilute in phosphate-buffered saline (PBS)).
9. PBS (Invitrogen).

2.3. Measurement

1. ELISA Kits for IL-1 and TNF as required (R&D, BD, or eBioscience).
2. NUNC MaxiSorb ELISA plates.
3. ELISA plate reader and analysis software.
4. Western blot equipment (Protean mini 1 mm, BioRad).
5. Nitrocellulose membrane (GE Healthcare, Hybond 0.45 μm).
6. 5% sodium azide in water (Sigma).
7. Skim milk powder (Sigma, or your local grocer).
8. PBS.
9. ECL solution (Pierce or GE).
10. Film, developer, dark room or equivalent development equipment.
11.

3× Western blot sample buffer	(For 500 ml)
187.5 mM Tris–HCl (pH 6.8)	178 ml of a 0.5-M Tris–HCl pH 6.8 stock solution
6% w/v SDS	150 ml of a 20% stock solution
0.03% w/v Phenol Red	150 mg
30% w/v Glycerol	172 ml of 87% stock solution

Adjust to pH 6.8 with 0.1N HCl, if necessary.

12.

15% separation gel	**(For 12 gels = 120 ml):**
Tris–HCl pH 8.8	30 ml of a 1.5-M Tris–HCl pH 8.8 stock solution
SDS	0.6 ml of a 20% stock solution
Acrylamide (37.5:1)	60 ml of a 30% stock solution
dH_2O	30 ml
TEMED	75 µl, mix well
APS	600 µl of a 10% stock solution, mix well, and cast immediately

13.

6% stacking gel	**(For 12 gels = 50 ml)**
Tris–HCl pH 6.8	12.5 ml of a 0.5-M Tris–HCl pH 6.8 stock solution
SDS	0.25 ml of a 20% stock solution
Acrylamide (37.5:1)	10.5 ml of a 30% stock solution
dH_2O	27 ml
TEMED	62.5 µl
Pyronine Y (red)	150 µl of a 1% w/v stock solution, mix well
APS	500 µl of a 10% stock solution, mix well, and cast immediately

14.

Running buffer	**(For 5 l)**
Tris base	75 g
Glycine	360 g
SDS (20%)	125 ml
dH_2O	To 5 l; mix well

15.

Blotting buffer	**(For 20 l)**
Tris base	50 g
Glycine	238 g
Ethanol	3.3 l
dH_2O	To 20 l; mix well

16.

Ponceau staining solution	(For 500 ml)
0.05% Ponceau S	250 mg
3% Trichloroacetic acid	15 g
dH_2O	To 500 ml

17. Wash buffer: PBS containing 0.5% Tween 20.
18. Blocking buffer: wash buffer containing 2% skim milk powder.

3. Methods

3.1. Differentiation of Murine BMDCs (see Note 1)

1. Prepare a 12-well plate with 2 ml of flushing medium per well in as many wells as you have mice, and keep it on ice.
2. Kill a mouse by cervical dislocation or CO_2 inhalation.
3. Lay the mouse down on its back (on some paper towels), pull up the fur at the center of the belly with two fingers, and make a small cut into the skin caudal of your fingers, without cutting into the peritoneum.
4. Grab the fur above (cranial) and below (caudal) the cut and pull it open, approximately half way around the body of the mouse.
5. Pull the fur over the hind legs and over the ankle (see Note 2).
6. Dislodge the ankle joint by repeatedly moving and twisting the foot into unnatural directions.
7. Put the knee on the table with the foot standing up and pull the foot with the attached muscles down along the tibia (see Note 3).
8. Remove the tibia by hyperextending it over the knee. The tibia should come off the knee cleanly and without muscles or the meniscus (cartilage) attached. Put the bone in a well of the 12-well plate.
9. With your thumbnail flip the cartilage off of the femur, and push the muscle surrounding the bone down toward the hip joint (see Note 4).
10. Grab the end of the femur in two fingers and hold the mouse up with it. Using sharp, pointy scissors, place several cuts in the muscles of the hip area until the hip-end of the femur comes loose from the rest of the mouse (see Note 5).
11. Remove the remaining muscle tissue at the end of the bone first with scissors, and then by rubbing the bone with paper towel between your fingers. Put it in the 12-well plate in the

same well as the tibia. Repeat the process with the other leg, and the other mice.

12. Move to a sterile work place, such as a laminar flow hood.
13. Put a pair of scissors and forceps into a 50-ml tube filled with 70% ethanol.
14. Using a new, sterile 12-well plate containing 1 ml of 70% ethanol per well, place the bones of each mouse in a separate well of for 1 min. Afterward, put them into cold, sterile flushing medium in a third, sterile 12-well plate.
15. Place a 100-μm cell strainer into a 50-ml tube.
16. Pour flushing medium into a 50-ml tube and use it to fill a 20-ml syringe. Attach a 22-G needle to the syringe.
17. Take one bone with the forceps and cut open the hip- or foot-side of the bone (see Note 6).
18. Insert the needle from knee-side of the bone and carefully flush the marrow into the cell strainer (see Note 7). Repeat the process with the other three bones of the same mouse.
19. Using the plunger of a 2-ml syringe, press the marrow through the mesh of the cell strainer. Use the remaining medium in the syringe to flush remaining cells (red clumps) out of the mesh. Put the tube on ice and repeat the process with the remaining bones.
20. Spin the cells down at $300 \times g$, resuspend them in 1 ml of RBC lysis buffer, and incubate them for 5 min at room temperature.
21. Add 9 ml of flushing medium to quench the RBC lysis buffer, and mix by inversion. Remove a 10-μl sample, dilute it in 40 μl trypan blue solution, and count the cells using the hemocytometer (see Note 8).
22. Spin the cells down and resuspend them in DC medium to a density of 10^6 per ml (see Note 9).
23. Plate the cells in 10 ml medium per plate, using 10-cm non-tissue culture treated Petri dishes (see Note 10).
24. Move plates to a 37°C, 5% CO_2 incubator and allow them to differentiate for 5–9 days (see Note 11).
25. On days 3 and 6, add 5 ml of fresh DC medium to the culture.

3.2. Stimulation

1. Harvest BMDCs between days 5 and 9 of culture by transferring them into 50-ml Falcon tubes (see Note 12).
2. Count the cells and adjust them to 10^6 per ml (see Note 13). For the experiment outlined in Fig. 4, put 5.5 ml in a 15-ml tube and 14 ml in another 15-ml tube.
3. Add 2.8 μl of LPS (100 μg/ml) to 14 ml of cells (final LPS concentration of 20 ng/ml) in order to prime them, and mix

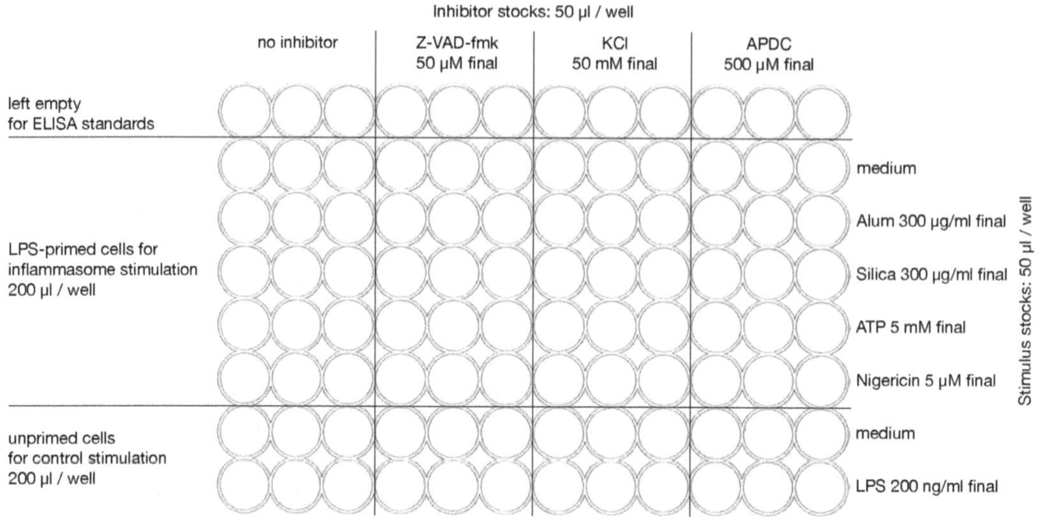

Fig. 4. **Overview of the example experiment outlined in the text.** First, cells are added to the plate (*left label*) and primed for 3 h. Second, inhibitors (*top label*) are added for 30 min. Third, inflammasome activators (*right label*) are added for a final incubation of 5 h. For details *see* Subheading 2 in Chapter 3.

by inverting. Leave 5.5 ml of cells unprimed for testing toxicity and side effects of the inhibitors (see Note 14).

4. Plate 200 µl of cells per well in a 96-well flat-bottom plate as indicated in Fig. 4 to perform triplicates of each condition (see Note 15).
5. Incubate the plates for 3 h at 37°C.
6. Prepare sixfold concentrated inhibitors: prepare four 1.5-ml tubes containing 1.2 ml of DC medium and add: (1) nothing; (2) 7.2 µl Z-VAD-fmk stock solution; (3) 164 µl KCl stock solution; (4) 3.6 µl APDC stock solution; vortex (see Note 16).
7. Add 50 µl of sixfold concentrated inhibitors (or medium) to the cells as indicated in Fig. 4 and incubate for 30 min to 1 h at 37°C (see Note 17).
8. Prepare sixfold concentrated stimuli: prepare seven 1.5-ml tubes containing 0.8 ml of DC medium and add: (1) nothing; (2) 36 µl of alum stock suspension; (3) 36 µl of silica stock suspension; (4) 24 µl of ATP stock solution; (5) 2.4 µl of nigericin stock solution; (6) nothing; (7) 9.6 µl of LPS stock solution; vortex and sonicate particles (see Note 18).
9. Add 50 µl of sixfold concentrated stimulus to the cells and incubate for 5 h at 37°C (see Note 19 for details about optimal incubation times for different activators).
10. In the meanwhile, coat an appropriate number of IL-1β and/or IL-1α and TNF ELISA plates according to manufacturer's instructions (see Note 20).

11. At the end of the incubation period, spin the inflammasome-stimulated plate(s) down at 300×*g* and carefully transfer the uppermost 200 μl of the supernatants to a fresh 96-well plate using a multichannel pipette (see Note 21). Store the fresh plate with the supernatants at 4°C if you plan to run ELISAs or western blots on the samples the next day, or at −20°C or −80°C until use.
12. Discard the remaining supernatant in the plate by decanting the contents into the sink and then, without flipping the plate back, tap the plate once firmly on a stack of paper towels.
13. Add 200 μl of cold PBS to each well, spin down, and repeat step 12.
14. Add 40 μl of 1× SDS sample buffer to each well and store the plate(s) at −20°C until use (see Note 22). See alternative method of analyzing cell lysates by ELISA in Subheading 4.4.3.

3.3. Measurement

1. Subject 50–100 μl of each well of the supernatants to ELISA for IL-1β and/or IL-1α according to manufacturer's recommendations (see Note 23).
2. Choose supernatant samples for western blot.
3. Using a multichannel pipette, pool 30 μl of each triplet into one well on a fresh 96-well plate.
4. Add 45 μl of 3× SDS sample buffer to each well, transfer the content of each well to a 1.5-ml tube, and incubate in a 95°C heat block for 5 min. Pool and boil the cell lysates accordingly (see Note 24).
5. Subject the samples to polyacrylamide gel electrophoresis using a 15% gel (see Note 25).
6. Blot the proteins on a nitrocellulose membrane (see Note 26).
7. Put the membrane for 2 min in Ponceau red solution with mild shaking. Wash for 2 min with distilled water. Make a photocopy or scan of the blot for your documentation, and note any loading differences between samples. Figure 3a shows an example (see Note 27).
8. Block the membranes by incubating in blocking buffer for 1 h at room temperature with mild shaking.
9. Add your antibody (1:1,000 to 1:2,000 (0.5 μg/ml final concentration) in blocking buffer containing 0.05% azide) and incubate over night at 4°C with mild shaking (see Note 28).
10. Wash 4×5 min with wash buffer at room temperature with mild shaking.
11. Dilute your HRP-conjugated secondary antibody 1:3,000 to 1:10,000 in blocking buffer (no azide!), add to the blots, and incubate for 2 h at room temperature with mild shaking.

12. Wash for 2 h in wash buffer (changing the buffer at least five times) at room temperature with mild shaking.
13. Tap the membrane dry using a tissue, immediately lay it over a 1-ml drop of regular or high fidelity ECL solution (on a piece of Parafilm or plastic) for 1 min, and tap dry again (see Note 29).
14. Develop your blot using standard techniques and equipments (see Note 30).
15. Repeat steps 9–14 with different primary and secondary antibodies. Example blots are provided in Fig. 3 (see Note 31).

4. Notes

4.1. Differentiation of Murine BMDCs

1. There are several published protocols on how to make BMDCs (29). They usually aim at a maximum yield of immature (but differentiated) BMDCs to use them, for example, for antigen presentation and T cell activation. As we prime the cells for inflammasome activation anyway, we are not as concerned about their preactivation status. If you want to use BMDCs for other purposes than inflammasome activation, you might want to consider using a different protocol.

   The method described here employs scissors to remove muscle from the bones only where absolutely necessary. It relies on the fact that the muscles are not attached directly to the bones, but rather to the knee and ankle joint, except at the proximal end of the femur. The method has the advantage that it is fast and yields bones with very little muscle attached. The procedure is, however, a bit hard to describe without actually demonstrating what you have to do. If you know someone who routinely takes bones from mice, you might want to ask for a demonstration of their method.
2. As soon as the knee is exposed, pinch the area under the knee and pull the fur further off. If the fur rips off before it is pulled clear over the ankle, make a cut into the fur over the ankle and pull it further.
3. The foot and lower leg muscles should come off with ease. If they do not, turn and twist the foot a bit more to completely dislodge the ankle joint. If you do not do this properly, the tibia bone might brake. Should this happen, continue with the protocol nonetheless. The bone usually breaks in the position where the fibula meets the tibia, and you will cut the bone at this point later in the protocol. I have never experienced a significant loss in yield or contamination due to breaking of a bone.

4. This requires a bit of fumbling around until you have gained some experience. A lip of cartilage projects off of the dorsal side of the femur. After you flipped if off, the femur ends bluntly, showing an X-like shape at its end. The older the mouse, the more tightly the cartilage is attached to the femur and the more difficult it is to remove. If you break open the bone, just continue anyway. Alternatively, you can use scissors to flip (but not cut!) the cartilage off.
5. Holding the mouse up by the exposed end of the femur allows you, by moving your hand in different angles with respect to the hanging mouse, to estimate where the bone is attached to the hip. It is in this area where you want to cut. The mouse will fall down once the femur is sufficiently dislodged from the hip.
6. For the tibia, cut at the site where tibia and fibula meet. For the femur, cut off just a little bit more than just the spherical femur-head.
7. The needle should easily enter the bone from this side. If it does not, the cartilage has not been properly removed. You can try to twist the needle to bore through the remaining cartilage, or carefully remove the remainder of the cartilage using scissors.
8. You can expect about 100 cells in one of the nine squares of the hemocytometer (=10^6 cells/ml in the trypan blue diluted cells and 5×10^7 cells per mouse).
9. There are various recipes for DC medium (29). Some laboratories add sodium pyruvate or nonessential amino acids, others use regular RPMI and add glutamine or leave out the HEPES buffer. These variables are relatively unimportant; as long as the medium contains GM-CSF and FCS, you should be able to generate DCs. However, the quality of the FCS used is critical. It might require testing several lots of FCS to find one that gives good yields of differentiated DCs. Those that contain measurable amounts of endotoxin according to the data sheet should not be considered. BMDCs in general do not proliferate much throughout the differentiation process. However, depending on the quality of the FCS, the cells will proliferate somewhat. Accordingly, you might have to alter the initial cell density to accommodate for that. The cell density should not exceed two million per ml.
10. The usual amount of bone marrow cells obtained per mouse is about 50 million, so you can expect to end up with five dishes of DCs per mouse.
11. During the differentiation process, the cells start to form floating clusters (after 2–3 days), and more cells become adherent and develop dendritic projections.

4.2. Stimulation

12. The longer you leave the DC culture, the higher is the amount of IL-1 and Caspase-1 secreted upon inflammasome stimulation, presumably, at least in part, due to the higher proportion of differentiated (CD11c$^+$) cells. However, the longer the DCs are in culture, the more cells that will become adherent. Classically, these cells would be deemed either mis-differentiated cells that drifted into a macrophage lineage or preactivated, mature dendritic cells. Both cell types are not useful for studying DC biology and would therefore be discarded. However, in terms of inflammasome activation, we have found that the adherent cells in a GM-CSF culture respond in the same way as the floating cells. Therefore, if you are short on cells or want to use a late DC culture in which a lot of cells have become adherent, these cells can also be used for the experiment. To dislodge the cells, add 10 ml of HBSS containing 5 mM EDTA to the plate after removing the floating cells in the medium. Incubate them for 10 min at 37°C, rinse the plate carefully, spin the cells down, discard the supernatant, and resuspend them in the medium containing the floating cells.
13. Counting and adjusting BMDCs is subject to inaccuracy. To avoid cells aggregating or sticking to the walls of the tube, it is best to count for each mouse/genotype immediately after pooling and resuspending the cells in one 50-ml tube. Do not put an aliquot in a 96-well plate for later counting, they might become adherent. Always carefully resuspend them by pipetting up and down several times with a 25-ml serological pipette before you take a sample. Count the cells again after readjusting, allowing a tolerance of ±10%. As outlined in the following note, a control stimulation will help to estimate the error derived from any differences in cell density.
14. If inhibitors are to be tested, it is important to include controls for off-target effects or toxicity. A good option is to do a parallel experiment in which a different signaling pathway is activated. I recommend using an aliquot of the cells you want to use for inflammasome stimulation but leave them unprimed. Plate and treat them with the inhibitor(s) in parallel with the primed cells. Instead of inflammasome activators, stimulate the control cells with LPS for the same duration as the inflammasome stimuli (4–6 h). Measuring TNF production from this stimulation gives an indication as to whether the compound is toxic to the cells, or whether it inhibits pathways unrelated to inflammasome activation.
15. A general formula to calculate the amount of cells *per genotype* you need is (# of stimuli + 1) × (# of inhibitors + 1) × 3 wells × 200 μl; the "+1" representing the negative controls. In the example used here, the numbers are (4 stimuli + 1 control) × (3 inhibitors + 1 control) × 3 wells × 200 μl = 12 ml; and for

the control stimulation (1 stimulus + 1 control) × (3 inhibitors + 1 control) × 3 wells × 200 μl = 4.8 ml.

16. Note that cell culture medium contains 5 mM KCl, while cytoplasmic concentrations are around 130 mM. Therefore, concentrations tested should stay within these boundaries. APDC and other ROS scavengers might become spontaneously oxidized over time, so stock solutions should not be stored for extended periods.
17. The time required for an inhibitor to be effective should be evaluated for each inhibitor individually, but should generally be kept as short as possible. It might be worth doing a time course experiment (after you have found the optimal time point for the stimulus first, see Note 19) to find the optimal preincubation period for optimal inhibition AND minimal off-target effects/toxicity.
18. Particulate stimuli tend to aggregate, but a short sonication (with either a bath or probe sonicator) will help to dissociate them.
19. The duration of the stimulation depends on the stimulus. ATP or nigericin induce robust inflammasome activation after only 30 min to 1 h, whereas most particulate stimuli, but also many pathogens, require at least 3–4 h. If you are interested in a specific stimulus, it is worth doing a time course experiment in order to identify the minimum stimulation period for full activation. Stimulation for more than 6 h should be avoided, in order to minimize contribution of feedback loops, cell death, or overgrowth of a pathogen. It might be useful to harvest fast-acting stimuli earlier than slow ones. (However, it does not change the signal if you leave fast activators on for longer, as we suggest in this protocol in order to reduce the experiment to one 96-well plate.) In a larger experiment, plate stimuli that will be harvested at the same time point on the same plate. If you do not use inhibitors, use 100 μl of a 3× stimulus instead.
20. Several companies produce good but rather expensive ELISA kits for these cytokines. In order to make a kit go further, use 50 μl instead of the recommended 100 μl per well for all steps except the substrate. This is mentioned here, as IL-1β is usually measured from undiluted supernatant. If 100 μl of sample per well is used, the total amount of sample might not be enough for all measurements intended or for repeating a measurement should the ELISA not work. In contrast to the recommendations, I use a highest standard of 10 ng/ml and an 11-point standard curve of 1/2 dilutions with a blank.

    In order to save time, coat your ELISA plates the day before the stimulation, so that you can transfer the supernatants to the plate immediately after the experiment.

21. Be extra careful at this step not to disturb the pelleted cells. 200 μl per replicate should be enough to do the measurements. If the control experiment (e.g., LPS stimulation) was done on a separate plate, it can be moved to storage immediately, without separating the supernatants from the cells.
22. Some stimuli like nigericin or ATP are quite toxic to the cells, which might reduce the total protein content of some of the samples.

4.3. Measurement

23. Most stimuli will not induce the production of more than 10 ng/ml of IL-1, so you can measure your supernatants undiluted. For TNF, dilute the supernatants 1:10. One well of the plate holding the supernatants corresponds to one well on the ELISA plates. Therefore, the final triplets measured consist of replicates of the whole experimental procedure, not only of the ELISA method.

    The sensitivity of the ELISA can be increased by using lower standard values than recommended (e.g., a 16-point standard curve ranging from 10 ng/ml to 0.6 pg/ml with a blank) and by incubating the supernatants overnight at 4°C rather than 2 h at room temperature. The last washing step before adding the substrate is critical in reducing the background and minimizing variation between replicate samples.
24. The cell lysates serve two purposes: They represent a control by which it can be demonstrated that equal amounts of cells (e.g., from different genotypes) were used. In addition, the pool of intracellular pro-IL-1 and pro-Caspase-1 and potential intracellular cleavage can be monitored. Pool the lysates from each triplet in one 1.5 ml tube. The samples are viscous until they have been boiled. Note that, as these are whole cell lysates, you might get more nonspecific bands than in samples prepared by methods that do not lyse organelles. In the case that the focus of a project lies in intracellular events that shall be monitored by western blot, you may wish to modify this protocol, do your stimulation in a larger plate format, and use a milder extraction buffer that solubilizes the cytoplasm but not the organelles.
25. We use the Mini-PROTEAN gel system from Bio-Rad with 1-mm thick gels and 15 lanes per gel and the related wet blot system. I let my gels run at 160 V for 1 h or until the dye front, but not the lowest marker in the protein molecular weight standard has left the gel. We have also obtained good results with precast gradient gels from Invitrogen, using the associated equipment and wet blot system.
26. A wet blot system may lead to better transfer than a semidry system, and thus a stronger signal. I blot at 100 V for 1 h–1.5 h. The current should be no higher than 300 mA per tank. The blotting buffer might get warm, but one should avoid letting

it get hot. Make sure that the blot buffer is thoroughly mixed, as undissolved salts and high ethanol concentrations will increase the resistance. If you do not frequently perform western blots, make the blot buffer without ethanol and add it freshly each time to avoid evaporation of the ethanol.

27. The most prominent bands are around 60 kDa, and correspond to albumin and the immunoglobulin heavy chain in the FCS.
28. Primary antibody solutions in blocking buffer with azide can be used until they are exhausted (i.e., when the signal becomes weaker and requires the use of stronger ECL solutions for detection).
29. Do not let your blots air-dry completely after tapping them but quickly move them again into ECL or your developing cassette. Use tissues or paper towels that have no pattern pressed into them, as this might lead to uneven removal of wash buffer or ECL solution. In case normal ECL is too weak, but high sensitivity ("femto") is too strong, you can dilute femto 1:3 in normal ECL solution.
30. Several methods for minimizing background are described here. First, you can try to wash longer or more intensely. Using a different wash buffer, containing 0.1% Trition X-100 and 1% skim milk powder in TBS or PBS seems to help for some antibodies, especially if used for the whole process from the first blocking of the membrane up to the second-last wash step before adding ECL solution (the last wash before ECL should be done without milk in the buffer). You can also try to use a lower concentration of primary and/or secondary antibodies, or a different secondary antibody.
31. As the common antibodies used (Fig. 3) were generated in different species (Caspase-1: rabbit, IL-1β: goat or sheep, IL-1α: hamster), the same membrane can be probed for all three proteins.

4.4. Supplementary and Alternative Methods

Below are methods for storage and shipment of bone marrow, as well as alternative methods for measuring inflammasome activation.

4.4.1. Storage of Bone Marrow in Liquid Nitrogen

The method described here uses a 96-well format and therefore, the amount of bone marrow obtained from one mouse might be more than sufficient for doing the intended experiment. Freezing the leftover bone marrow for long-term storage in liquid nitrogen is a good method to reduce mouse consumption and increase experimental flexibility. The whole freezing process should be done quickly, as DMSO is toxic to cells at room temperature.

1. Before performing erythrocyte lysis (after step 19 in Subheading 3.1 of the main protocol), set a portion of the bone marrow aside for freezing. Keep on ice.

2. Spin down the bone marrow set aside.
3. Discard the supernatant and gently but quickly resuspend the bone marrow in freezing medium (ice-cold FCS containing 10% DMSO). Use 1–3 ml freezing medium per mouse.
4. Immediately transfer 1 ml of the suspension per tube to prechilled cryo-tubes (NUNC) and transfer them into a prechilled (on ice) “Mr Frosty” (Nalgene) freezing container. It is important to label the tubes with a pencil or an alcohol-resistant pen.
5. Immediately transfer to –80°C.
6. The next day, transfer for long-term storage to liquid nitrogen. Do not store bone marrow at –80°C for longer periods as the cells will eventually die under these conditions.
7. In order to thaw the cells, put them in a 37°C water bath until the contents of the tube is half liquid but there is still some ice left.
8. Sterilize the outside of the tube with ethanol and quickly but gently transfer the content to 10 ml of medium.
9. Spin the cells down and resuspend them in DC medium.
10. Count the living cells using trypan blue and a hemocytometer and adjust them to 10^6 cells/ml. Continue at step 23 in Subheading 3.1 of the main protocol.

4.4.2. Shipping Bone Marrow

The easiest way to ship bone marrow to collaborators is to send frozen bone marrow on dry ice. An alternative is to ship bones or legs in medium in a 15-ml tube on ice. The cells inside the bones easily survive 24 h at 0°C. If the shipping takes a longer time it might be safer to flush the bone marrow and send it in complete medium in a 15-ml tube on ice. In any case, make sure that the sample does not freeze during shipping, as freezing in regular medium will kill the bone marrow cells. For this reason, do not put –20°C cold packs into a parcel containing bones or cells in medium. Instead, use blue ice or 0°C cold packs.

For the shipping itself, make sure you include the necessary paperwork, especially the dry ice declaration, declaration of hazards, and commercial value. Include enough dry ice or cold packs so the package will remain cool even if shipping is delayed.

4.4.3. Quantification of Intracellular Pro-IL-1β by ELISA

Commercially available ELISA kits also measure pro-IL-1, and therefore can be used to monitor intracellular pro-IL-1 levels. In addition, a specific mouse pro-IL-1β ELISA kit is available from eBioscience. By subjecting the cells to repeated freeze–thaw cycles, the intracellular pro-IL-1 is released and can be quantified by an additional ELISA test. This allows direct comparison of the amount of pro-IL-1 inside the cells to the amount of secreted IL-1 in the supernatants.

1. After step 13 in Subheading 3.2, instead of adding SDS sample buffer to the cells in the plate, add 200 μl of medium containing 10% FCS.
2. Subject the cells to three freeze–thaw cycles by transferring the plate between a freezer (–20°C or –80°C) and a 37°C incubator. Make sure that the medium is completely frozen or thawed after each step.
3. After the last thaw, spin the plates down at 300 × *g* to pellet debris, transfer the cell lysate supernatant to a new plate, and perform an ELISA using the same method as was used to measure IL-1 in the cell culture supernatants (step 1 in Subheading 3.3).

4.4.4. Protein Precipitation from Supernatants

This protocol replaces complete medium with unsupplemented OptiMEM, which allows precipitation of the secreted proteins, and can be used to increase signal strength. Using the method presented here, the signal strength can be increased 20–30-fold as compared to unprecipitated samples. However, as cells primed in OptiMEM secrete 2–10-fold less IL-1β (Fig. 2b), the actual gain is reduced. Medium containing FCS should not be precipitated, as this will overload the gel. This protocol can be scaled up as much as necessary, by using more cells and precipitating from larger volumes of medium.

1. In contrast to the standard protocol, following step 2 in Subheading 3.2, spin down the cells after counting, and resuspend them in OptiMEM medium (Invitrogen) without any supplements at 10^6 cells/ml. Add LPS to 20 ng/ml final concentration for priming.
2. Seed 1 ml of cells 12-well plates without replicates (instead of 96-well plates as triplicates) and perform inflammasome stimulations analogous to the standard protocol using 10× inhibitors and stimuli in OptiMEM.
3. Harvest the supernatants (1.2 ml) into 1.5-ml tubes and spin for 5 min at 400 × *g*.
4. Transfer two aliquots of 500 μl into fresh 1.5-ml tubes, leave some medium with the cells in the original tube to make sure you do not carry over any cells or debris.
5. Store one of the two parallel tubes at –20°C as a backup.
6. To the other tubes, add 500 μl methanol and 150 μl chloroform, vortex, and spin down in a microcentrifuge at maximum speed for 10 min at room temperature. Spinning will separate the sample into three phases: an organic phase at the bottom containing chloroform, an aqueous phase containing water and methanol, and a protein-containing interphase.

7. Discard the aqueous phase (at the top) without touching the interphase. Leave some of the aqueous phase to ensure that the interphase is not disturbed or removed. If you do not see a lower (organic) phase, add 50 μl more of chloroform, mix, and spin again.
8. Add 800 μl of methanol, mix, and spin again. (Now, the chloroform is dissolved and the protein is pelleted).
9. Remove the supernatant carefully, without disturbing the brittle pellet.
10. Dry the pellets for 10 min at 37°C. If the methanol is not completely removed, the sample will float out of the well when loading the gel.
11. Add 25–50 μl of SDS sample buffer, mix, and incubate for 5 min at 95°C.
12. Spin down and perform western blot following the standard protocol (step 5 in Subheading 3.3).

4.4.5. Caspase-1 ELISA

Various companies offer ELISA kits, primarily for the detection of human Caspase-1. In principle, these kits suffer from the same limitation as those for IL-1β as they do not allow distinguishing pro-Caspase-1 from cleaved and active Caspase-1 p10 and p20 subunits. However, they can be useful for the detection of Caspase-1 in serum and other body fluids.

4.4.6. Fluorescent Caspase-1 Substrates

FLICA™ (ImmunoChemistry Technologies) is a fluorescent probe that binds to active Caspase-1, thereby labeling cells in which the inflammasome is active. It consists of carboxyfluorescein (FAM) bound to the irreversible Caspase-1 inhibitor Y-VAD-FMK. This reagent can be useful for the determination of inflammasome activation by fluorescence microscopy and flow cytometry-based assays.

Acknowledgments

I would like to thank Michael Bscheider, Manuel Ritter, and James Vince for helpful suggestions, and Christina Thomas for proofreading and correcting the manuscript. My work is funded by an EMBO long-term fellowship. I declare no competing financial interests.

References

1. Schroder, K., Tschopp, J. (2010) The inflammasomes, *Cell* **140**, 821–832.
2. Martinon, F., Petrilli, V., Mayor, A. et al. (2006) Gout-associated uric acid crystals activate the NALP3 inflammasome, *Nature* **440**, 237–241.
3. Dostert, C., Petrilli, V., Van Bruggen, R. et al. (2008) Innate immune activation through Nalp3 inflammasome sensing of asbestos and silica, *Science* **320**, 674–677.
4. Yazdi, A. S., Guarda, G., Riteau, N. et al. (2010) Nanoparticles activate the NLR pyrin domain containing 3 (Nlrp3) inflammasome and cause pulmonary inflammation through release of IL-1alpha and IL-1beta, *Proc Natl Acad Sci USA* **107**, 19449–19454.
5. Henderson, C., Goldbach-Mansky, R. (2010) Monogenic autoinflammatory diseases: new insights into clinical aspects and pathogenesis, *Curr Opin Rheumatol* **22**, 567–578.
6. Okamoto, M., Liu, W., Luo, Y. et al. (2010) Constitutively active inflammasome in human melanoma cells mediating autoinflammation via caspase-1 processing and secretion of interleukin-1beta, *J Biol Chem* **285**, 6477–6488.
7. Ghiringhelli, F., Apetoh, L., Tesniere, A. et al. (2009) Activation of the NLRP3 inflammasome in dendritic cells induces IL-1beta-dependent adaptive immunity against tumors, *Nat Med* **15**, 1170–1178.
8. Zhou, R., Yazdi, A. S., Menu, P. et al. (2010) A role for mitochondria in NLRP3 inflammasome activation, *Nature* doi:10.1038/nature09663
9. Masters, S. L., Dunne, A., Subramanian, S. L. et al. (2010) Activation of the NLRP3 inflammasome by islet amyloid polypeptide provides a mechanism for enhanced IL-1beta in type 2 diabetes, *Nat Immunol* **11**, 897–904.
10. Duewell, P., Kono, H., Rayner, K. J. et al. (2010) NLRP3 inflammasomes are required for atherogenesis and activated by cholesterol crystals, *Nature* **464**, 1357–1361.
11. Bauer, C., Duewell, P., Mayer, C. et al. (2010) Colitis induced in mice with dextran sulfate sodium (DSS) is mediated by the NLRP3 inflammasome, *Gut* **59**, 1192–1199.
12. Gross, O., Poeck, H., Bscheider, M. et al. (2009) Syk kinase signalling couples to the Nlrp3 inflammasome for anti-fungal host defence, *Nature* **459**, 433–436.
13. Hornung, V., Bauernfeind, F., Halle, A. et al. (2008) Silica crystals and aluminum salts activate the NALP3 inflammasome through phagosomal destabilization, *Nat Immunol* **9**, 847–856.
14. Dinarello, C. A. (2009) Immunological and inflammatory functions of the interleukin-1 family, *Annu Rev Immunol* **27**, 519–550.
15. Burckstummer, T., Baumann, C., Bluml, S. et al. (2009) An orthogonal proteomic-genomic screen identifies AIM2 as a cytoplasmic DNA sensor for the inflammasome, *Nat Immunol* 10, 266–272.
16. Fernandes-Alnemri, T., Yu, J. W., Datta, P. et al. (2009) AIM2 activates the inflammasome and cell death in response to cytoplasmic DNA, *Nature* **458**, 509–513.
17. Hornung, V., Ablasser, A., Charrel-Dennis, M. et al. (2009) AIM2 recognizes cytosolic dsDNA and forms a caspase-1-activating inflammasome with ASC, *Nature* **458**, 514–518.
18. Roberts, T. L., Idris, A., Dunn, J. A. et al. (2009) HIN-200 proteins regulate caspase activation in response to foreign cytoplasmic DNA, *Science* **323**, 1057–1060.
19. Poeck, H., Bscheider, M., Gross, O. et al. (2010) Recognition of RNA virus by RIG-I results in activation of CARD9 and inflammasome signaling for interleukin 1 beta production, *Nat Immunol* **11**, 63–69.
20. Boyden, E. D., and Dietrich, W. F. (2006) Nalp1b controls mouse macrophage susceptibility to anthrax lethal toxin, *Nat Genet* **38**, 240–244.
21. Mariathasan, S., Newton, K., Monack, D. M. et al. (2004) Differential activation of the inflammasome by caspase-1 adaptors ASC and Ipaf, *Nature* **430**, 213–218.
22. Martinon, F., Burns, K., and Tschopp, J. (2002) The inflammasome: a molecular platform triggering activation of inflammatory caspases and processing of proIL-beta, *Mol Cell* **10**, 417–426.
23. Mariathasan, S., Weiss, D. S., Newton, K. et al. (2006) Cryopyrin activates the inflammasome in response to toxins and ATP, *Nature* **440**, 228–232.
24. Bauernfeind, F. G., Horvath, G., Stutz, A. et al. (2009) Cutting edge: NF-kappaB activating pattern recognition and cytokine receptors license NLRP3 inflammasome activation by regulating NLRP3 expression, *J Immunol* **183**, 787–791.
25. Greten, F. R., Arkan, M. C., Bollrath, J. et al. (2007) NF-kappaB is a negative regulator of IL-1beta secretion as revealed by genetic and pharmacological inhibition of IKKbeta, *Cell* **130**, 918–931.

26. Bauernfeind, F., Ablasser, A., Bartok, E. et al. (2011) Inflammasomes: current understanding and open questions, *Cell Mol Life Sci* **68**, 765–83.
27. Bergsbaken, T., Fink, S. L., Cookson, B. T. (2009) Pyroptosis: host cell death and inflammation, *Nat Rev Microbiol* **7**, 99–109.
28. Keller, M., Ruegg, A., Werner, S. et al. (2008) Active caspase-1 is a regulator of unconventional protein secretion, *Cell* **132**, 818–831.
29. Lutz, M. B. (2004) IL-3 in dendritic cell development and function: a comparison with GM-CSF and IL-4, *Immunobiology* **209**, 79–87.

Chapter 16

Arginine and Macrophage Activation

Mònica Comalada, Andree Yeramian, Manuel Modolell, Jorge Lloberas, and Antonio Celada

Abstract

In order to perform their functions, macrophages must be activated either by Th1-type cytokines, such as interferon-gamma which is called classical activation or M1, or by Th2-type cytokines, such as IL-4, IL-10, IL-13, etc. referred as alternative activation or M2. In all of these conditions, macrophages require the uptake of exogenous arginine to meet their metabolic demands. Depending on the intracellular availability of this amino acid, the activities of these cells are differentially modulated. In this regard, macrophage activation requires this amino acid for the synthesis of proteins, production of nitric oxide via classical activation, and production of polyamines and proline through alternative activation. Therefore, the study of the arginine transport for amino acid system transporters may be a key regulatory step for physiological responses in macrophages. In this chapter, we present simple and direct methods to determine the mRNA expression and activity of arginine transporters. Moreover, we describe a direct method to measure the arginine catabolism using thin-layer chromatography.

Key words: Cationinc amino acid transporters, Heterodimeric amino acid transporters, Polyamines, Citruline, Alternative activation, Classical activation

1. Introduction

Inflammation is a tightly regulated process initiated following body injury caused by a large number of agents (physical, chemical, infection, trauma, etc.). The main function of inflammation is to eliminate the pathogenic insult and to remove damaged tissue, with the aim of restoring tissue homeostasis. The concerted action of phagocytes—neutrophils, monocytes, and macrophages—is crucial for the effective removal of intruders and cell debris. In this regard, macrophages are especially appreciated for their phagocytic and microbial capacity. The activation of macrophages implies that they undergo a serie of functional, morphological, and biochemical

Robert B. Ashman (ed.), *Leucocytes: Methods and Protocols*, Methods in Molecular Biology, vol. 844,
DOI 10.1007/978-1-61779-527-5_16, © Springer Science+Business Media, LLC 2012

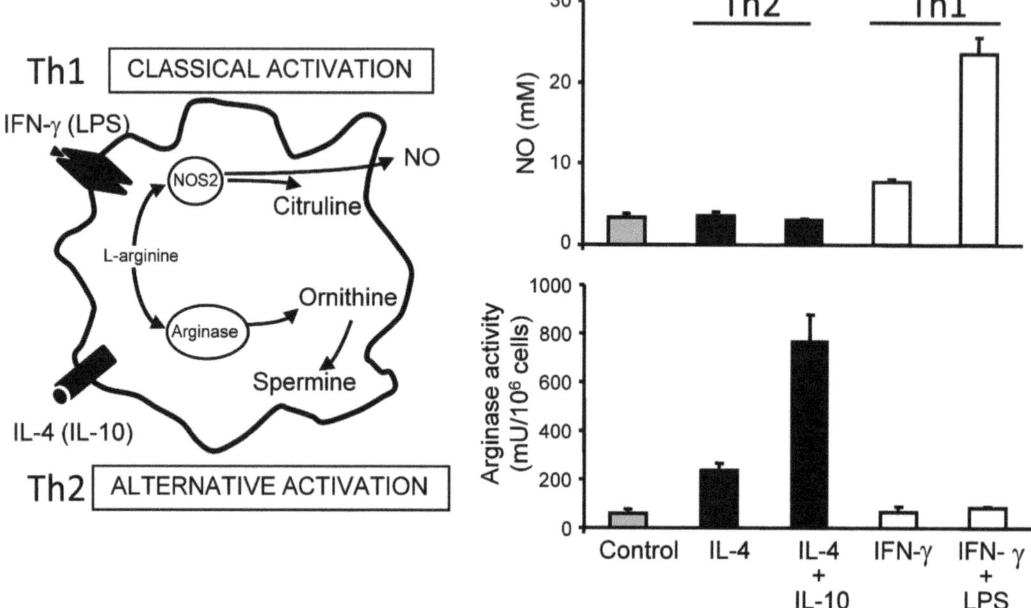

Fig. 1. IFN-γ and LPS induce classical (M1) activation of macrophages while IL-4 and IL-10 lead to alternative (M2) activation. On the left, Th1 activators induce the expression of NOS2, which catabolizes arginine in NO and citruline. Th2 cytokines induce the expression of arginase, which catabolizes arginine in ornitine, proline, glutamate, and polyamines. On the right, bone marrow-derived macrophages were cultured for 24 h in the presence of IL-4 (10 ng/ml), IL-10 (10 ng/ml), IFN-γ (10 ng/ml), or LPS (10 ng/ml). Nitrites (NO) or arginase activity were then determined. The values shown correspond to the mean ± SD of three independent experiments.

modifications produced by the regulation of a large number of genes. When these cells interact with Th1-type cytokines, such as interferon-gamma (IFN-γ), or microbial components, such as lipopolysaccharide (LPS), they produce a serie of products, such as nitric oxide (NO) and oxygen-free radicals, which destroy microorganisms (Fig. 1). This type of action is known as classical activation or M1. When the inciting stimulus is removed from the inflammatory loci, a period of reconstruction ensues, with the removal of apoptotic cells, production of collagen, etc. During this period, macrophages become activated by Th2-type cytokines, such as IL-4 or IL-13. This type of activation is called alternative activation or M2 (1–3) (Fig. 1).

Interestingly, although these phenotypes, M1 and M2, exhibit distinct properties, they are both involved in metabolism of the amino acid arginine through distinct biochemical pathways to yield their ultimate characteristics. IFN-γ and LPS induce macrophage NO synthase 2 (NOS2), which converts arginine into OH arginine and then into NO. In contrast, when macrophages are activated by IL-4 or IL-13, arginase 1 is induced. This molecule downregulates NOS2 expression and degrades arginine into urea and ornithine, which are then subsequently metabolized into proline, glutamate, and polyamines (putrescine, spermidine, and spermine). Proline mediates the production of collagen while polyamines induce cell

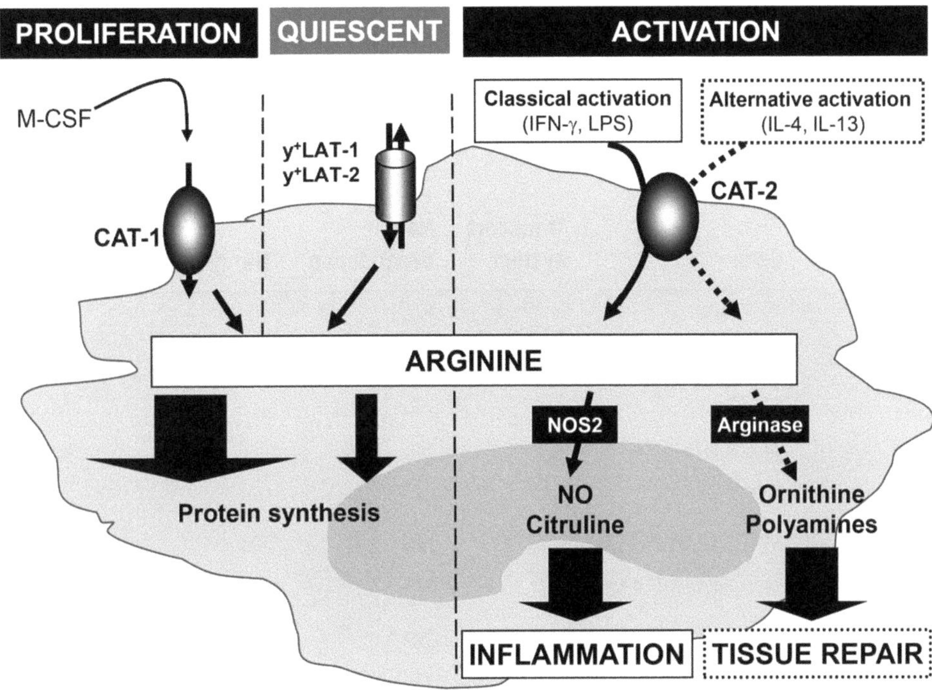

Fig 2. Arginine uptake and catabolism by quiescent, proliferative, and activated macrophages. In quiescent cells, under basal conditions, arginine is incorporated in the cell mainly through the y^+L transport system, which is due to y^+LAT-1 and y^+LAT-2. In M-CSF-dependent proliferation, arginine is required from the extracellular medium through the activation of CAT-1. Under these conditions, arginine is used for protein synthesis and only half is returned to the media; however, in quiescence, only a small amount of arginine is incorporated to the proteins and the rest returns to the media. The y^+ system is strongly induced in the classical (IFN-γ+LPS) and alternative (IL-4+IL-10) activation modes of macrophages and this increased arginine transport is caused mainly by CAT-2 activation.

proliferation, thereby contributing to the reconstitution of the damaged extracellular matrix, a process that occurs during the final phases of inflammation (1–3). Thus, macrophage activation requires high concentrations of arginine either for protein synthesis or for producing either NO during classical activation or polyamines and proline after alternative activation (Fig. 2).

The extracellular milieu is the main source of arginine and several transport systems are involved in carrying this amino acid across the plasma membrane. The transport systems that mediate arginine uptake (y^+, $B^{0,+}$, $b^{0,+}$, and y^+L) are well-characterized (4, 5) and depending on the cell type and stimuli, a number of transport activities are expressed (6). Briefly, system $B^{0,+}$ is a Na^+- and Cl^--dependent transporter for neutral and cationic amino acids; system $b^{0,+}$ handles both neutral and cationic amino acids in a Na^+-independent fashion; system y^+ interacts with cationic amino acids and only very weakly (Km > 10 mM) with neutral amino acids in either the presence or absence of Na^+; and finally, system y^+L handles cationic amino acids in a Na^+-independent fashion and neutral amino acids in the presence of Na^+ (4–7) (Table 1).

Table 1
Membrane transporters that accept L-arginine as a substrate

			Arginine transport	Leucine transport	
Protein	**Gene**	**Transport system**	**Na⁺ dependence**	**Na⁺ dependence**	**Trans-stimulation**
CAT-1	SLC7A1	y^+	No	–	Yes
CAT-2A	SLC7A2	ND	No	–	No
CAT-2B	SLC7A2	y^+	No	–	Moderate
CAT-3	SLC7A3	y^+	No	–	Moderate
CAT-4	SLC7A4	–	–	–	–
Y^+LAT-1 4F2hc	SLC7A7 SLC3A2	y^+L	No	Yes	Yes
Y^+LAT-2 4F2hc	SLC7A6 SLC3A2	y^+L	No	Yes	Yes
bo+AT rBAT	SLC7A9 SLC3A1	bo+	No	No	Yes
Bo+	SLC6A14	Bo+	Yes + Cl^-	Yes + Cl^-	No

AT amino acid transporter, *ND* not described. Adapted table from Closs et al. (6) and Verrey et al. (5)

In a structural manner, the transport systems y^+ and $B^{0,+}$ are constituted by a unique protein. System y^+ of cationic amino acid transporters (the CAT family) includes four members, CAT-1 to CAT-4, whose gene products are *SLC7A1* to *SLC7A4* (5–7). The gene product of *SLC6A14* is transport system $B^{0,+}$. Systems y^+L and $b^{0,+}$ belong to a family of heterodimeric transporters known as heterodimeric amino acid transporters (HATs) and the proteins that produce these transporters are codified by *SLC7A6-7,9* (4–7) (Table 1).

Although data on arginine transport in macrophages is limited, has been demonstrated that arginine transport through the y^+ or y^+L systems differs depending on the macrophage population and the induction stimuli (6, 8, 9). In bone marrow-derived macrophages (BMDMs) under basal conditions, arginine is incorporated mainly through y^+L (>75%), with a small contribution of y^+ (<10%) (6). The activity of y^+L can be due to y^+LAT-1 and y^+LAT-2 (6, 8). The activity of the y^+ is mediated by CAT-1, although CAT-2 has also been implicated depending on the mouse strain (6, 8, 9). However, when macrophages are grown with macrophage colony-stimulating factor (M-CSF), the arginine required for proliferation is obtained by increasing transport through the y^+ system, in which CAT-1 plays a crucial role without CAT-2 involvement (6) (Fig. 2). This observation is demonstrated by the finding that proliferation and arginine catabolism are not affected in macrophages from CAT-2 *knockout* mice (6). In this regard, CAT-1 transport is also increased in proliferating activated B and T cells and tumoral cells (10).

Interestingly, the y^+ system is strongly induced in both classical (IFN-γ + LPS) and alternative (IL-4 + IL-10) activation of BMDM and accounts for almost 90% of the transport system in these cells (6, 9). In these activation conditions, the increased arginine transport is due mainly to CAT-2 (Fig. 2). This is demonstrated because in the absence of CAT-2, although CAT-1 is functional, there is a decrease in NO production as well as in polyamines and proline (9). Similarly, it has also been described that in mouse peritoneal macrophages NO production requires CAT-2 activity (11). Finally, BMDMs stimulated with granulocyte–macrophage colony-stimulating factor (GM-CSF) are highly dependent on the induction of CAT-2 for the increase in L-arginine consumption through arginase 1 activity (8).

Consequently, the transport of arginine across the plasma membrane is an essential regulatory first step during proliferation and activation of the macrophage, both processes requiring the synthesis of a large number of proteins. Studies using macrophage-derived cell lines or primary differentiated macrophages have shown that these cells express CAT-1 under resting conditions and CAT-2 upon classical and alternative activation (6, 9). Therefore, the uptake of this amino acid may be a key regulatory step for physiological responses in macrophages. Here, we describe approaches to analyze arginine metabolism and transport activity. For this purpose, we use BMDMs, a population of non-transformed cells that, unlike peritoneal or alveolar macrophages, respond to both proliferative and activating stimuli.

2. Materials

2.1. Cell Culture

1. Dulbecco's modified Eagle's medium (DMEM).
2. Heat inactivate fetal bovine serum (FBS) at 56°C for 30 min; store aliquots at −20°C.
3. L cell-conditioned medium as a source of M-CSF.
4. Penicillin (10,000 U/ml)/streptomycin (10 mg/ml) (P/S).
5. *Stimulating agents*: IFN-γ, LPS, IL-4, IL-10.
6. PBS 1× (500 ml): 4 g NaCl; 0.1 g KCl; 0.38 g $Na_2HPO_4 \times 2H_2O$; 0.1 g KH_2PO_4 completed with ddH_2O to reach a volume of 500 ml.
7. Sterile surgical material (forceps and scissors); needle (25 gauge), syringe, cell scrapers, Petri dishes (150 mm).

2.2. Quantitative Real-Time PCR

1. TRI Reagent®.
2. Chloroform.

3. 2-Propanol.
4. 75% ethanol absolute diluted in RNase-free H_2O.
5. DNase- and RNase-free H_2O.
6. Moloney murine leukaemia virus (M-MLV) reverse transcriptase RNase H Minus reaction: M-MLV RT 5× reaction buffer; oligo-(dT) 15 primer; M-MLV reverse transcriptase; dNTP mix. Store these reagents at –20°C.
7. Power SYBR® Green PCR Master Mix.
8. Primers (Table 2).
9. RNase- and DNase-free material: 1.5-ml Eppendorf tubes; Petri dishes (60 mm), 384-well PCR plates; PCR ultraclear film.

2.3. Arginine Transport

1. *Transport buffer 1*×. 10 mM HEPES; 5.4 mM KCl; 1.2 mM $MgSO_4 \cdot 7H_2O$; 2.8 mM $CaCl_2 \cdot 2H_2O$; 1 mM KH_2PO_4; and 137 mM *N*-methyl-D-glucamine (MGA). In the transport measurement, when the presence of NaCl is required, MGA is replaced by 137 mM NaCl. The solution is brought to pH 7.4, can be made some days before the experiment, and stored at 4°C. Pre-warm the solution at 37°C just before use.

Table 2
Arginine transporters and L14 and β-actin primer pair sequences used by quantitative real-time PCR (5, 8, 9)

Protein	Gene	PCR primer pair sequences
CAT-1	Slc7A1	FW: GTTTCCCATGCCCCGAGTTATCTAT RV: GTTTCCCATGCCCCGAGTTATCTAT
CAT-2	Slc7A2	FW: GTTATGGCCGGCCTTTGCTATG RV: TCCGACCGTGACGTAAGTGTAT
CAT-3	Slc7A3	FW: TTGCAATTTCTGGGGTCATC RV: GGGAGTGCGGTTCTGTG
y+LAT-1	Slc7A7	FW: CTGCCCTTCTACTTCTTCATCATCA RV: CTCTCCATCTTCCAAGTCCATTTCT
y+LAT-2	Slc7A6	FW: CCTTGGCCATTGGGATTTCTAT RV: ACAGCCACAGCGTCACTCTTATG
β-ACTIN	β-actin	FW: ACTATTGGCAACGAGCGGTTC RV: AAGGAAGGCTGGAAAAGAGCC
60S ribosomal protein	L14	FW: TCCCAGGCTGTTAACGCGGT RV: GCGCTGGCTGAATGCTCTG

2. *Stock of arginine (50 mM).* The stock of arginine is prepared with MGA solution or water depending on the concentration used and aliquots are maintained at –20°C.
3. *Radioactive transport media.* Add the amount of arginine (from stock) to the transport buffer until a concentration of 50 μM is achieved. Pre-warm the solution at 37°C just before use and add the radioactive isotope. Usually, add 1 μCi/well of L-[3H) arginine (5 μCi/ml; 200 μl/well).
4. *Stop solution.* Ice-cold transport buffer 1× with 5 mM of non-radioactive L-arginine is used to remove the unspecific binding of labelled amino acid. Buffer is stored at 4°C.
5. *Cell lysis buffer.* 0.1% SDS; 100 mM NaOH. Buffer is stored at RT.
6. Specific material: 24-well plates; incubator at 37°C; water bath at 37°C; scintillation liquid EcoLite and β-scintillation counter.

2.4. Arginine Catabolism

1. *Radioactive incubation media.* Arginine-free DMEM plus 2% FBS with 0.1 μCi of L-[U-^{14}C) arginine (100,000 cpm/well in 10 μl arginine-free medium).
2. *TLC standard solution.* Mix a solution containing citruline, glutamate, proline, ornithine, L-arginine, putrescein, spermidine, and spermine (2.5 mg/ml) (10 μl/well).
3. *Solvent solution.* Chloroform/methanol/ammonium hydroxide 33%/water (0.5:4.5:2:1 v/v).
4. Ninhydrin spray solution (100 ml).
5. Specific material. Sterile microplate; TLC plates (Cromatoplates TLC 20×20 cm, Silica Gel 60F254); filter paper; heater; scintillation tubes; EcoscintA™ Scintillation liquid; scintillation counter.

3. Methods

3.1. Cell Culture (12)

1. Prepare sterile surgical materials, needle and syringe.
2. Kill mice by cervical dislocation. Remove the legs and separate the femur and tibiae. Cut the bones open and flush them with DMEM using an injection needle.
3. Separate cells by pipetting up and down and culture them in DMEM supplemented with 30% L cell-conditioned medium, 20% FBS, and 1% P/S. The bone marrow extracted from one mouse (2 tibiae + 2 femurs) is usually divided into four Petri dishes (40 ml/plate).

4. To obtain an almost homogenous population of macrophages, culture cells for 6 days at 37°C in 5% CO_2 without changing media.
5. After 6 days, detach the adherent cells by scraping in the same media. Centrifuge them at 500 × *g* at 4°C for 5 min, and resuspend them in 1 ml DMEM plus 10% FBS and 1% P/S.
6. Count cells using a Neubauer cell chamber.
7. Replate (or dilute) macrophages in several plates as established in the experiment. If required, to obtain quiescent macrophages, culture cells at 37°C in 5% $CO_2$for 18 h in DMEM plus 10% FBS and 1% P/S without M-CSF. Under these conditions, macrophages stop proliferating and are arrested in the G0 phase of the cell cycle.

3.2. Determination of Arginine Transporters by mRNA Expression (13)

All reagents and materials must be RNase free. Work with gloves!

1. Harvest day-6 bone marrow-derived macrophages, as described in Subheading 3.1, and seed approximately 2×10^6 cells per small Petri dish. Allow cells to adhere and stimulate them the next day.
2. Treat the quiescent cells with the desired stimuli for the established times (e.g. 24 h of stimuli with LPS or IFN-γ to check the M1 pathway or with IL-4 for M2 or with specific growth factors, such as M-CSF or GM-CSF).
3. Remove the supernatant and add 300 μl of TRI Reagent® per plate and lyse cells directly on the culture dish by scraping. Transfer the cell suspension to an RNase-free 1.5-ml tube. In this step, it is possible to freeze the samples at −20°C until use.
4. Allow samples to stand for 5 min at RT to ensure complete dissociation of nucleoprotein complexes.
5. Add 60 μl of chloroform. Cover the sample tightly, shake vigorously for 15 s, and allow to stand for 5–15 min at RT.
6. Centrifuge the samples at 12,000 × *g*, 4°C, for 15 min. Centrifugation separates the mixture into three phases: a colourless upper aqueous phase contains the RNA.
7. Transfer the aqueous phase to a new tube, add 150 μl 2-propanol, mix, and incubate for 10 min at RT. It is possible to leave samples at −20°C overnight.
8. Centrifuge the samples at 12, 000 × *g* for 10 min at 4°C. The RNA precipitate forms a pellet on the side and bottom of the tube.
9. Carefully discard the supernatant and add 300 μl of 75% ethanol.
10. Centrifuge the tubes at 7,500 × *g* for 5 min at 4°C.
11. Discard the supernatant and dry RNA pellet.

12. Dissolve the pellet in approximately 20 μl RNase-free water. To facilitate dissolution, mix by repeated pipetting and incubate the RNA samples at 65°C for 10 min. Store the samples at -20°C for short-term storage and at -80°C for long-term storage.
13. Measure RNA concentration: 1 μl RNA in a NanoDrop Spectrophotometer ND-1000.
14. Mix 1 μg RNA with 1 μl oligo-dT primer mix and fill with ddH_2O to a final volume of 14.5 μl.
15. Incubate tubes at 65°C for 5 min.
16. Incubate tubes at 4°C for 5 min to prevent the formation of secondary structures.
17. Meanwhile, prepare a master mix: 4 μl M-MLV RT 5× reaction buffer; 1 μl dNTP mix (10 mM each); 0.5 μl M-MLV reverse transcriptase.
18. Add 5.5 μl of the master mix to each tube and spin quickly.
19. Incubate the samples for 50 min at 40°C.
20. Incubate the samples for 15 min at 70°C.
21. Spin for a short time and prepare dilutions of each sample in a ratio 1:4 (sample:H_2O) and store samples at -20°C.
22. For quantitative real-time PCR, design primers using the software Primer3 in the laboratory. For normalization, use a gene that is constitutively expressed, like β-Actin or L14.
23. Prepare a standard curve that allows evaluation of the efficiency of the reaction: Dilution 1 (1/2) = 1 μl of sample + 1μl H_2O; dilution 2 (1/4) = 1 μl of dilution 1 + 1μl H_2O; dilution 3 (1/8) = 1 μl of dilution 2 + 1 μl H_2O; dilution 4 (1/16) = 1 μl of dilution 3 + 1 μl H_2O. These values are multiplied by the number of samples evaluated. The standard curve is used to calculate the dilutions of the cycle threshold (CT).
24. Next, prepare a master mix for each pair of primers: 6.25 μl SYBR® Green PCR Master Mix; 4.875 μl ddH_2O; 0.1875 μl forward primer 20 μM; 0.1875 μl reverse primer 20 μM. This is the volume required per well. Calculate the volume for triplicates of each sample and include one well without sample. The standards should be done in duplicate for each gene.
25. Add 11.5 μl of master mix and 1 μl sample per well, except for the negative control.
26. Cover the plate with transparent film.
27. Perform the PCR using the software SDS 2.3. Dissociation stage: 10 min at 95°C. Program the following 35 repeats cycles: 30 s at 95°C; 1 min at 60°C; and 1 min at 72°C. Stop the reaction at 4°C.
28. Analyze data using the software SDS 2.3.

3.3. Arginine Uptake (6, 9)

The functional analysis of the transporters consists of measuring the amount of radioactive amino acid influx into cells (uptake) after their incubation with the transport media.

1. Harvest day-6 bone marrow-derived macrophages as described in Subheading 3.1 above and seed approximately 1×10^6 cells per well in a 24-well plate. Allow cells to adhere and stimulate the next day.
2. Treat the quiescent cells with the desired stimuli for the established times (e.g. 24 h of stimuli with LPS or IFN-γ to check the M1 pathway or with IL-4 for M2 or with specific growth factors, such as M-CSF or GM-CSF). Each point of the 24-well plate is a point of transport. Each transport value is determined by 4 wells ± standard deviation (SD).
3. Aspirate culture media and wash cells twice with 1 ml of warm (37°C) transport buffer 1× (see Note 1).
4. Add 200 µl of radioactive transport media (37°C) for 1 min (see Note 2).
5. Remove the media and wash cells three times with 1 ml of the stop solution (4°C).
6. After washes, lyse the cells by adding 200 µl of cell lysis buffer to each well.
7. From this preparation, add 100 µl from each well to a vial that contains 3 ml of scintillation liquid. The β-scintillation counter measures the radioactivity (counts per min: c.p.m) in each vial.
8. In counterpart, add 10 µl of radioactive transport media to one vial in order to count the specific radioactivity of the transport media.
9. Moreover, from the wells of point 7, remove 20 µl to determine the protein content per well using a conventional protein assay method (Bradford, Pierce bicinchoninic acid (BCA), etc.).
10. The amino acid uptake measure is expressed in picomoles of amino acid transported per mg of protein in a set time (pmol/mg protein/min) (see Note 3).

3.4. Arginine Catabolism (6, 9) (Fig. 3)

To determine arginine catabolism, macrophages must be incubated with ^{14}C-radiolabelled arginine. The distinct products of degradation are then measured simultaneously in the cell and in the supernatant using thin-layer chromatography (TLC).

1. Harvest day-6 bone marrow-derived macrophages as described in Subheading 3.1 and seed approximately 1×10^5 cells per well in a microplate. Allow cells to adhere to the plate.
2. Treat the quiescent cells with the desired stimuli for the established times (e.g. 24 h of stimuli with LPS or IFN-γ to check the M1 pathway or with IL-4 for M2 or with specific growth

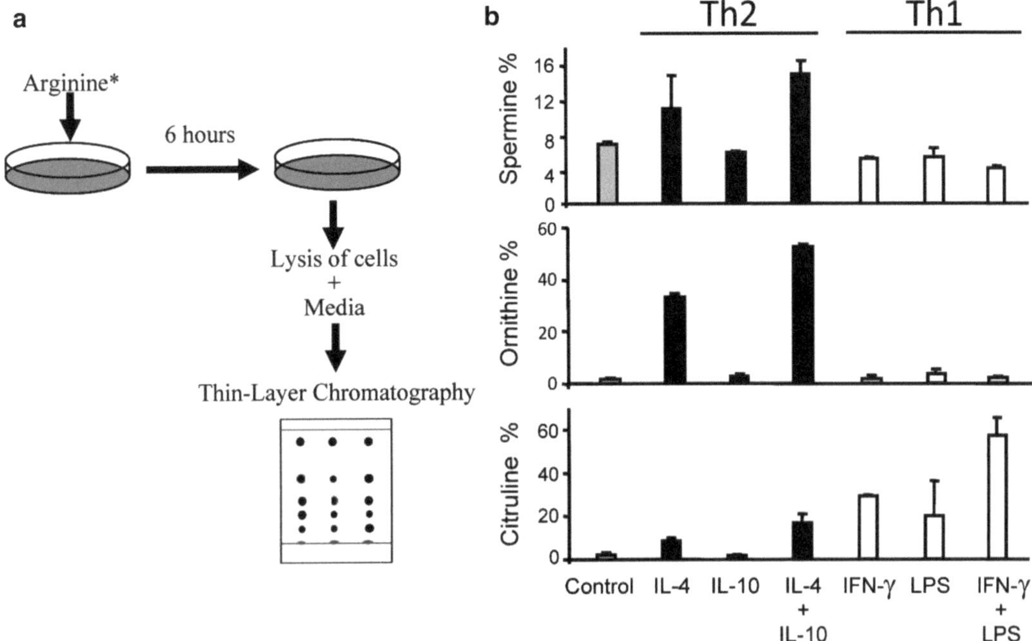

Fig. 3. Measurement of arginine catabolism. (**a**) Schematic representation of the method used. (**b**) Bone marrow-derived macrophages were cultured for 24 h as in Fig. 1. Macrophages were then incubated for 6 h with radiolabelled arginine. The products of catabolized arginine were then separated by TLC. The results are indicated as percentage of the arginine added at the beginning of the assay. The values shown correspond to the mean ± SD of three independent experiments.

factors, such as M-CSF or GM-CSF). Each point value is determined by 3 wells ± SD.

3. After 24 h, aspirate culture media and wash cells with PBS 1× (200 μl/well).
4. Add 100 μl of arginine-free DMEM containing 2% FBS.
5. Add 10 μl/well (100,000 cpm) of *radioactive incubation media* and incubate the microplate for 2 or 6 h at 37°C (see Note 4).
6. Lyse cells with a subsequently two freeze–thaw cycles in the same radioactive incubation media. Freeze the plate at −80°C for at least 30 min.
7. Add 10 μl/well of a *TLC standard solution* to the cell lysates to identify the spots.
8. The remaining arginine and the catabolic products (citruline, glutamate, proline, ornithine, L-arginine, putrescein, spermidine, and spermine) are evaluated by TLC. Add 20 μl of the samples onto TLC reference band in the plate and 30 μl of a standard solution (dilution 1:10; orange–green) (application width: 2 cm, 4 samples + standard/plate).
9. Dry the TLC plates for 1 h at 50°C.
10. Develop the plates in the solvent solution (2.5–3 h).

11. After development of the plate, allow to dry it for 1 h at 50°C.
12. To reveal the spots, spray the TLC with ninhydrin spray solution. Once the spray has been applied, heat the plate at 120°C for 5 min. In this step, spots become visible. Mark them with a pencil.
13. Scrape each spot of interest into a scintillation tube containing 5–6 ml Ecoscint ATM. Maintain the samples 30 min before measurement.
14. The scintillation counter measures radioactivity. The values for each compound are expressed as percentage of the total radioactivity measured in triplicate cultures ± SD.

4. Notes

1. Total arginine transport in cells is measured using the transport buffer 1× in the presence of 137 mM NaCl. To indirectly evaluate the involvement of the $B^{0,+}$ transport system in arginine uptake, 137 mM of NaCl should be replaced by MGA.
2. To determine the involvement of the y^+ (CAT-1-4) or y^+L (y^+LAT-1, y^+LAT-2) transport systems in arginine uptake, the radioactive transport media can be pre-incubated for 5 min with transport inhibitors, such as L-leucine (5 mM) or *N*-ethyl maleimide (NEM) (5 mM). The y^+L component is inhibited by L-leucine in a Na^+-dependent fashion. The NEM-sensitive component corresponds to system y^+. It is possible to make a combination of L-leucine and NEM to determine residual L-arginine transport cause by low-affinity carriers or passive diffusion.
3. Arginine uptake can also be expressed in pmol/10^6 cells/min.
4. The incorporation time of L-[U-^{14}C] arginine varies depending on the experimental approach. The optimal incorporation time is between 4 and 8 h.

Acknowledgments

The authors thank Dr. Susana Bodoy and Dr. Lorena Martin for their help in developing some of the techniques presented here. This work was supported by a grant from the *Ministerio de Ciencia e Innovación* MEC-FEDER BFU2007-63712/BMC to A.C. M Comalada is recipient of a "Ramon y Cajal" Program from Spanish *Ministerio de Ciencia e Innovación*. We thank Tanya Yates for editing the manuscript.

References

1. Martinez, F.O., Helming, L., Gordon, S. (2009) Alternative activation of macrophages: an immunologic functional perspective, *Annu Rev Immunol.* **27**, 451–83.
2. Nathan, C., Ding, A. (2010) Nonresolving inflammation, *Cell* **140**, 871–82.
3. Medzhitov, R. (2008) Origin and physiological roles of inflammation, *Nature* **454**, 428–35.
4. Closs, E.I., Simon, A., Vékony, N. et al. (2004) Plasma membrane transporters for arginine, *J Nutr,* **134**, 2752 S–2759 S.
5. Verrey, F., Closs, E.I., Wagner, C.A. et al. (2004) CATs and HATs: the SLC7 family of amino acid transporters, *Pflugers Arch* **447**, 532–42.
6. Yeramian, A., Martin, L., Arpa, L. et al. (2006) Macrophages require distinct arginine catabolism and transport systems for proliferation and for activation, *Eur J Immunol* **36**, 1516–26.
7. Closs, E.I. (2002) Expression, regulation and function of carrier proteins for cationic amino acids, *Curr Opin Nephrol Hypertens* **11**, 99–107.
8. Martin, L., Comalada, M., Marti, L. et al. (2006) Granulocyte-macrophage colony-stimulating factor increases L-arginine transport through the induction of CAT2 in bone marrow-derived macrophages, *Am J Physiol Cell Physiol* **290**, C1364–72.
9. Yeramian, A., Martin, L., Serrat, N. et al. (2006) Arginine transport via cationic amino acid transporter 2 plays a critical regulatory role in classical or alternative activation of macrophages, *J Immunol* **176**, 5918–24.
10. Yoshimoto, T., Yoshimoto, E., Meruelo, D. (1992) Enhanced gene expression of the murine ecotropic retroviral receptor and its human homolog in proliferating cells, *J Virol* **66**, 4377–81.
11. Nicholson, B., Manner, C.K., Kleeman, J. et al. (2001) Sustained nitric oxide production in macrophages requires the arginine transporter CAT2, *J Biol Chem* **276**, 15881–5.
12. Celada, A., Gray, P. W., Rinderknecht, E. et al. (1984) Evidence for a -interferon receptor that regulates macrophage tumoricidal activity, *J Exp Med* **160**, 55–74.
13. Valledor, A.F., Arpa, L., Sánchez-Tilló, E. et al. (2008) IFN-{gamma}-mediated inhibition of MAPK phosphatase expression results in prolonged MAPK activity in response to M-CSF and inhibition of proliferation, *Blood* **112**, 3274–82.

Chapter 17

Immunodetection of Granzyme B Tissue Distribution and Cellular Localisation

Catherina H. Bird, Corrine Hitchen, Mark Prescott, Ian Harper, and Phillip I. Bird

Abstract

Investigation of Granzyme B (GrB) function and pathophysiology in both human settings and rodent models increasingly involve the use of indirect immunofluorescence imaging and fluorescence-activated cell sorting, which requires reliable GrB antibodies that do not recognise other closely related granzymes. Here, we describe the validation (using a set of recombinant granzymes, and GrB-deficient cells) and application of widely available monoclonal antibodies to specifically monitor GrB in human or mouse cells.

Key words: Granzyme B, Immunoblotting, Confocal immunofluorescence, Fluorescence-activated cell sorting, Antibody specificity, ELISA

1. Introduction

Granzyme B (GrB) is a member of a family of serine proteases produced by cytotoxic lymphocytes (CD8$^+$ T cells and natural killer (NK) cells) (1). It is stored in lysosome-related organelles (granules) following synthesis in the endoplasmic reticulum and trafficking through the Golgi apparatus, and it is released from granules and secreted from the cytotoxic cell during the perforin-mediated destruction of abnormal cells. GrB is regulated by the nucleocytoplasmic protease inhibitor, SERPINB9, which serves to protect cells from the lethal effects of stress-induced escape of GrB from granules (2–4). The inhibitor binds to GrB forming a stable complex which is sodium dodecyl sulphate (SDS) resistant and detectable as a higher molecular weight species following SDS polyacrylamide gel electrophoresis (PAGE). The complex contains both serpin and protease epitopes which are evident via immunoblotting.

Robert B. Ashman (ed.), *Leucocytes: Methods and Protocols*, Methods in Molecular Biology, vol. 844,
DOI 10.1007/978-1-61779-527-5_17, © Springer Science+Business Media, LLC 2012

Although GrB is a commonly used activation marker of cytotoxic lymphocytes, emerging evidence suggests that it has a broader distribution and role than previously appreciated (1, 5, 6), and investigators thus require reliable and specific tools for its detection.

We have tested many anti-GrB antibodies, both monoclonal and polyclonal. Cross-reactivity with other family members is a problem, especially with polyclonal antibodies. Moreover, monoclonal antibodies, although usually more specific, often have limited utility due to changes in epitope structure induced by sample preparation. Thus, no single GrB antibody is suitable for all applications. Here, we have focused on three commonly used monoclonal antibodies readily available through a number of suppliers. 2C5 (7) and GB11 (8) were raised to human GrB, and 16G6 to mouse GrB.

To characterise the specificity of the antibodies, we have performed ELISA and immunoblotting experiments using a panel of purified recombinant human and mouse granzymes produced in our laboratory (see Fig. 1). This was followed by an assessment of immunostaining via fluorescence-activated cell sorting (FACS) and confocal microscopy of fixed cells. The 2C5 antibody detects human GrB (hGrB) by ELISA and immunoblotting but does not detect granular GrB by immunostaining. It does not cross-react with any other human or mouse granzyme tested. By contrast, GB11 detects both hGrB and mouse GrB (mGrB) by ELISA but fails to detect them by immunoblotting. It is, however, suitable for immunostaining (see Figs. 3 and 4). It detects granular GrB in both human *and* mouse lymphocytes and is GrB specific in mouse cells as indicated by its failure to stain GrB-null cells. It cross-reacts with human GrH (hGrH) by ELISA, but we have not determined if GB11 will also detect hGrH by immunostaining. Finally, 16G6 detects mGrB well by ELISA and immunoblotting, but it is not suitable for immunostaining. It binds much less efficiently to mGrM by ELISA and immunoblotting, and to hGrB, mGrA, and mGrK by ELISA only.

2. Materials

2.1. Culture of YT Cells and Purification of Human and Mouse Lymphocytes

1. Culture medium: YT cells (9) are cultured in RPMI 1640 supplemented with 10% (v/v) heat-inactivated fetal bovine serum (FBS), 2 mM glutamine, 50 μg/ml streptomycin, 50 U/ml penicillin, and 55 μM 2-mercaptoethanol. Human or mouse lymphocytes are grown in RPMI 1640 supplemented with 10% (v/v) heat-inactivated FBS, 2 mM glutamine, 50 μg/ml streptomycin, 50 U/ml penicillin, 1 mM sodium pyruvate, 0.1 MEM nonessential amino acids, and 55 μM 2-mercaptoethanol (complete medium).

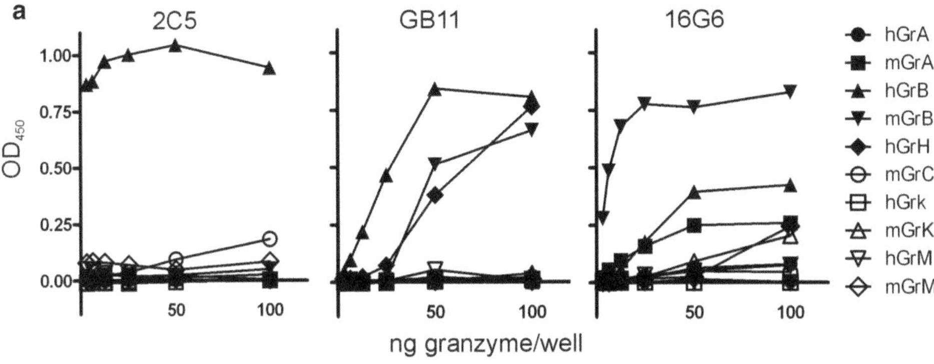

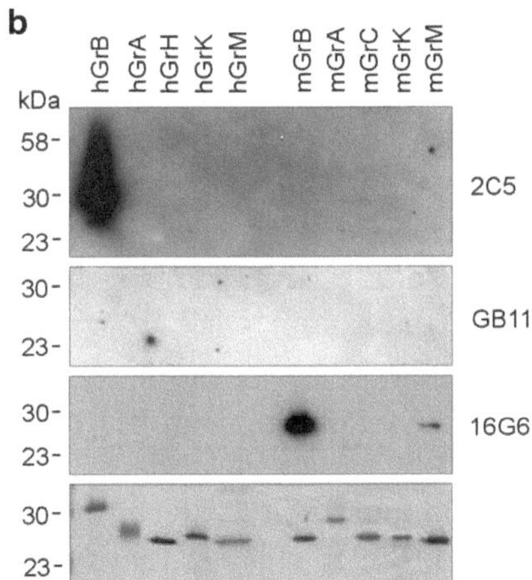

Fig. 1. Validation of antibody specificity by ELISA and immunoblotting. Recombinant mouse and human granzymes were produced in either yeast or bacterial expression systems. (**a**) Microtitre plates were coated overnight at 4°C with varying amounts of each recombinant granzyme. After extensive washing, granzyme B antibodies (1:100 2C5 or GB11 hybridoma culture supernatants, or 1:1,000 16G6) were applied for 2 h at room temperature. Non-bound antibody was removed and replaced with appropriate secondary antibodies conjugated to horseradish peroxidase. Bound antibodies were detected using the 3,3′,5,5′-tetramethylbenzidine (TMB)/H_2O_2 colorimetric assay. (**b**) Approximately 100 ng of each granzyme was loaded onto 12.5% acrylamide gels and run under denaturing conditions. After transfer to nitrocellulose, membranes were incubated with 1:100 2 C5 or GB11 hybridoma culture supernatants, or 1:1,000 16G6 as indicated. The *bottom panel* is a silver-stained gel of 150 ng of each granzyme. Note that hGrB, hGrA, and mGrA are produced in yeast and so are glycosylated resulting in higher molecular weight. The other granzymes are produced in bacteria which do not glycosylate proteins.

2. Anti-coagulant: Make a fresh solution of 3.8% (w/v) tri-sodium citrate in water. Sterilise by filtration. Dilute 1:10 into freshly drawn blood.
3. LeucoSep™ centrifuge tubes (with porous membrane frit) are purchased from Greiner Bio-One, Longwood, FL.
4. Lymphocyte separation medium: Ficoll-Paque™ PLUS is purchased from GE Healthcare, UK. Long-term storage is at 4°C, but use it at room temperature.

5. Phosphate-buffered saline (PBS): Dissolve 9.0 g NaCl, 0.726 g $Na_2HPO_4 \cdot 7H_2O$ (or 0.385 g anhydrous Na_2HPO_4), and 0.21 g KH_2PO_4 in 1 L water. Check the pH (should be about 7.2; adjust with HCl if necessary) and then autoclave to sterilize.
6. "MACS" buffer: PBS + 0.5% heat-inactivated FBS + 2 mM ethylenediamine tetraacetic acid (EDTA). Prepare aseptically from sterile solutions and store at 4°C.
7. Lymphocyte purification: Magnetic beads coupled to either CD56 or CD8 (from Miltenyi Biotec, Germany) are used to purify NK and $CD8^+$ T cells from blood. We follow the manufacturer's instructions, except that the buffer used is "MACS" buffer.
8. Concanavalin A: A 5 mg/ml stock solution of concanavalin A (Sigma–Aldrich) is made by adding sterile water to the aseptically prepared powder. Once dissolved, it is stored at –20°C in small aliquots.
9. Cytokines: Recombinant mouse IL-7 and IL-2 are widely available commercially.
10. Antibodies for activation: Purified anti-mouse CD28 and anti-mouse CD3e are from BD Biosciences.

2.2. Preparation of Cell Lysates and Immunoblotting

1. Laemmli sample buffer (LSB): 20 mM Tris–HCl, pH 6.8, 2% (w/v) SDS, 10% (v/v) glycerol, 0.02% (w/v) bromophenol blue. Store at room temperature.
2. NP-40 lysis buffer: 50 mM Tris–HCl, pH 8.0, 10 mM EDTA, 1% (v/v) Nonidet P-40 (see Note 1). Store at room temperature.
3. 10× 2-mercaptoethanol (2ME) (β-mercaptoethanol): 2ME is usually obtained as a 14.4 M solution. Prepare a 10× stock solution by diluting to 0.5 M in water. Store frozen.
4. Tris-buffered saline (TBS): A 10× stock is prepared with 200 mM Tris–HCl, pH 7.4, and 1.5 M NaCl. Dilute 100 ml with 900 ml water for use.
5. "Blotto": 5% (w/v) skim milk powder in TBS with 0.02% (w/v) sodium azide (see Note 2). Store at 4°C.
6. TBS/Tween: Add 0.1% (v/v) Tween 20 to 1× TBS.
7. GrB antibodies: The rat monoclonal antibody to mGrB, 16G6 (eBioscience), is diluted 1:1,000 in TBS. The mouse monoclonal antibody to hGrB, 2C5 hybridoma culture supernatant, is diluted 1:100 in TBS (see Note 3).
8. Secondary antibodies: Affinity-purified goat anti-mouse IgG conjugated to horseradish peroxidase (Rockland Immuno-chemicals, Gilbertsville, PA.) is diluted 1:5,000 in TBS.

This antibody is mouse specific and shows little cross-reactivity with immunoglobulins from other species. Anti-rat IgG conjugated to horseradish peroxidase (Sigma–Aldrich) is also diluted 1:5,000 in TBS.

9. Enhanced chemiluminescent (ECL) reagents: Western Lightning® Plus, Perkin Elmer, Waltham, MA, USA.

2.3. Detection of Granzyme B by FACS Analysis

1. FACS tubes: 5 ml Falcon polystyrene round-bottom tubes (12 × 75 mM) are from BD Biosciences.
2. PBS: See Subheading 2.1.
3. Fix solution: Dissolve 4.0 g paraformaldehyde (Sigma–Aldrich) in 100 ml PBS by heating the solution at 56°C for about an hour (all solids must be dissolved). Cool, then aliquot, and freeze at −20°C for long-term storage. Once thawed, store the solution at 4°C, protected from light, and use within a few weeks.
4. Saponin buffer: Prepare a 10% (w/v) stock solution of saponin (Sigma–Aldrich) in water. Aliquot and store at −20°C for up to 1 year. Prepare saponin buffer by diluting the stock solution to a final concentration of 0.1% (w/v) in sterile Hank's balanced salt solution (GIBCO, Invitrogen). Add 0.05% (w/v) sodium azide (see Note 2) and store the buffer at room temperature for no longer than 1 month. Note that saponin powder is an irritant; use a face mask when preparing stock solutions. Do not use old, discoloured powder.
5. Primary antibodies: Either Alexa Fluor® 647 mouse anti-human GrB (clone GB11, mouse IgG1, κ) (BD Biosciences) diluted 1:100 or 1:100 of unconjugated GB11 antibody (see Note 3).
6. Secondary antibody: Anti-mouse IgG conjugated to Alexa Fluor® 568 diluted 1:800 (from Molecular Probes, Invitrogen).
7. Isotype control: MOPC-31 (BD Biosciences) diluted 1:100 (see Note 4).

2.4. Preparation of Glass Slides for Immunofluorescence

1. Multi-well glass slides: We routinely use teflon printed glass slides; ten wells, 6 mm in diameter (ProSci Tech).
2. BD Cell-Tak™ Cell and Tissue Adhesive (BD Biosciences).
3. Bicarbonate buffer: 0.1 M $NaHCO_3$, pH 8.0; filter sterilize and store at 4°C.
4. NaOH: 1 M NaOH; filter sterilize or autoclave.

2.5. Confocal Immunofluorescence

1. PBS: See Subheading 2.1.
2. Fix solution: see Subheading 2.3.
3. Quench solution: 20 mM NH_4Cl in PBS.

4. Permeabilization solution: 0.5% (v/v) Triton X-100 in PBS.
5. Primary antibodies: Antibodies are diluted in PBS. GB11 directly conjugated to Alexa Fluor® 647, GB11 hybridoma culture supernatant (see Subheading 2.3), or rat anti-mouse CD3 (clone 17A2: BD Biosciences) are diluted 1:100. The sheep polyclonal antibody raised to human LAMP-1 (J. Hopwood, Women's and Children's Hospital, North Adelaide, Australia) is used at 0.75 μg/ml.
6. Secondary antibodies: FITC-conjugated goat anti-rat IgG (Sigma) is diluted 1:200. The anti-mouse IgG conjugated to Alexa Fluor® 488 and anti-sheep IgG conjugated to Alexa Fluor® 647 are from Molecular Probes (Invitrogen) and diluted 1:800 prior to use. Antibodies are diluted in PBS.
7. Mounting medium: PermaFluor™ aqueous mounting medium (Thermo Scientific).

3. Methods

Here, we focus on cultivation, activation, and preparation of cells to enable the researcher to look at GrB in situ or in cell extracts, and to assess its state (either free or complexed with SERPINB9). We assume that the reader is familiar with standard techniques of SDS/PAGE and immunoblotting.

3.1. Culture of YT Cells and Purification of Human and Mouse Lymphocytes

1. YT cells are passaged twice a week by transferring approximately 5×10^5 cells in 10 ml fresh medium to new 100-mm tissue culture dishes.
2. Human cytotoxic lymphocytes cells are purified and activated as follows.
   (a) Fresh blood from informed healthy volunteers is collected into 50-ml tubes containing the anti-coagulant tri-sodium citrate.
   (b) Mononuclear blood cells are isolated from the citrated blood using LeucoSep™ centrifuge tubes as described by the manufacturer and Ficoll-Paque™ PLUS. The only change to the protocol is that we use "MACS" buffer instead of PBS.
   (c) Cytotoxic T lymphocytes are purified from mononuclear cells using MACS magnetic cell separation based on CD8 expression following the manufacturer's instructions. The $CD8^+$ T cells are activated by culturing the cells at 2×10^6/ml in complete medium supplemented with 5 μg/ml concanavalin A and 100 U/ml IL-2 for 3 days (see Note 5).

(d) NK cells are also purified from the mononuclear cells by MACS magnetic cell separation. We use CD56-micro beads and follow the manufacturer's instructions. Cells are activated by culturing 2×10^6 cells/ml in complete medium containing 100 U/ml IL-2 for 4 days.

3. Splenocytes are purified and activated as follows.
   (a) Spleen is collected aseptically into 5 ml of complete medium in a 10-cm dish.
   (b) Sterile curved forceps are used to hold the spleen. A 1-ml syringe fitted with a 23-G needle is filled with medium which is then injected into the spleen. This process is repeated until most of the cells are released from the capsule. The remaining bits of spleen are teased apart with the needle and forceps.
   (c) Cells are transferred to a 10-ml tube and left for about 10 min to allow large chunks to settle. The supernatant is transferred to a fresh 10-ml tube and cells are collected by centrifugation at $420 \times g$ for 5 min.
   (d) To lyse red blood cells, the supernatant is removed and cells are resuspended in 1 ml of sterile 0.9% NH_4Cl per spleen followed by incubation at 37°C for 5 min.
   (e) The tube is filled with medium and cells collected by centrifugation. Cells are washed once in 10 ml medium.
   (f) Splenocytes are resuspended at 5×10^6 cells/ml in complete medium and activated by adding 100 U/ml IL-2, 2 ng/ml IL-7, 0.1 μg/ml anti-CD3, and 0.1 μg/ml anti-CD28.
   (g) Cultures are split 1:2 after 2 days (only need to add more IL-2) and harvested on day 3 (see Notes 6 and 7).

3.2. Preparation of Cell Lysates and Immunoblotting

1. LSB lysis: Lysis in LSB traps preformed complexes and prevents the formation of new complexes post lysis. Cells are washed twice with PBS and resuspended in 1× LSB at a density of 5×10^7 cells/ml. Lysates are sonicated briefly to reduce viscosity (see Note 8) and stored at −20°C until needed.
2. NP-40 lysis: Lysis in NP-40 disrupts cellular membranes, releasing granule contents and allowing interaction of serpin and protease. Incubation at 37°C promotes complex formation. Cells are washed twice with PBS, resuspended in NP-40 lysis buffer at a density of 1×10^8 cells/ml, and incubated in a 37°C water bath for 10 min. An equal volume of 2× LSB is added and lysates are sonicated briefly to reduce viscosity. Because nuclei are not removed following lysis, this yields total cell protein. Lysates can be stored at −20°C until needed.

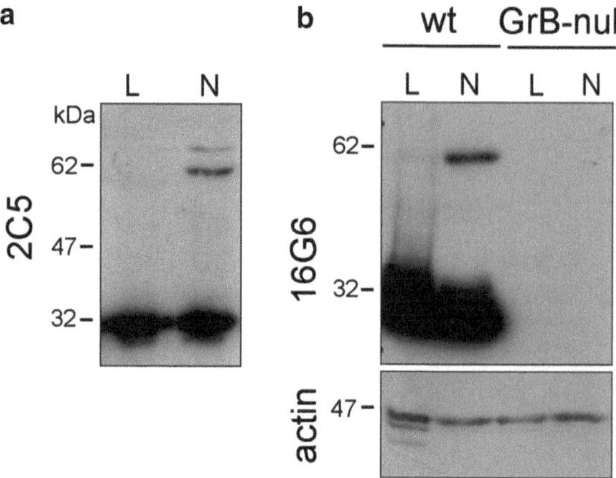

Fig. 2. Detection of GrB in its free state or complexed to SERPINB9 by immunoblotting. Human NK cells and mouse splenocytes were purified and activated as described. Cells were lysed in either NP-40 lysis buffer (N) or LSB (L) and a volume equivalent to 1×10^6 cells (20 μl) was resolved by SDS/PAGE and transferred to nitrocellulose. (**a**) 2C5 detects both free (32 kDa) and SERPINB9-bound (62 and 84 kDa) GrB in lysates prepared from human NK cells. (**b**) 16G6 is specific for mouse GrB since no other proteins are detected in lysates from GrB-null cells derived from knockout mice (11). The antibody detects both free and complexed mGrB. Note that we generally find that only a small fraction of total SERPINB9 interacts with mGrB, so longer exposure of the immunoblot may be necessary to clearly see the 62 kDa complex.

3. 50 mM 2ME is added to cell lysates, the samples heated to 100°C for 5 min, and then resolved by SDS/PAGE on 12.5% acrylamide gels.
4. Following transfer to nitrocellulose, membranes are incubated with "Blotto" for at least 1 h at room temperature.
5. Membranes are rinsed with TBS and then incubated with primary antibodies overnight at 4°C.
6. Following 4 × 15-min washes in TBS/Tween, secondary antibody diluted in TBS is applied for 1 h at room temperature.
7. The membranes are again washed in TBS/Tween four times, 15 min each, and then developed using ECL following the manufacturer's instructions (see Fig. 2).

3.3. Detection of Granzyme B by FACS Analysis

1. Aliquot 0.5–1×10^6 cells per FACS tube.
2. Wash twice in PBS. Sediment primary lymphocytes at $360 \times g$ and cell lines at $120 \times g$ for 5 min.
3. Fix cells by resuspending in 0.5 ml fix solution for 10 min at room temperature.
4. Add 2 ml PBS and sediment cells. Wash once more with 2 ml of PBS.
5. Resuspend the cells in 1 ml saponin buffer to permeabilise and incubate at room temperature for about 30 min.

6. Collect cells by spinning at about 740 × *g* for primary lymphocytes or 250 × *g* for cell lines (see Note 9).
7. Resuspend cells in 50–100 μl primary antibody (diluted in saponin buffer) and incubate for 30 min at room temperature in the dark (see Note 10).
8. Wash twice in 1 ml saponin buffer.
9. Resuspend cells in 50–100 μl secondary antibody (if used) diluted in saponin buffer and incubate at room temperature for 30 min in the dark.
10. Wash twice in saponin buffer as above.
11. Resuspend cells in approximately 300 μl of PBS and analyse by FACS (see Note 11). Representative FACS data are shown in Fig. 3.

3.4. Preparation of Glass Slides for Indirect Immunofluorescence

1. Calculate the surface area of the chamber or well to be used. This protocol is for 10-well glass slides, where each well is 6 mm in diameter.
2. Calculate how much Cell-Tak is required to coat that area—use about 1 μg Cell-Tak/cm^2 (see Note 12).
3. Prepare a mixture of 30 μl bicarbonate buffer + required volume of Cell-Tak (xμl) + ½ xμl of NaOH for each well. Mix and add to wells immediately as the Cell-Tak spontaneously adsorbs to surfaces once it has been added to a neutral buffer.
4. Incubate at either 37°C or room temperature for at least 30 min.
5. Wash wells three times with sterile PBS.

3.5. Indirect Immunofluorescence for Confocal Microscopy

1. Collect primary lymphocytes by centrifugation at 360 × *g*. Wash twice with PBS.
2. Resuspend the cells at a density of 2 × 10^6/ml in PBS and add 25–30 μl to each well. Incubate at 37°C for 20 min (see Note 13).
3. Gently aspirate the PBS (see Note 14), add 100 μl fix solution to each well, and incubate for 20 min at 37°C (see Note 15). Wash three times with PBS.
4. Add 100 μl quench solution to each well for at least 5 min (fixed cells can be stored at 4°C for several days at this stage). This reduces non-specific fluorescence due to residual formaldehyde.
5. Remove the quench solution, wash once with PBS, and permeabilise by adding 0.5% (v/v) Triton X-100 in PBS for 5 min at room temperature.
6. Wash three times with PBS. Dry between wells with a rubber-tipped Pasteur pipette (see Note 16).

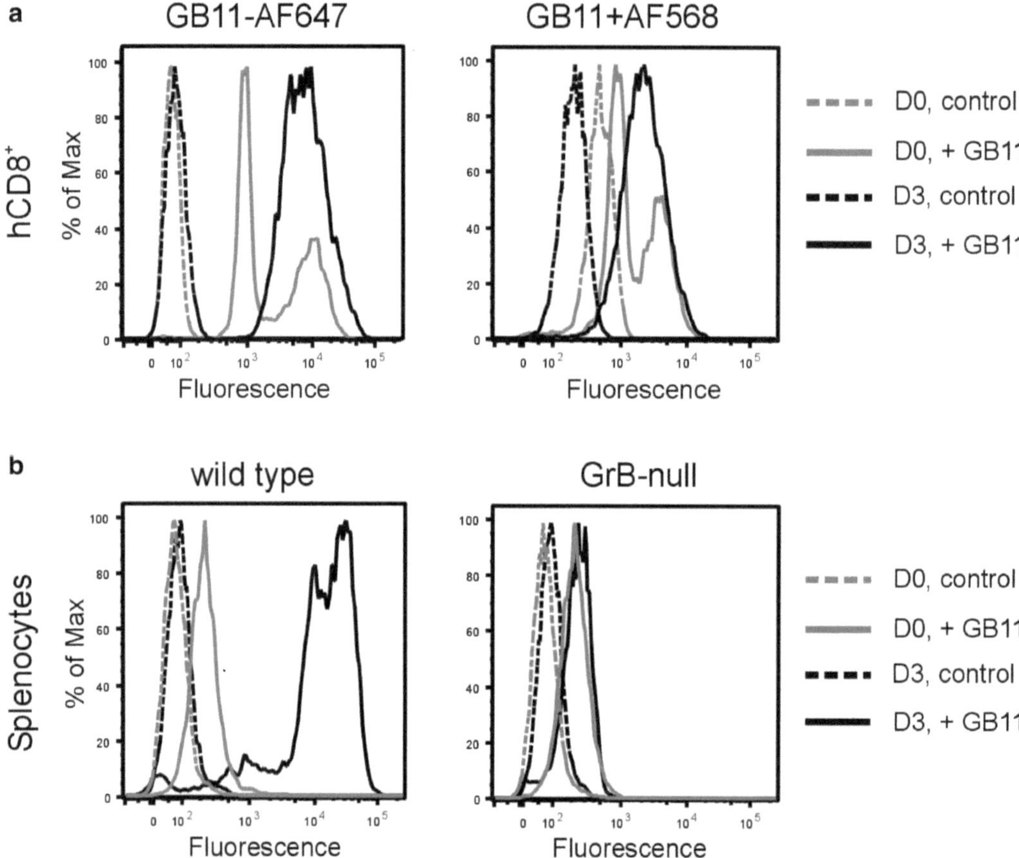

Fig. 3. Detection of GrB in purified human and mouse lymphocytes by FACS. (**a**) Both direct and indirect methods of staining were used to compare GrB levels in freshly purified human CD8+ T cells (D0) and cells which had been activated with concanavalin A and IL-2 (D3). In experiments using directly conjugated GB11 (GB11−AF647), the control was unstained cells. Where unconjugated GB11 was used (GB11+AF568), the control was cells stained first with an isotype-control antibody (MOPC-31) followed by the secondary antibody. Although the directly conjugated antibody gives clearer results, both methods show similar trends. In freshly isolated cells, there are two peaks, presumably corresponding to cells expressing different levels of GrB. Once activated, all cells express similar levels of the granzyme. (**b**) Freshly isolated (D0) and activated (D3) splenocytes from wild-type and GrB-null mice were stained with Alexa Fluor® 647-conjugated GB11. The data indicate that freshly isolated cells do not express GrB and that the antibody is specific for GrB.

7. Add 15–20 μl of primary antibody for 30 min at room temperature. Do three 5-min washes in PBS and then again dry between the wells.
8. Add 15–20 μl of secondary antibody for 30 min at room temperature, protected from light. Do three 5-min washes in PBS.
9. Mount samples in a suitable mounting fluid (see Note 17). Store slides in the dark to prevent “fading” of the fluorophores.
10. Cells are then viewed by confocal microscopy. Examples of images are shown in Fig. 4. The GB11 antibody clearly detects both mouse and human granzyme B and is specific since no signal is evident in cells isolated from GrB-null mice (see Note 18).

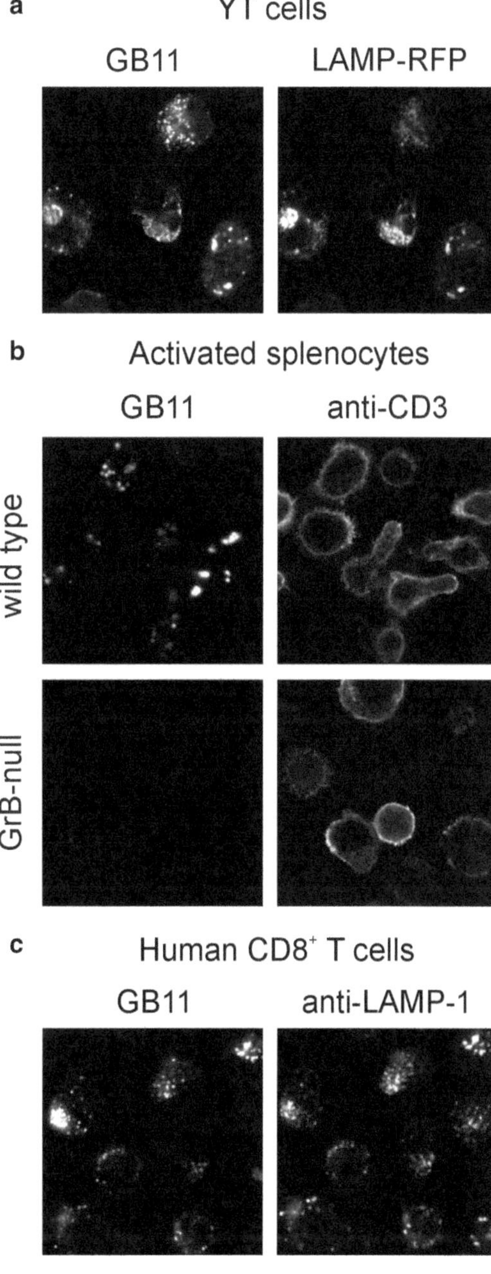

Fig. 4. Cells of different origin were prepared for staining as described in Subheading 3. Both direct and indirect methods of immunofluorescence were used to demonstrate effectiveness of the GB11 antibody which clearly detects mature GrB present in granules (LAMP-1$^+$ vesicles). (**a**) YT cells were transduced with lentivirus encoding human LAMP-1 fused to red fluorescent protein (RFP). Cells were fixed and stained using GB11 directly conjugated to Alex Fluor® 647. (**b**) Splenocytes were purified from wild-type or GrB-null mice and activated as described. They were first surface stained with anti-CD3 followed by anti-rat IgG conjugated to FITC to allow visualisation of individual cells. Following permeabilisation, cells were stained with Alexa Fluor® 647-conjugated GB11. (**c**) Human CD8$^+$ T cells were purified from peripheral blood and activated as described. Indirect immunofluorescence was performed using GB11 followed by anti-mouse Alexa Fluor® 488 and anti-LAMP-1 followed by anti-sheep Alexa Fluor® 647.

4. Notes

1. Nonidet P-40 is no longer sold by Sigma–Aldrich. It has replaced it with IGEPAL CA-630 which is described as "chemically indistinguishable from Nonidet-P40".
2. It is most convenient to prepare a 10% (w/v) stock solution of sodium azide in water. Exercise caution as sodium azide is toxic.
3. 2C5 and GB11, either unconjugated or directly conjugated to a number of different fluorophores, are available from several commercial sources. 16G6 is from eBioscience. Required dilutions should be empirically determined. In general hybridoma culture, supernatants require less dilution than ascites fluid or affinity-purified antibodies. Start with the supplier's recommended dilution.
4. Isotype-control antibodies are generally used for FACS experiments, where cells are surface stained. We have found that in some cases the isotype-control antibody detects intracellular targets when cells are permeabilised. The antibody used here does not detect either surface or intracellular targets. An alternative negative control for GrB expression is the staining of naïve cells since GrB is up-regulated in both NK and T cells upon activation.
5. Activated lymphocytes can be expanded by washing out the concanavalin A and culturing the cells in complete medium containing only 100 U/ml IL-2. These cells need to be split 1:2 daily as they expand rapidly for several days. Eventually, the growth rate slows and the level of GrB decreases.
6. Splenocytes can also be expanded. Wash cells and culture in complete medium with 100 U/ml IL-2 alone as for human lymphocytes.
7. This method yields a mixed lymphocyte population which is about 50–70% $CD8^+$ and 20–40% $CD4^+$. However, FACS analysis indicates that all cells express GrB.
8. If a sonicator is not available, viscosity can be reduced by mechanically shearing the DNA using a 26-gauge needle and syringe.
9. Cells are more buoyant after permeabilisation and need to be spun harder to avoid cell loss.
10. Due to the reversible nature of membrane permeabilisation by saponin, all antibody dilutions and washes need to be carried out in the presence of saponin.

11. Use PBS + 0.02% sodium azide if stained cells are to be stored prior to FACS analysis. We have stored cells for up to 2 weeks at 4°C, protected from light.
12. Each batch of Cell-Tak is at a different concentration which is noted on the tube. BD recommends 3.5 μg/cm^2, but we find less works fine.
13. Cells attach to the Cell-Tak-treated glass in medium containing FBS, but it seems to be much more efficient in the absence of serum. Also, the plating density and time required for cells to attach vary depending on the cell type. The protocol described here is for primary human or mouse lymphocytes. Cell lines, such as YT, are much larger; therefore, fewer cells need to be plated and 10–15 min is sufficient time for attachment.
14. Do not allow wells to dry out during the staining procedure. If you are doing a number of slides, work on one at a time. Cells which have dried stain non-specifically.
15. Experiments using YT cells have indicated that the fixation temperature has a significant impact on the quality of staining using GB11. At 37°C, there is little apparent cytosolic staining and GrB is localised to discrete granules. At room temperature, cytosolic staining increases and fewer granules are evident. If the slides are chilled and fixed on ice, then no granules are detected. Therefore, we fix all cells for microscopy at 37°C.
16. It is necessary to wash the slides very well after permeabilisation as residual detergent allows solutions to spread between the wells. If surface tension is not restored after washing, cover the entire surface with PBS and leave for 10 min. Aspirate and dry between the wells with a Pasteur pipette fitted with a short length of silicon rubber tubing cut to give an angled tip and test surface tension by adding a bit of PBS to each well.
17. There are many mounting fluids available, both commercial and home-made. We use Permafluor which solidifies after overnight incubation at room temperature. If using a non-setting mounting fluid, coverslips will need to be sealed with a couple of thin coats of nail polish. In both cases, it is important to avoid air bubbles under the coverslips. Once mounted, store slides in Petri dishes covered with aluminium foil preferably at 4°C.
18. 2C5 and 16G6 do not detect mature granular GrB using this method of sample preparation, although they do detect immature GrB in other compartments, such as the endoplasmic reticulum. Permeabilisation of cells with 0.1% (w/v) SDS in PBS for 10 min at room temperature instead of Triton X-100 sufficiently denatures granular GrB and unmasks the epitopes recognised by these antibodies (10).

Acknowledgements

The authors would like to thank Dr. J. Sun, Ms. S. Stewart, Mr. A. Matthews, and Dr. D. Kaiserman for preparing recombinant granzymes. We are grateful to Dr. J. Trapani (Peter MacCallum Cancer Institute, Melbourne, Australia), Dr. C. Froelich (Northwestern University, Chicago, USA), and Dr J. Hopwood (Women's and Children's Hospital, Adelaide, Australia) for providing 2C5, GB11, and anti-LAMP-1 antibodies, respectively. Support was provided by the National Health and Medical Research Council, Australia.

References

1. Afonina, I. S., Cullen, S. P., Martin, S. J. (2010) Cytotoxic and non-cytotoxic roles of the CTL/NK protease granzyme B, *Immunol Rev* **235**, 105–116.
2. Bird, C. H., Sutton, V. R., Sun, J. et al. (1998) Selective regulation of apoptosis: the cytotoxic lymphocyte serpin proteinase inhibitor 9 protects against granzyme B-mediated apoptosis without perturbing the Fas cell death pathway, *Mol Cell Biol* **18**, 6387–6398.
3. Bird, P. I. (1998) Serpins and regulation of cell death, *Results Probl Cell Differ* **24**, 63–89.
4. Hirst, C. E., Buzza, M. S., Bird, C. H. et al. (2003) The intracellular granzyme B inhibitor, proteinase inhibitor 9, is up-regulated during accessory cell maturation and effector cell degranulation, and its overexpression enhances CTL potency, *J Immunol* **170**, 805–815.
5. Buzza, M. S., Hirst, C. E., Bird, C. H. et al. (2001) The granzyme B inhibitor, PI-9, is present in endothelial and mesothelial cells, suggesting that it protects bystander cells during immune responses, *Cell Immunol* **210**, 21–29.
6. Hirst, C. E., Buzza, M. S., Sutton, V. R. et al. (2001) Perforin-independent expression of granzyme B and proteinase inhibitor 9 in human testis and placenta suggests a role for granzyme B-mediated proteolysis in reproduction, *Mol Hum Reprod* **7**, 1133–1142.
7. Trapani, J. A., Browne, K. A., Dawson, M. et al. (1993) Immunopurification of functional Asp-ase (natural killer cell granzyme B) using a monoclonal antibody, *Biochem Biophys Res Commun* **195**, 910–920.
8. Spaeny-Dekking, E. H., Hanna, W. L., Wolbink, A. M. et al. (1998) Extracellular granzymes A and B in humans: detection of native species during CTL responses in vitro and in vivo, *J Immunol* **160**, 3610–3616.
9. Yodoi, J., Teshigawara, K., Nikaido, T. et al. (1985) TCGF (IL 2)-receptor inducing factor(s). I. Regulation of IL 2 receptor on a natural killer-like cell line (YT cells), *J Immunol* **134**, 1623–1630.
10. Robinson, J. M., and Vandre, D. D. (2001) Antigen retrieval in cells and tissues: enhancement with sodium dodecyl sulfate, *Histochem Cell Biol* **116**, 119–130.
11. Heusel, J. W., Wesselschmidt, R. L., Shresta, S. et al. (1994) Cytotoxic lymphocytes require granzyme B for the rapid induction of DNA fragmentation and apoptosis in allogeneic target cells, *Cell* **76**, 977–987.

Chapter 18

Detection of Human and Mouse Granzyme B Activity in Cell Extracts

Sarah Elizabeth Stewart, Matthew Stephen James Mangan, Phillip Ian Bird, and Dion Kaiserman

Abstract

The serine protease granzyme B (GrB) is a key effector molecule in cell-mediated immunity, released by cytotoxic lymphocytes (CLs) to induce cell death in neoplastic or virus-infected cells. The ability to detect and measure GrB activity is important for understanding CLs. Unfortunately, such analyses are complicated by significant differences in the substrate specificities of human and mouse GrB, which is reflected by their different activities on commonly used peptide substrates. Here, we present methods for the detection of active human and mouse GrB in extracts from primary cells, and evaluate the sensitivity of the various substrates and inhibitors. Mouse splenocytes produce approximately 120-fold more GrB than similarly activated human cells, which allows the use of the hGrB substrate IETD-AFC to follow mouse GrB activity despite its unfavourable kinetic properties.

Key words: Granzyme B, Serine protease, Activity assay

1. Introduction

One of the major functions of cytotoxic lymphocytes (CLs) is the induction of apoptosis in target cells. The effector molecules necessary for this function are a family of granule-associated serine proteases called granzymes. They are able to activate a number of pro-apoptotic pathways following delivery into the target cell cytoplasm by the pore forming protein, perforin. Of the five human and ten mouse granzymes, granzyme B (GrB) is the most potent known inducer of apoptosis.

GrB cleaves substrates after acidic residues, in particular aspartic acid (1), however human GrB (hGrB) and mouse GrB (mGrB) have different extended substrate specificities (2). We have previously shown that the residues N-terminal of the aspartic acid are important for discriminating between hGrB and mGrB substrates. Where

Robert B. Ashman (ed.), *Leucocytes: Methods and Protocols*, Methods in Molecular Biology, vol. 844, DOI 10.1007/978-1-61779-527-5_18,

hGrB shows a preference for the sequence [I/V]-[G/E]-A-D, mGrB has a preference for [I/L]-X-[F/Y]-D (2). These differences, particularly at the P2 position immediately upstream of the aspartate, lead to hGrB activating apoptosis primarily through cleavage of Bid (3–6), while mGrB activates the caspase cascade (2, 3).

Moreover, $CD8^+$ T cells do not express GrB until they have been activated through interactions with antigen-presenting cells. Thus, GrB is a good marker for activation in human and mouse CLs and plays a very important role in their function. There are several commercially available substrates for detection of GrB activity; however, due to the differences in substrate specificity between human and mouse GrB these are not always ideal for detection. Here, we describe methods for detection of both hGrB and mGrB activity in extracts from primary cells and assess the sensitivity and specificity of the most commonly used substrates and inhibitors. We show that the level of GrB can vary dramatically between the human and mouse systems, with mouse splenocytes producing 120-fold more GrB per cell than human $CD8^+$ T cells.

2. Materials

2.1. Buffers

Lysis Buffer: 50 mM Tris-base, 10 mM EDTA, 1% (v/v) IGEPAL CA-630, pH 7.5. The following protease inhibitors should be added to the lysis buffer immediately prior to use: 150 μM aprotinin (Sigma), 10.5 μM leupeptin (Sigma), 1 μM pepstatin (Merck), and 0.2 μM AEBSF (4-(2-Aminoethyl) benzene sulfonyl fluoride hydrochloride) (Sigma).

Tris-buffered saline (TBS): 20 mM Tris-base, 150 mM NaCl, pH 7.4

2.2. Plate Readers

For detection of fluorescent signals, we use a BMG FLUOstar Galaxy microplate reader with the following filter sets: excitation 380 ± 10 nm, emission 510 ± 10 nm (detection of AFC) and excitation 320 ± 20 nm, emission 420 ± 12 nm (detection of Abz).

Chromogenic substrates were detected by monitoring absorbance at 405 nm on a Molecular Devices Thermo max microplate reader.

2.3. Recombinant Protein

Human GrB was produced in *Pichia pastoris* as previously described (7) and mouse GrB was expressed in inclusion bodies in *Escherichia coli* and refolded. Each was purified to homogeneity and active site titrated.

2.4. Substrates

The following substrates are described in this paper:
AAD-sBzl (carboxybenzyl-Ala-Ala-Asp-thiobenzylester) (Bachem), IETD-AFC (carboxybenzyl-Ile-Glu-Thr-Asp-7-amido-4-trifluo-

romethylcoumarin) (CalBiochem), Abz-IEPDSSMES(K-dnp) (2-aminobenzoyl-Ile-Glu-Pro-Asp-Ser-Ser-Met-Glu-Ser-dinitrophenol lysine) (synthesised by Mimotopes).

Stock solutions of all substrates were prepared by dissolving powder in dimethyl sulfoxide (DMSO) at 10 mM, and then stored at –20°C until use. Working solutions were prepared by dilution into TBS to a final concentration of 100 μM.

A stock solution of 5,5′-dithio-bis-(2-nitrobenzoic acid) (DTNB) was produced by dissolving the powder in DMSO at 250 mM and then stored at –20°C until use. The working solution was prepared by dilution into TBS to a final concentration of 2 mM.

2.5. Inhibitors

The GrB inhibitor compound 20 was synthesised according to ref. 8 and dissolved in DMSO at 10 mM, then stored at –80°C until use.

3. Methods

Peptide substrates can be bought or synthesised with a variety of N-terminal protection groups (which usually have very little effect on activity detection) and C-terminal "indicator" groups. Although fluorescent, colourimetric, and chemical indicator groups are available, all of them utilise the same principle to detect protease activity. When conjugated via a peptide or ester bond to the greater peptide, they are quiescent. Cleavage between the C-terminal amino acid and the indicator liberates the indicator group, which can then be detected by its intrinsic fluorescent, colour, or chemical properties.

Activity assays typically monitor liberation of the indicator group over time. Initially, the concentration of substrate greatly exceeds that of the protease and a linear increase in free indicator over time is observed that is solely dependant on the concentration of protease. As the concentration of intact substrate falls, access to substrate becomes rate limiting and the rate of turn over decreases. It is for this reason that continuous measurement of free indicator is preferable to single point assays, since the latter can only be quantitative if the end-point lies within the time frame of substrate in large excess to protease.

The two commonly used substrates for detection of human or mouse GrB are AAD-sBzl and IETD-AFC. It is important to determine whether these are appropriate for detection of GrB in cell lysates by looking at their sensitivity and specificity. For both mouse and human GrB detection in lysates, we compare the sensitivity of these substrates and investigate whether they are able to detect the levels of GrB expressed. In the case of mouse, splenocyte lysates from *gzmB*$^{-/-}$ mice provide a means of establishing a background measurement when mGrB is not present, thus providing information on how specific a substrate is for mGrB. However, in the human system to investigate the specificity of the substrate we are reliant on commercially available GrB inhibitors, in this case we have used compound 20.

3.1. Detection of Recombinant GrB Activity

As the progress curve method is only accurate over the period in which substrate turnover is linear (i.e. concentration of protease, not substrate, is limiting), a titration of recombinant human and mouse GrB must be performed to ensure this period is long enough to be analysed.

1. In a 96-well microtitre plate (Falcon) add recombinant protease to the first well to give a final concentration of 800 nM mGrB or 100 nM hGrB in a final volume of 100 μl.
2. Add 50 μl TBS to wells 2–8.
3. Perform serial twofold dilutions of GrB by taking 50 μl from well 1 and mixing with the TBS in well 2. Repeat to well 7 and then discard 50 μl of GrB solution from well 7. Well 8 serves as a negative control.
4. Add substrate as described below.

3.1.1. AAD-sBzl

AAD-sBzl is a colourometric substrate. GrB cleavage causes release of the thiobenzyl ester group which reacts with DTNB to produce a chromophore that absorbs at 412 nM. This substrate only tests for asp-ase activity, and hence is less specific for GrB compared to IETD-AFC.

1. Dilute substrate stock in TBS to a final concentration of 100 μM.
2. Make up fresh dilution of DTNB from stock in TBS to a final concentration of 2 mM.
3. Add 50 μl 2 mM DTNB to each well and pre-heat the plate in plate reader at 37°C to ensure there are no temperature variations (see Note 1).
4. To each well, add 100 μl substrate and measure the absorbance at 405 nm continuously for 10–20 min (see Note 2).
5. Plot the change in absorbance at 405 nm over time. Restrict the curves to the linear portion and analyse by linear regression (Fig. 1). One unit of activity is defined as a gradient of 1 (see Note 3).

3.1.2. IETD-AFC or Abz-IEPDSSMES(K-dnp)

IETD-AFC is a fluorogenic substrate, when the AFC is conjugated to a peptide it has an optimal emission max of 400 nm however once liberated, by GrB cleavage after the Asp, it has an optimal emission max of 505 nm and excitation max 400 nm. Due to this

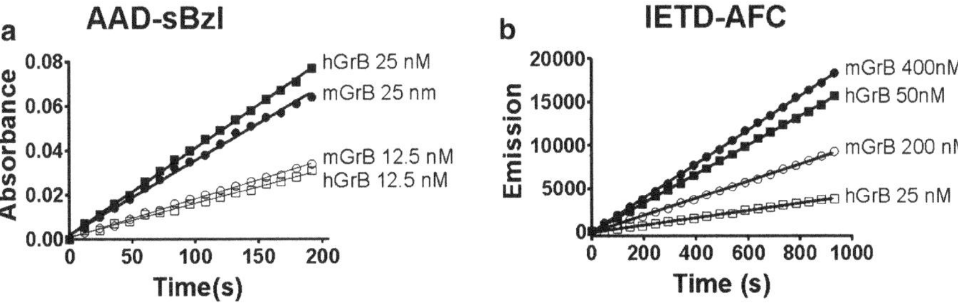

Fig. 1. Detection of recombinant mouse and human granzyme B activity. Detection of recombinant mouse and human GrB activity. Recombinant mouse and human GrB were incubated with AAD-sBzl (**a**) or IETD-AFC (**b**) at 37°C. Progress curves were restricted to the linear portion and analysed by linear regression to determine activity.

property, we can measure only the emission at 505 nm and thus measure cleavage of the substrate over time. This is the most commonly used GrB substrate and matches the preferred substrate sequence for hGrB. Mouse GrB will cleave this substrate; however, differences in preference at P2 and P4 lead to a significant drop in sensitivity compared to hGrB (2). This can be offset by increasing the amount of mGrB in the assay. Indeed the much larger amounts of GrB in mouse splenocytes explains why this poor mGrB substrate has been successfully used to follow GrB activity in mouse model systems (9).

Abz-IEPDSSMES(K-dnp) is a quenched fluorescence substrate and its detection is slightly different to that of the AFC and sBzl substrates. Rather than a quiescent reporter group that only becomes active when liberated from the peptide, the Abz group is always fluorescent (excitation maximum of 320 nm, emission maximum of 420 nm). However, the emission wavelength of Abz matches the absorbtion wavelength of dnp so that when they are in close proximity (i.e. held together by the intervening peptide), no fluorescence is observed. Cleavage anywhere within the peptide separates the fluorescence and quencher groups so that fluorescence can be measured. This substrate is less specific, but far more sensitive than IETD-AFC or AAD-sBzl (10).

1. Dilute substrate in to TBS to a final concentration of 100 μM.
2. Add 50 μl TBS to each well and pre-heat the plate in plate reader at 37°C so there are no temperature variations.
3. To each well add 100 μl substrate and measure fluorescence over 10–20 min.
4. Plot the change in fluorescence over time. Restrict the curves to the linear portion and analyse by linear regression (Fig. 1). One unit of activity is defined as a gradient of 1.

3.2. Preparation of Mouse Splenocyte Lysates

When preparing cell lysates a weak non-denaturing detergent is used that will permeabilise membranes without perturbing protein fold. Thus, GrB is released from its storage compartment but remains active. It is important to note that naive splenocytes contain no detectable GrB expression and therefore must be activated. The method of activation is not critical, although the level of GrB expression in splenocytes will vary with different methods. We have activated mouse splenocytes by stimulation with IL-2, IL-7, soluble anti-CD3, and soluble anti-CD28 for 3 days, then washed and rested them in IL-2 for 24 h. Post-activation the population contained 60% $CD8^+$ T cells and 30% $CD4^+$ T cells, 90% of the population was GrB positive.

1. Count cells in a haemocytometer.
2. Collect cells by centrifugation at $600 \times g$ for 5 min.
3. Wash with 1 ml TBS.
4. Resuspend cells in lysis buffer at 2×10^8 cells/ml (see Note 4) and incubate on ice for 10 min.
5. Separate debris by centrifugation at $16{,}000 \times g$ for 5 min at 4°C.
6. Collect supernatant (lysate) and keep on ice at all times (see Note 5).

3.3. Detection of Mouse GrB Activity in Splenocytes

Neither AAD-sBzl nor IETD-AFC are totally specific for GrB, as both can be cleaved by caspases. As such, an appropriate negative control is required to prove the specificity. In the mouse system, the best control are activated splenocytes from the GrB null (genotype: *gzmB*$^{-/-}$/*ΔPGK-neo)* mouse (11), although lysates of naïve splenocytes can also be used. Again, a titration must be performed to ascertain the number of cells that will give a progress curve with a linear portion for analysis.

1. Mix 7.5 μl cells with 92.5 μl TBS in a 96-well microtitre plate (Falcon). This will yield 1.5×10^6 cells in the first well in a final volume of 100 μl.
2. Add 50 μl TBS to wells 2–8.
3. Perform serial twofold dilutions of lysate by taking 50 μl from well 1 and mixing with the TBS in well 2. Repeat to well 7 and then discard 50 μl of lysate dilution from well 7. Well 8 serves as a negative control. This will give a final starting cell density of 7.5×10^5 cells in the first well down to 1.2×10^4 cells in well 7.
4. Add substrate as described in Subheadings 3.1.1 and 3.1.2, respectively. Example progress curves for naive, wild type, and *gzmB*$^{-/-}$ splenocyte lysates on each substrate are shown in Fig. 2.
5. The sensitivity and specificity of each substrate on the lysates can be determined relative to the *gzmB*$^{-/-}$ is summarised in Table 1.

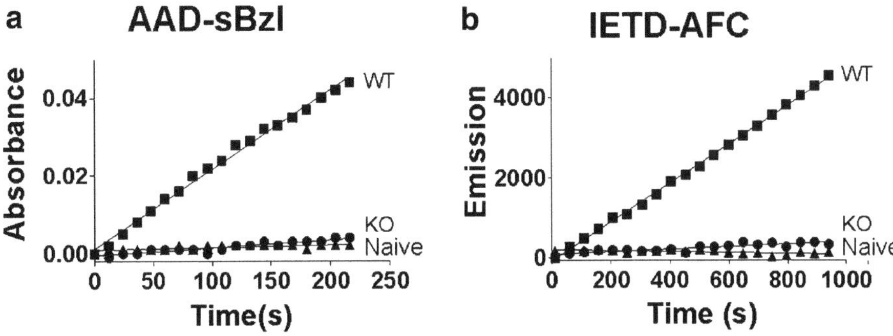

Fig. 2. Detection of mouse granzyme B activity in splenocyte lysates. Lysate equivalent to 3.75×10^5 cells was incubated with IETD-AFC (**a**) or AAD-sBzl (**b**) at 37°C. Progress curves were restricted to the linear portion and analysed by linear regression to determine activity. Naive wild type splenocytes have undetectable levels of activity. By contrast, activated wild type lysates give the greatest activity (indicated by the steepest gradient) and activity in the *gzmB*$^{-/-}$ lysates indicate background activity not due to GrB.

Table 1
Specificity of GrB substrates in primary cell lysates

Species	Substrate	Granzyme B per cell	Specificity (WT/KO or inhibitor)
Mouse	AAD-sBzl	0.19 pg	7.64
	IETD-AFC	1.48 pg	14.45
Human	Abz-IEPDSSMES(K-dnp)	6.8 fg	1.5

Activated mouse wild type or *gzmB*$^{-/-}$ splenocyte lysates and human CD8$^+$ T cell lysates with or without compound 20 were incubated with the indicated substrate. Progress curves were restricted to the linear portion and analysed by linear regression to determine activity. Activity in units was then related back to the amount of recombinant mouse or human GrB required to give this reading. This amount of GrB was divided by the number of cells to calculate GrB per cell. For the mouse lysates, the activity in the *gzmB*$^{-/-}$ lysates was subtracted from that detected in the wild type, while for the human lysates the activity in the presence of compound 20 was subtracted from that with no treatment. Specificity of each substrate was defined as the ratio of activity between the wild type and *gzmB*$^{-/-}$ lysates (mouse) or compound 20 treated and untreated lysates (human)

3.4. Preparation of Human Primary CD8$^+$ T Cell Lysates

As with the mouse system, the method of activation will have a great effect on expression of GrB in human cells. Here, we have purified and activated human CD8$^+$ T cells as described in ref. 12. Briefly, mononuclear cells from peripheral blood of healthy volunteers were obtained by density gradient centrifugation, and then further purified by MACS magnetic cell separation based on CD8 expression. Cells were activated by culture in medium supplemented with 5 μg/ml concanavalin A and 100 U/ml recombinant IL-2 for 3 days.

1. Count cells in a haemocytometer.
2. Collect cells by centrifugation at $300 \times g$ for 5 min.

3. Wash with 1 ml TBS.
4. Resuspend cells in lysis buffer at 2×10^8 cells/ml (see Note 4) and incubate on ice for 10 min.
5. Separate debris by centrifugation at $16{,}000\times g$ for 5 min at 4°C.
6. Collect supernatant (lysate) and keep on ice at all times (see Note 5).

3.5. Detection of Human GrB Activity in Primary CD8+ T Cells

1. Mix 20 μl lysate with 80 μl TBS in a 96-well microtitre plate (Falcon). This will yield 4.0×10^6 cells in a final volume of 100 μl.
2. Add 50 μl TBS to wells 2–8.
3. Perform serial twofold dilutions of the lysate by taking 50 μl from well 1 and mixing with the TBS in well 2.
4. Repeat to well 7 and then discard 50 μl of lysate dilution from well 7. Well 8 serves as a negative control. This will give a final starting cell density of 2.0×10^6 cells in the first well down to 3.1×10^4 cells in well 7.
5. Add substrate as described in Subheadings 3.1.1 and 3.1.2, respectively. Example progress curves for human lysates on each substrate are shown in Fig. 3.

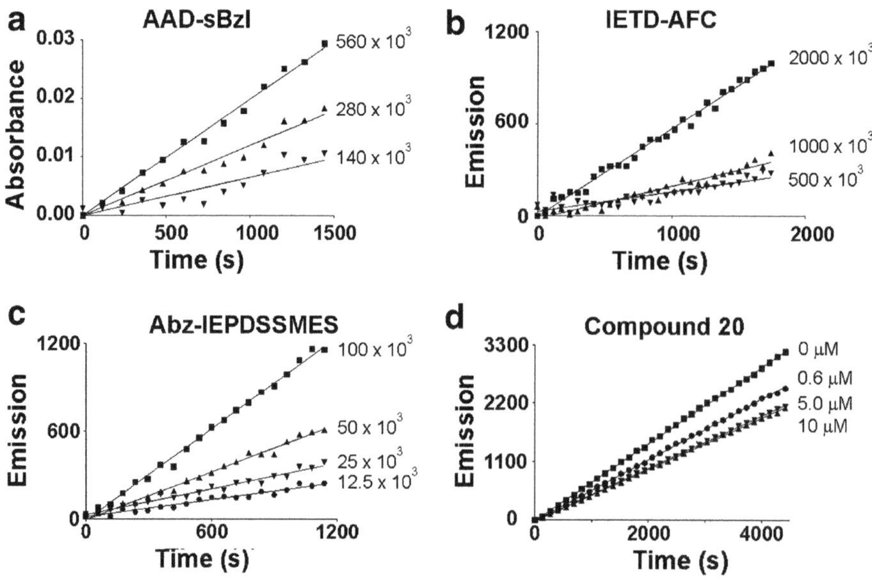

Fig. 3. Detection and inhibition of human granzyme B activity in CD8+ T cell lysates. Serial dilution of T cell lysates were incubated with AAD-sBzl (**a**), IETD-AFC (**b**), or Abz-IEPDSSMES(K-dnp) (**c**) at 37°C. Progress curves were restricted to the linear portion and analysed by linear regression to determine activity. GrB activity was inhibited in 1×10^5 cells with serial dilutions of compound 20 (**d**) and residual non-GrB activity measured with Abz-IEPDSSMES(K-dnp).

3.6. Inhibition of Human GrB Activity

In the absence of a null mutant control for background activity in the human system, we must rely instead on the use of chemical inhibitors. Two common commercial inhibitors of GrB are Ala-Ala-Asp-chloromethyl ketone (AAD-cmk) and Ile-Glu-Thr-Asp aldehyde (IETD-CHO). Unfortunately, IETD-CHO also inhibits caspase-8, which is a major source of non-GrB activity in lysates. AAD-cmk is an irreversible inhibitor, but is relatively weak and as such requires long pre-incubation times to fully inhibit GrB, which can lead to loss of GrB activity within the lysate. By contrast, the recently reported GrB inhibitor compound 20, is a strong inhibitor of GrB that does not cross-react with caspases (8).

1. Dilute compound 20 1:10 with TBS.
2. Mix 5 μl diluted compound 20 with 100 μl TBS in a 96-well tray. This will give a concentration of 20 μM.
3. Add 50 μl TBS to wells 2–8.
4. Perform serial twofold dilutions of compound 20 by taking 50 μl from well 1 and mixing with the TBS in well 2.
5. Repeat to well 7 and then discard 50 μl of solution from well 7. Well 8 serves as the no inhibitor control.
6. Dilute lysate 1:100 with TBS to give 4×10^6 cells/ml. Add 50 μl of dilute lysate (100,000 cells) to well 1–8 and incubate at 37°C for 15 min (see Note 6).
7. Add Abz-IEPDSSMES(K-dnp) as described in Subheading 3.1.2. An example progress curve showing inhibition by compound 20 is shown in Fig. 18.3d.

4. Notes

1. Ensure that all reagents (substrate, 96-well plate, etc.) are at 37°C before starting the assay as the activity is dependent on the temperature. Failure to do so will result in an upward curving, rather than linear, progress curve. This is caused by GrB activity increasing until the temperature stabilises.
2. When adding substrate, ensure no air bubbles are introduced to the wells as this will affect the readings.
3. Progress curves can only be analysed along the linear portion. Data beyond this point must be removed from the calculation. To ensure the data are linear, the r^2 value of the line must be greater than 0.8.
4. Lyse cells in a small volume to keep the number of cells as concentrated as possible. A final concentration of lysate of 2×10^8 cells/ml is generally sufficient; however, this is dependent on the method of activation.

5. Once the lysate is made it can be stored at –80°C with only a marginal drop in the activity of GrB. Do not freeze thaw lysates multiple times or keep on ice at 4°C for an extended period of time (over 4 h) as GrB will degrade.
6. Begin titration of compound 20 at 10 μM. This should give at least two concentrations with identical amounts of inhibition, indicating that maximal inhibition has been reached. Any resultant activity is therefore not due to GrB. If the two highest concentrations of inhibitor do not show equivalent inhibition, then repeat the titration beginning with a higher concentration or less lysate.

References

1. Odake, S., Kam, C.M., Narasimhan, L. et al. (1991) Human and murine cytotoxic T lymphocyte serine proteases: subsite mapping with peptide thioester substrates and inhibition of enzyme activity and cytolysis by isocoumarins, *Biochemistry* **30**, 2217–2227.
2. Kaiserman, D., Bird, C.H., Sun, J. et al. (2006) The major human and mouse granzymes are structurally and functionally divergent, *J Cell Biol* **175**, 619–630.
3. Adrain, C., Murphy, B.M., Martin, S.J. (2005) Molecular ordering of the caspase activation cascade initiated by the cytotoxic T lymphocyte/natural killer (CTL/NK) protease granzyme B, *J Biol Chem* **280**, 4663–4673.
4. Barry, M., Heibein, J.A., Pinkoski, M.J. et al. (2000) Granzyme B short-circuits the need for caspase 8 activity during granule-mediated cytotoxic T-lymphocyte killing by directly cleaving Bid, *Mol Cell Biol* **20**, 3781–3794.
5. Davis, J.E., Sutton, V.R., Smyth, M.J. et al. (2000) Dependence of granzyme B-mediated cell death on a pathway regulated by Bcl-2 or its viral homolog, BHRF1, *Cell Death Differ* **7**, 973–983.
6. Sutton, V.R., Davis, J.E., Cancilla, M. et al. (2000) Initiation of apoptosis by granzyme B requires direct cleavage of bid, but not direct granzyme B-mediated caspase activation, *J Exp Med* **192**, 1403–1414.
7. Sun, J., Bird, C.H., Buzza, M.S. et al. (1999) Expression and purification of recombinant human granzyme B from Pichia pastoris, *Biochem Biophys Res Commun* **261**, 251–255.
8. Willoughby, C.A., Bull, H.G., Garcia-Calvo, M. et al. (2002) Discovery of potent, selective human granzyme B inhibitors that inhibit CTL mediated apoptosis, *Bioorg Med Chem Lett* **12**, 2197–2200.
9. Pardo, J., Wallich, R., Ebnet, K. et al. (2007) Granzyme B is expressed in mouse mast cells in vivo and in vitro and causes delayed cell death independent of perforin, *Cell Death Differ* **14**, 1768–1779.
10. Sun, J., Whisstock, J.C., Harriott, P. et al. (2001) Importance of the P4' residue in human granzyme B inhibitors and substrates revealed by scanning mutagenesis of the proteinase inhibitor 9 reactive center loop, *J Biol Chem* **276**, 15177–15184.
11. Heusel, J.W., Wesselschmidt, R.L., Shresta, S. et al. (1994) Cytotoxic lymphocytes require granzyme B for the rapid induction of DNA fragmentation and apoptosis in allogeneic target cells, *Cell* **76**, 977–987.
12. Prakash, M.D., Bird, C.H., and Bird, P.I. (2009) Active and zymogen forms of granzyme B are constitutively released from cytotoxic lymphocytes in the absence of target cell engagement, *Immunol Cell Biol* **87**, 249–254.

Chapter 19

T Cell Transfer Model of Colitis: A Great Tool to Assess the Contribution of T Cells in Chronic Intestinal Inflammation

Rajaraman Eri, Michael A. McGuckin, and Robert Wadley

Abstract

Inflammatory bowel diseases (IBD) consist of Crohn's disease (CD) and ulcerative colitis (UC) affecting about 0.1% of the western population. These two chronic gut diseases affect youth at their prime of life causing diarrhoea, intestinal bleeding, and severe gut discomfort. Mouse models of colitis have been major tools in understanding the pathogenesis of IBD. A number of mouse models are available to assess the contribution of T cells in the pathogenesis of CD and UC. Among these, the T cell transfer model of colitis is the most widely used model to dissect the initiation, induction, and regulation of immunopathology in chronic colitis mediated by T cells. The methodology below describes the classification of various animal models and explains the T cell transfer model in detail, including flow cytometry-based isolation of naïve T cells that are used in the transfer, immunological concepts, detailed immune-pathological assessment, shortcomings of the model, and the latest improvements to this colitis model. A special focus is paid to the utilisation of the T cell transfer model in delineating the immunopathology in a primary epithelial defect model of colitis, namely *Winnie*.

Key words: Inflammatory bowel disease, Mouse, T cell, Pathogenesis, Flow cytometry

1. Introduction

Inflammatory bowel diseases (IBD)—Crohn's disease (CD) and ulcerative colitis (UC) are chronic intestinal disorders. These two diseases affect millions of people worldwide and IBD patients, usually young people suffer from severe intestinal discomfort, weight loss, diarrhoea, and rectal bleeding (1). Despite four decades of intense research, pathogenesis of IBD is still elusive. Recent GWA studies have shed light on multiple genetic components that predispose to IBD. There are at least 70 genes that have been implicated in IBD pathogenesis (2). Yet, these genes could only account for around 20% of IBD disease susceptibility. The current working hypothesis revolves around the idea that there is an inappropriate

Robert B. Ashman (ed.), *Leucocytes: Methods and Protocols*, Methods in Molecular Biology, vol. 844,
DOI 10.1007/978-1-61779-527-5_19, © Springer Science+Business Media, LLC 2012

host reaction to one's own microbiota leading to IBD in a genetically susceptible individual. The major research advances in IBD came about because of the use of animal models of IBD. There are over 50 animal models of IBD mainly mouse models of small and large intestinal inflammation.

Major IBD therapy advances were derived from the analysis of immune responses in CD and UC. The immune responses in IBD involve both innate and adaptive immunity manifested in the form of multiple cytokines propagating intestinal inflammation (3). Among the cell types, epithelial cells, myeloid innate cells, T effector cells, regulatory cells, and B cells to name a few have been implicated in IBD pathogenesis. The most widely studied and understood out of all are the T cell responses; thanks to one mouse model that revolutionised the understanding of that area, namely the transfer model of colitis.

In this chapter, we briefly discuss the available mouse models to study IBD and discuss transfer model of colitis in detail.

2. Animal Models of IBD

Animal models of IBD can be broadly grouped as (1) chemically induced models of colitis; (2) altered/defective innate/adaptive immune system; and (3) Models with intestinal epithelial defects.

2.1. Chemical Induction Models of IBD

Chemicals such as dextran sulphate sodium (DSS), 2, 4, 6-trinitrobenzenesulfonic acid (TNBS), and oxalazone are administered to cause acute and chronic colitis in this type of mouse model. DSS administration in drinking water is one of the most widely used models of IBD.

2.2. Innate Immune System-Based Models of IBD

Innate immune cells function through the expression of pattern recognition receptors that include toll-like receptors (TLRs) and nod-like receptors (NLRs) that are able to detect and react to pathogen-associated molecular patterns. Mouse models that have aided in dissecting innate immune pathways include mice deficient for IL-10, STAT3, A20, NFκB, NOD2, and TGFβ. Please see attached Table 1 for more details. The most widely characterised models in IBD involve disease propagation mainly through effector T cells. T cell-based models include either excessive effector T cell cytokine secretion or defective regulatory T cell function.

2.3. Epithelial Cell Defective Models of IBD

Only recently, a series of papers have described the importance of the intestinal epithelium and its function as a primary factor in IBD pathogenesis. Important pathways described include (1) NFκB pathway: epithelial-specific deficiency of IKKβ, NEMO; (2) secretory cell defects/ER stress pathways: *Winnie*, XBP-1 epithelial cell

Table 1
Mouse models of colitis

Chemical induction

Chemical	Action	References
Dextran sodium sulphate (DSS)	Luminal toxin, widely used, both acute and chronic types	(5, 6)
TNBS	Hapten, delayed-type hypersensitive response	(7, 8)
Oxazalone	Hapten, mainly Th2 type response	(9)

Altered innate/adaptive immune responses

Deficient/transgenic mice	Manifestation	References
TGFβ1 deficiency	Severe colitis, multiorgan disease	(10, 11)
IL-10 deficiency	Spontaneous colitis	(12, 13)
Stat 3 deficiency	Transmural colitis	(14)
Stat 4 transgenic	Transmural colitis	(15)
Nod 2	Reduced defensins, uncontrolled NFκB activation	(16)
Samp/YitFc (Samp)	Spontaneous ileitis	(17)
WASP	T cell-dependent colitis	(18)
A20 deficiency	Spontaneous multi-organ inflammation	(19)
TCRα deficiency		(20)

Defective intestinal epithelial responses

Deficiency/strain	Manifestation	References
Mdr1	Spontaneous colitis	(21)
IKK-γ (NEMO) in IEC	Severe chronic pancolitis	(22)
IKKβ (IEC specific)	Defective pathogen response	(23)
Muc 2	Spontaneous colitis	(24)
Winnie/Eyore	ER stress, spontaneous colitis	(25, 26)
Agr 2	Spontaneous colitis	(27)
Xbp 1 (IEC specific)	Paneth/GC pathology, ileitis	(28)
Atg5/Atg16l1	Hypomorphic for autophagy proteins. Paneth cell pathology	(29)

deficiency, Muc2 deficiency, and Agr2 deficiency; (3) multi-drug resistance MDR1 deficiency; (4) autophagy pathway: epithelial-specific deficiency of Atg16l1 gene. Table 1 describes all these models in detail.

3. Transfer Model of Colitis

One mouse model that is extremely suited to studying $CD4^+$ T cell-based pathogenesis is the adoptive transfer model of colitis. This model works on two important concepts that have been well established.

1. T cells that express high CD45RB (mainly naïve T cells) from the spleen or lymph nodes possess T cells capable of trafficking to the intestine and causing severe intestinal disease.
2. The effect of regulatory T cells (contained in $CD45RB^{lo}$ cells) can overcome the effect of the above mentioned $CD4^+$ effector T cells.

Briefly, the method originally described by Powrie (4) involves adoptive transfer of naïve T cells ($CD4^+CD45RB^{high}$) into immune-deficient Rag KO mice. These naïve T cells react in a severe fashion on experiencing the gut antigens, become activated forming colitogenic T cells secreting cytokines that causing severe gut inflammation involving both small and large intestines. It usually takes around 6–8 weeks, depending on the microbial populations present in the animal facility, to develop severe disease manifested by hunching, progressive weight loss, and diarrhoea. Histopathology reveals transmural inflammation, dense infiltrates involving neutrophils, and crypt abscessation. This model is well suited for studying multiple drug targets that work via modulation of T cell-mediated cytokines or are antibiotics.

3.1. Methods

3.1.1. Animals and Reagents

C57BL/6 mice: Both males and females can be used as donors. If female donors are used, recipients have to be female mice but male donors can be used to transfer cells to both male and female mice. These specificities are due to the problems associated with graft rejection.

Rag $1^{-/-}$ mice: Immune-deficient $RAG^{-/-}$ mice are used as the recipients of naïve T cells from donor splenocytes.

CD4 isolation kit (Miltenyi, Auburn, CA).

CD4-FITC (BD Biosciences, Franklin Lakes, NJ).

CD45RB-PE (BD, 553101).

Fc Block (purified CD16/32, eBiosciences, San Diego, CA. Catalogue No. 14-0161).

7 AAD (Amino-Actinomycin D) (Calbiochem, San Diego, CA): 1 mg/ml in 1× PBS.

FACS buffer: 1× PBS, pH 7.4 without Ca^{2+} and Mg^{2+} plus 1% BSA.

ACK lysing buffer (Lonzabio, Walkersville, MD).

3.1.2. Equipment/Materials

AutoMACS (Miltenyi).

Flow cytometer.

100-mm Petri dish.

5- and 10-ml Syringes.

Glass slides.

15-ml and 50-ml Falcon tubes.

Eppendorf tubes.

Polystyrene flow cytometer tubes.

Needles 27 G.

Cold centrifuge.

3.1.3. Isolation of Naïve T Cells from Donor Wild Type BL/6 Mice Spleens

1. Euthanise donor mice by approved method.
2. Open up the abdomen and collect spleens.
3. Place the collected spleens in 50-ml tube containing FACS buffer. Place the spleens on a Petri dish and cut away any pieces of fat left on the spleens.
4. Using two sterile glass slides with frosted ends, crush the spleen until all the cells are teased out into a Petri dish containing FACS buffer. Make sure the glass slides are wet with medium at all times during the process. Remove any visible pieces of connective tissue in the Petri dish.
5. Aspirate all the splenocytes in Petri dish using a pipette and place in a 50-ml tube. Rinse the Petri dish with a further 10–15 ml of buffer and collect into 50-ml conical tube. Keep repeating this process until all the cells are collected by filling the 50-ml tube.
6. Centrifuge the collected cells at 400×*g* for 5 min at 4°C. Discard the supernatant and gently tap to disrupt the cell pellet and resuspend in 2 ml FACS buffer and keep on ice.
7. Accurate counting of cells for further $CD4^+$ enrichment: Take 20 μl of the cell suspension, add 80 μl (1:5 dilution) of ACK hypotonic cell lysing buffer (aiding in RBC lysis). To 10 μl of these cells, add 90 μl trypan blue (0.4% in PBS) to count viable cells in a haemocytometer. Accurately record the number of cells.

3.1.4. Enrichment for $CD4^+$ T Cells (Miltenyi Kit)

The methodology described here is for using the Miltenyi MACS isolation and autoMACS usage. One could also use other magnetic separation techniques like Dynal kits as well.

1. Centrifuge the cells at 400×*g* for 5 min at 4°C. Discard the supernatant and completely resuspend the cells in 40 μl FACS buffer per 10^7 cells.
2. Add 10 μl biotin-antibody cocktail per 10^7 cells.

3. Mix well and incubate for 10 min at 4°C.
4. Add 30 μl FACS buffer and 20 μl anti-biotin microbeads per 10^7 cells.
5. Mix well and incubate for 15 min at 4°C.
6. Wash cells with FACS buffer by adding 10–20 times labelling volumes and centrifuge at 400 × *g* for 5 min at 4°C.
7. Discard the supernatant. Resuspend the cells in 500 μl FACS buffer per 10^8 cells. Place on ice.

3.1.5. AutoMACS Isolation of CD4⁺ T Cells

1. Prepare and prime the autoMACS separator. Apply the magnetically labelled cells and choose the separation program "Deplete".
2. Collect the negative ($CD4^+$) and positive fractions ($CD4^-$).
3. Count the cells as described before.

3.1.6. Fluorescent Labelling of $CD4^+$ T Cells for Cell Sorting

1. Pellet the negative fraction by centrifugation at 400 × *g* for 5 min at 4°C. Discard the supernatant.
2. Resuspend the cells ($CD4^+$ fraction) with antibody cocktail containing CD4-FITC and CD45RB-PE to the pre-optimised concentrations. 1 ml of soup stains 5×10^7 cells.
3. Incubate the cells on a rocking platform with gentle tilting for 15 min at 4°C.
4. Wash the cells with FACS buffer by adding 10–20 times labelling volume and centrifuge at 400 × *g* for 5 min at 4°C.
5. Resuspend the cells at 5×10^7 cells/ml, place on ice. (Take 0.5×10^6 cells and stain with 1/1,000 7AAD in 200 μl FACS buffer for viability check).
6. Staining the single colour control and unstained control at the same time:
   (a) Cells from step 5, put into 96-well plate 10^6/well, centrifuge 300 × *g*, 3 min, 4°C. Aspirate and discard supernatant.
   (b) Add 100 μl Fc block soup (2 μl Fc block + 98 μl FACS buffer) per well. Mix well and incubate on a rocking platform for 15 min at 4°C.
   (c) Fill each well to 200 μl with FACS buffer, centrifuge 300 × *g*, 3 min, 4°C. Aspirate and discard supernatant.
   (d) Add 100 μl appropriate antibody soup (single colour control) or FACS buffer (unstained). Mix well and incubate on a rocking platform for 15 min at 4°C.
   (e) Fill each well with 100 μl with FACS buffer, centrifuge 300 × *g*, 3 min, 4°C. Aspirate and discard supernatant.
   (f) Add 200 μl FACS buffer to each well, centrifuge 300 × *g*, 3 min, 4°C. Aspirate and discard supernatant.

(g) Add 200 μl FACS buffer to each well, transfer to FACS tubes and place on ice for 10 min before going to the cell sorting machine.

7. On a cell sorting flow cytometer (BD FACSAria IIu; Conversion date (from Aria 1): August 2009; DiVa version 6.1.1.) use unstained control cells to establish baseline auto-fluorescence and for proper gating of lymphoid cells and voltage setting for each parameter.
8. Use single colour controls to establish appropriate compensation for different fluorochromes.
9. The gating protocol starts with FSC-A vs. SSC-A; then to assist in reducing doublets, exclude outliers in FSC-H vs. FSC-W and SSC-H vs. SSC-W; dead cells are excluded using a 7AAD histogram (or an FSC vs. 7AAD dot plot if preferred); and finally the cells of interest are defined in an FITC vs. PE contour plot. The CD45RB^{high} population is identified as the 35–45% of cells exhibiting the brightest CD45RB staining as shown in Fig. 1.

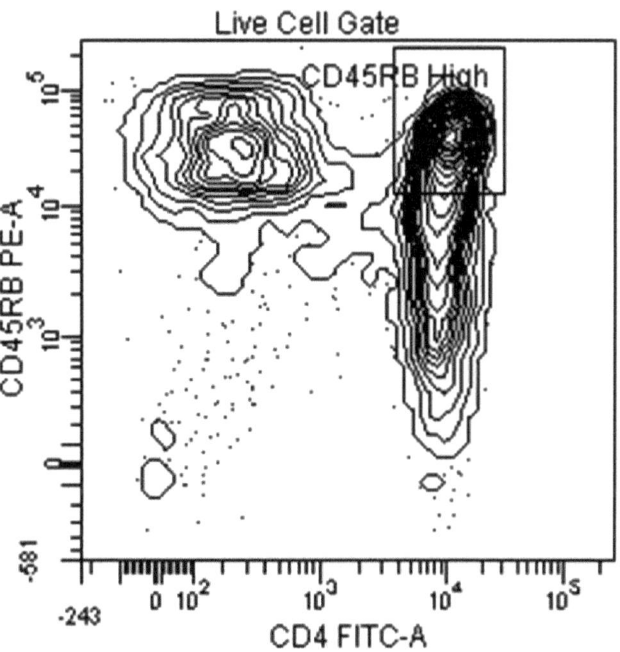

Fig. 1. Sorting of CD4^{+}CD45RB^{high} cells: single live cells are gated on their percentage and concentration. The contour plot is much more informative as to distribution than a dot plot. The PE MFI of the sorted cells will tend to be lower on post sort analysis, this is thought to be due to Ab shedding through the sort process, CD4 appears unaffected.

10. The purity of the sort can be determined by redoing the run with a small proportion of the sorted cells. One should expect >95% purity of these cells from the sort. Note: The sorting process tends to sheer cytoplasm and antigen from the cell, so the sorted population will tend to be slightly dimmer in CD45RB than was originally gated.

3.1.7. Injection of Immune-Deficient Mice with Purified Naïve CD4$^+$CD45RB^{high} T Cells

1. Centrifuge the sorted cells in FACS buffer at 400 × *g* for 5 min at 4°C. Discard the supernatant.
2. Resuspend sorted cells with cold PBS to a final cell concentration of 5×10^6/ml. Keep cells on ice prior to injection.
3. Weigh recipient mice and record it. Very slowly draw 0.1 ml cells into 1-ml syringe with 27-G needle attached.
4. Using proper intra-peritoneal injection technique, slowly inject the cells (IP) into recipient mice. Injecting the cells improperly into the subcutaneous area will result in no induction of colitis.

3.1.8. Disease Progression and Analysis

1. Typically, mice will gain weight in the first 3–4 weeks. Hence, body weight measurements can be taken thrice weekly for the first 3–4 weeks.
2. Once any mice start showing some weight loss, all mice need to be weighed three times a week. There are a proportion of mice that will show severe body weight loss in 4–6 weeks. Try to weigh the mice at the same time every time weighing is done. It is important to be consistent. Use the same weighing balance throughout the procedure.
3. Development of colitis in this model is dependent on multiple factors like the microbial flora present in the animal facility, intact intestinal flora, good i/p techniques, viable naïve T cell preparation, etc.
4. Apart from weighing the mice, record all the clinical parameters as well. Please see Table 2 for detailed clinical analysis. A typical "hunched-back" position is seen in colitic mice in this model. As described in the score sheet in Table 2, clinical parameters also include diarrhoea and rectal bleeding. All the clinical parameters are subjective and hence the person carrying out scoring for these clinical symptoms should preferably be blinded to the identity of the experimental groups.
5. By week 8 and in some settings even later, you will euthanise mice with combined clinical scores necessitating euthanasia (typically mandated by an institutional animal welfare committee).

Table 2
Score sheet for clinical evaluation in-transfer colitis

This score sheet includes clinical observations like the appearance and posture. A typical sick mouse has a hunched back appearance with crusty eyes and ruffled fur. Diarrhoea and rectal bleeding are other important clinical signs. It is better done when you weigh the mice. Observe the stools while weighing. Alternatively, the mice can be kept in separate divided plastic boxes and observed for tool consistency.

DAILY OBSERVATIONS

♀/♂

Experiment number:_______________
Mouse number:_____________
Date and time of Transfer :______________________
Pre-challenge weight (g):__________________

DATE												
Time post-challenge												
Days post-DSS	Day 1	Day 2	Day 3	Day 4	Day 5	Day 6	Day 7	Day 8	Day 9	Day 10	Day 11	Day12
Observations from a distance												
Inactive												
Hunched posture												
Ruffled fur												
Rate of breathing												
Crusty Eyes												
Shivering												
Diarrhoea												
Rectal bleeding												
On handling												
Not inquisitive or alert												
Bodyweight (% charge from start/score)*												
Any other abnormal behaviour or signs noted												
ACTION TAKEN^												
NOTES												
TOTAL SCORE												

Scoring Details:
0 = normal, 1 = equivocal symptoms, 2 = mild symptoms, 4 = severe symptoms. **Total of 4 or over = mouse culled.**
0= no weight loss, 2 = 5-15% weight loss, 4 = >15% weight loss

3.1.9. Tissue Collection, Sampling, and Scoring for Colitis

1. Depending on the aims of the experiment, tissue sample collection and analysis may vary. Use ethically approved euthanasia method and open up the abdomen using aseptic techniques (ethanol spray and open completely to keep away from hair).
2. Identify the small and large intestine by following from the stomach. As transfer colitis also exhibits small intestinal inflammation, one may need to collect a stretch of small intestine as well. Typically, hold the ileocecal junction. Leave 2–3 cm of ileum and disengage the small intestine. Care has to be taken in case you are collecting the draining mesenteric lymph nodes (MLN). First, collect MLN for further processing in cold PBS or medium. When colitis has developed, the MLN are easily visible (see Fig. 3 for an example).
3. Tracing from the ileocecal junction, move towards the entire cecum and colon removing the entire colon to the rectum. Remove the mesenteric fat and make sure the colon is free of all mesentery and adipose tissue connections.
4. Measure the colon length by taking the measurement from ileocecal junction to the rectum. Record the data and clean the faecal contents by a gentle wash.
5. Cut open the entire colon using fine scissors and gently remove the remaining faecal contents. Weigh the colon and record. Colon weight is a good non-subjective parameter of the severity of colitis.
6. For histology, either sampling from different regions of the colon (Cecum, proximal colon, mid colon, distal colon) is done or the “Swiss-roll” method is employed where the entire colon is rolled. We believe Swiss-roll technique enables a panoramic analysis of the severity of colitis and avoids sampling error. The colon material is then placed in 10% buffered formalin and changed into 70% ethanol after 24 h of fixation.
7. If samples are desired for RNA or protein analysis, collect and place the colon tissue as soon as possible and free on dry ice or liquid nitrogen.
8. For immunochemistry methods where paraffin embedding destroys epitopes, samples may need to be mounted in optimum cutting temperature (OCT) and frozen.
9. Scoring for histological colitis using the H&E stained sections of colon is described in Table 3.

Table 3
Score sheet for chronic transfer model of colitis: histology (H&E sections)

H&E sections from either Swiss-roll or individual samples from various regions will be scored for every region of the intestine. All the important parameters defining colitis are graded and scores are given accordingly. A ruler with μm divisions can be used for crypt length measurements.

Animal ID No. ☐ **Regions of Intestine**

	SI	Ce	PC	MC	DC
Crypt Architecture 0 = normal 1 = irregular 2 = moderate crypt loss (10-50%) 3 = severe crypt loss (50-90%) 4 = small/medium sized ulcers (<10 crypt widths) 5 = large ulcers (>10 crypt widths)					
Crypt Abscesses 0 = none 1 = 1-5 2 = 6-10 3 = >10					
Crypt Length **Caecum** – 0 = < 130 uM, 1 = 130-150 um, 2 = 150-200, 3 = 200-250, 4 = >250 **PC** – 0 = < 150 uM, 1 = 150-200 um, 2 = 200-250, 3 = 250-300, 4 = >300 **MC** – 0 = < 250 uM, 1 = 250-300 um, 2 = 300-400, 3 = 350-400, 4 = >400 **DC** – 0 = < 200uM, 1 = 200-250 um, 2 = 250-300, 3 = 300-350, 4 = >350					
Tissue Damage 0 = no damage 1 = discrete lesions 2 = mucosal erosions 3 = extensive mucosal damage					
Goblet Cell Loss 0 = normal <10% loss 1 = 10-25% 2 = 25-50% 3 = >50%					
Inflammatory Cell Infiltration 0 = occasional infiltration 1 = increasing leukocytes in lamina propria 2= confluence of leukocytes extending to submucosa 3 = transmural extension of inflammatory infiltrates					
Lamina Propria Neutrophils (PMN) 0 = 0-5 PMNs/HPF 1 = 6-10 2 = 11-20 3 = >20					

4. Troubleshooting

4.1. Transfer Results in No Colitis Development

It is important to remember that this model depends on many additional factors apart from the viability of naïve T cells, sufficient naïve T cells, and good animal procedures. These include the cleanliness of the animal house and the intact intestinal flora in recipient mice.

Possible causes:

1. Low naïve T cell numbers transferred: it is critical that 0.5×10^6 naïve T cells per mouse are given and make sure the intraperitoneal injection technique is correct. Injecting low numbers or subcutaneously causes little or no disease induction.
2. Cell viability is another critical factor. As the procedure takes around 5–6 h to be completed, make sure the cells are always maintained at 4°C. It is important to assess the cell viability after the sorting is completed and only >90% viability is acceptable for transfer. Cell viability is affected by sorting. T cells are generally considered to be robust in regard to sorting. However, sort times of <30–45 minutes are highly recommended, the cell sample should be in a quality media, and the collected cells in the best possible media to assist in post sort recovery. If possible the collection tube holder in the sorter should be chilled.

4.2. Low Yield of $CD4^+$ $CD45RB^{high}$ Cells

Planning the number of spleens required for transfer is an aspect to avoid this problem. Usually, 6–8 mice spleens are sufficient to yield enough naïve T cells for 13–15 recipients.

Possible causes:

1. Insufficient spleens or poor gating of $CD45RB^{high}$ cells: Donor mice are usually 8–12 weeks of age and the gating of the $CD45RB^{high}$ cells must be strictly adhered to. Figure 3 represents a typical gating of $CD45RB^{high}$ cells. Around 35–40% of the gated $CD4^+$ cells (very bright for CD45RB) are termed naïve T cells.
2. Low cell viability: Attention must be paid to every step to ensure the cells are processed at 4°C and also make sure the cells are not treated harshly. For example, vigorous vortexing, pipetting, and shaking should be avoided.

5. Transfer Colitis in *Winnie*

We recently characterised *Winnie* mice which carry a single missense mutation in the *Muc2* mucin gene leading to severe endoplasmic reticulum stress in intestinal goblet cells, premature epithelial apoptosis, increased intestinal permeability, and spontaneous colitis (25). We further characterised the genesis and nature of the immune response due to this intestinal epithelial dysfunction using the transfer model of colitis (26). We generated Rag $1^{-/-}$ X Winnie (RaW) and adoptively transferred naïve T cells from spleens of BL/6 mice. Rag $1^{-/-}$ mice were controls for this experiment. Figure 2 illustrates the survival of mice after transfer. The disease manifested very severely in RaW mice compared to Rag $1^{-/-}$ control mice. The weight loss post-transfer of naïve T cells is shown in Fig. 2. Rag $1^{-/-}$ mice normally exhibit weight loss between 8 and 12 weeks depending on the hygiene condition of the animal house. In this example, RaW mice lost weight much earlier. RaW mice also displayed very severe clinical symptoms and more severe histological colitis. Tables 2 and 3 describe the scoring methodology used for clinical symptoms

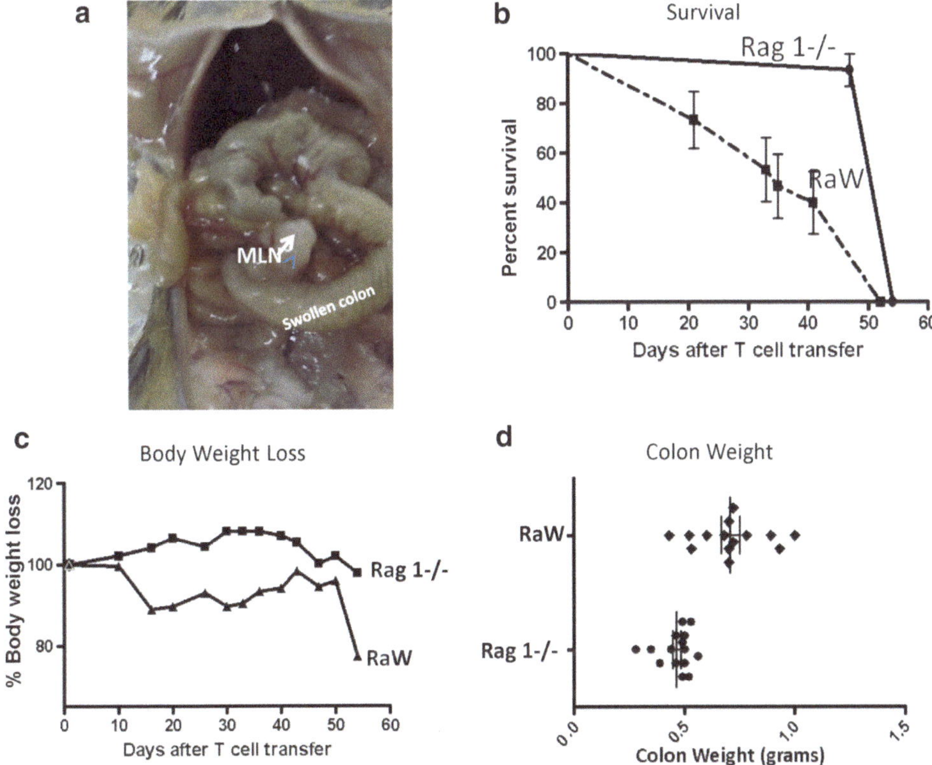

Fig. 2. An example for transfer model of colitis: age and sex matched immune-deficient Rag $1^{-/-}$ controls and Rag $1^{-/-}$ X Winnie (RaW) were transferred with 0.5×10^6 naïve T cells from C57BL/6 donor mice. (**a**) RaW mouse abdominal cavity opened after euthanasia. Duration: 5 weeks after transfer of $CD4^+CD45RB$ high cells. The *photograph* shows the swollen large intestine from cecum to the distal colon and also the mesenteric lymph nodes (*shiny white* in *colour*). (**b**) Survival graph showing the extreme susceptibility of RaW mice to transfer colitis. (**c**) Body weight loss following naïve T cell transfer shown for duration of 8 weeks after transfer. (**d**) Colon weight measurements after culling. Data taken from experiments detailed in ref. 26.

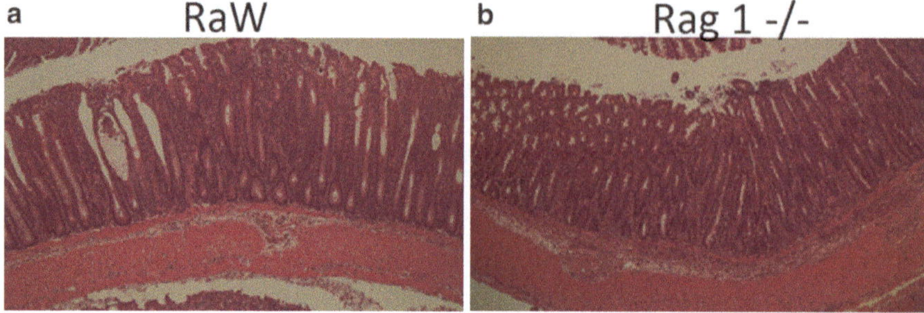

Fig. 3. Histological appearance of colitis in transfer model: Typical colitis features in histology includes transmural inflammation, increased proliferation, crypt abscesses, and severe infiltration as seen in both (**a**) and (**b**). (**a**) RaW mice colon 5 weeks after transfer of naïve T cells. ×10 magnification illustrating a more severe colitis phenotype with multiple crypt abscesses and severe infiltration. (**b**) Rag $1^{-/-}$ mice colon after 8 weeks of naïve T cell transfer. Colitis is a bit less severe than RaW but still has features of colitis described above.

and histological assessment, respectively. A typical dissection image to visualise severe colitis and swollen mesenteric lymph nodes is presented in Fig. 2. Representative images of histology in RaW and Rag $1^{-/-}$ mice are shown in Fig. 3.

References

1. Xavier, R.J., Podolsky, D.K. (2007) Unravelling the pathogenesis of inflammatory bowel disease, *Nature* **448**, 427–34.
2. Franke, A., McGovern, D.P., Barrett, J. C. et al. (2010) Genome-wide meta-analysis increases to 71 the number of confirmed Crohn's disease susceptibility loci, *Nat Genet* **42**, 1118–25.
3. Strober, W., Fuss, I., Mannon, P. (2007) The fundamental basis of inflammatory bowel disease, J Clin Invest **117**, 514–21.
4. Powrie, F., Leach, M.W., Mauze, S. et al. (1994) Inhibition of Thl responses prevents inflammatory bowel disease in scid mice reconstituted with CD45RBhi CD4+ T cells, *Immunity* **1**, 553–62.
5. Wirtz, S., Neufert, C., Weigmann, B. et al. (2007) Chemically induced mouse models of intestinal inflammation, *Nat Protocols* **2**, 541–546.
6. Strober, W., Fuss, I.J., Blumberg, R.S. (2002) The immunology of mucosal models of inflammation, *Annu Rev Immunol* **20**, 495–549.
7. Qiu, B.S., Vallance, B.A., Blennerhassett, P.A. et al. (1999) The role of CD4+ lymphocytes in the susceptibility of mice to stress-induced reactivation of experimental colitis, *Nat Med* **5**, 1178–82.
8. Scheiffele, F., Fuss, I.J. (2002) Induction of TNBS colitis in mice, *Curr Protoc Immunol* **Unit 15.19** doi: 10.1002/0471142735.im1519s49.
9. Heller, F., Fuss, I. J. Nieuwenhuis, E.E. et al. (2002) Oxazolone colitis, a Th2 colitis model resembling ulcerative colitis, is mediated by IL-13-producing NK-T cells, *Immunity* **17**, 629–38.
10. Gorelik, L., Fields, P.E., Flavell, R.A. (2000) Cutting edge: TGF-beta inhibits Th type 2 development through inhibition of GATA-3 expression, *J Immunol* **165**, 4773–7.
11. Shull, M.M., Ormsby, I., Kier, A.B. et al. (1992) Targeted disruption of the mouse transforming growth factor-beta 1 gene results in multifocal inflammatory disease, *Nature* **359**, 693–9.
12. Kuhn, R., Lohler, J., Rennick, D. et al. (1993) Interleukin-10-deficient mice develop chronic enterocolitis, *Cell* **75**, 263–74.
13. Rennick, D., Davidson, N., Berg, D. (1995) Interleukin-10 gene knock-out mice: a model of chronic inflammation, *Clin Immunol Immunopathol* **76**, S174–8.
14. Takeda, K., Clausen, B. E., Kaisho, T. et al. (1999) Enhanced Thl activity and development of chronic enterocolitis in mice devoid of Stat3 in macrophages and neutrophils, *Immunity* **10**, 39–49.
15. Simpson, S.J., Shah, S., Comiskey, M. et al. (1998) T cell-mediated pathology in two

models of experimental colitis depends predominantly on the interleukin 12/Signal transducer and activator of transcription (Stat)-4 pathway, but is not conditional on interferon gamma expression by T cells, *J Exp Med* **187**, 1225–34.

16. Kobayashi, K.S., Chamaillard, M., Ogura, Y. et al. (2005) Nod2-dependent regulation of innate and adaptive immunity in the intestinal tract, *Science* **307**, 731–4.
17. Matsumoto, S., Okabe, Y., Setoyama, H. et al. (1998) Inflammatory bowel disease-like enteritis and caecitis in a senescence accelerated mouse P1/Yit strain, Gut **43**, 71–8.
18. Snapper, S.B., Rosen, F.S., Mizoguchi, E. et al. (1998) Wiskott-Aldrich syndrome protein-deficient mice reveal a role for WASP in T but not B cell activation, *Immunity* **9**, 81–91.
19. Lee, E.G., Boone, D.L., Chai, S. et al. (2000) Failure to regulate TNF-induced NF-kappaB and cell death responses in A20-deficient mice, *Science* **289**, 2350–4.
20. Mizoguchi, A., Mizoguchi, E., Bhan, A.K. (1999) The critical role of interleukin 4 but not interferon gamma in the pathogenesis of colitis in T-cell receptor alpha mutant mice, Gastroenterology **116**, 320–6.
21. Panwala, C.M., Jones, J.C., Viney J.L. (1998) A novel model of inflammatory bowel disease: mice deficient for the multiple drug resistance gene, mdr1a, spontaneously develop colitis, *J Immunol* **161**, 5733–44.
22. Nenci, A., Becker, C., Wullaert, A. et al. (2007) Epithelial NEMO links innate immunity to chronic intestinal inflammation, *Nature* **446**, 557–61.
23. Zaph, C., Troy, A.E., Taylor, B.C. et al. (2007) Epithelial-cell-intrinsic IKK-beta expression regulates intestinal immune homeostasis, *Nature* **446**, 552–6.
24. Van der Sluis, M., De Koning, B.A., De Bruijn, A.C. et al. (2006) Muc2-deficient mice spontaneously develop colitis, indicating that MUC2 Is critical for colonic protection, *Gastroenterology* **131**, 117–129.
25. Heazlewood, C.K., Cook, M.C., Eri, R. et al. (2008) Aberrant mucin assembly in mice causes endoplasmic reticulum stress and spontaneous inflammation resembling ulcerative colitis, *PLoS Med* **5**, e54. doi: 10.1371/journal.pmed.0050054.
26. Eri, R.D., Adams, R.J., Tran, T.V. et al. (2010) An intestinal epithelial defect conferring ER stress results in inflammation involving both innate and adaptive immunity, *Mucosal Immunol* **4**, 354–64.
27. Park, S.W., Zhen, G., Verhaeghe, C. et al. (2009) The protein disulfide isomerase AGR2 is essential for production of intestinal mucus, *Proc Natl Acad Sci USA* **106**, 6950–5.
28. Kaser, A., Lee, A.H., Franke, A. et al. (2008) XBP1 links ER stress to intestinal inflammation and confers genetic risk for human inflammatory bowel disease, *Cell* **134**, 743–56.
29. Cadwell, K., Liu, J.Y., Brown, S. L. et al. (2008) A key role for autophagy and the autophagy gene Atg16l1 in mouse and human intestinal Paneth cells, *Nature* **456**, 259–63.

Chapter 20

Measurement of Nitrite in Urine by Gas Chromatography-Mass Spectrometry

Dimitrios Tsikas, Maria-Theresia Suchy, Anja Mitschke, Bibiana Beckmann, and Frank-Mathias Gutzki

Abstract

Nitric oxide (NO) is enzymatically produced from L-arginine and has a variety of biological functions. Autoxidation of NO in aqueous media yields nitrite (O = N–O⁻). NO and nitrite are oxidized in erythrocytes by oxyhemoglobin to nitrate (NO_3^-). Nitrate reductases from bacteria reduce nitrate to nitrite. Nitrite and nitrate are ubiquitous in nature, they are present throughout the body and they are excreted in the urine. Nitrite in urine has been used for several decades as an indicator and measure of bacteriuria. Since the identification of nitrite as a metabolite of NO, circulating nitrite is also used as an indicator of NO synthesis and is considered an NO storage form. In contrast to plasma nitrite, the significance of nitrite in the urine beyond bacteriuria is poorly investigated and understood. This chapter describes a gas chromatography-mass spectrometry (GC-MS) protocol for the quantitative determination of nitrite in urine of humans. Although the method is useful for detection and quantification of bacteriuria, the procedures described herein are optimum for urinary nitrite in conditions other than urinary tract infection. The method uses [^{15}N]nitrite as internal standard and pentafluorobenzyl bromide as the derivatization agent. Derivatization is performed on 100-μL aliquots and quantification of toluene extracts by selected-ion monitoring of *m/z* 46 for urinary nitrite and *m/z* 47 for the internal standard in the electron-capture negative-ion chemical ionization mode.

Key words: Nitric oxide, Nitrite, [^{15}N]nitrite, Pentafluorobenzyl bromide, Gas chromatography-mass spectrometry, Electron-capture negative-ion chemical ionization, Selected-ion monitoring

1. Introduction

1.1. Origin and Significance of Nitrite in Urine

The guanidine group of L-arginine is converted by nitric oxide synthases (NOS) in various types of cells to nitric oxide (NO) and L-citrulline (see Fig. 1) (1). Upon arrival in erythrocytes, NO is rapidly oxidized to nitrate (ONO_2^-) by oxyhemoglobin. The half-life of NO in the human circulation is less than 0.1 s (2) so that

Robert B. Ashman (ed.), *Leucocytes: Methods and Protocols*, Methods in Molecular Biology, vol. 844,
DOI 10.1007/978-1-61779-527-5_20, © Springer Science+Business Media, LLC 2012

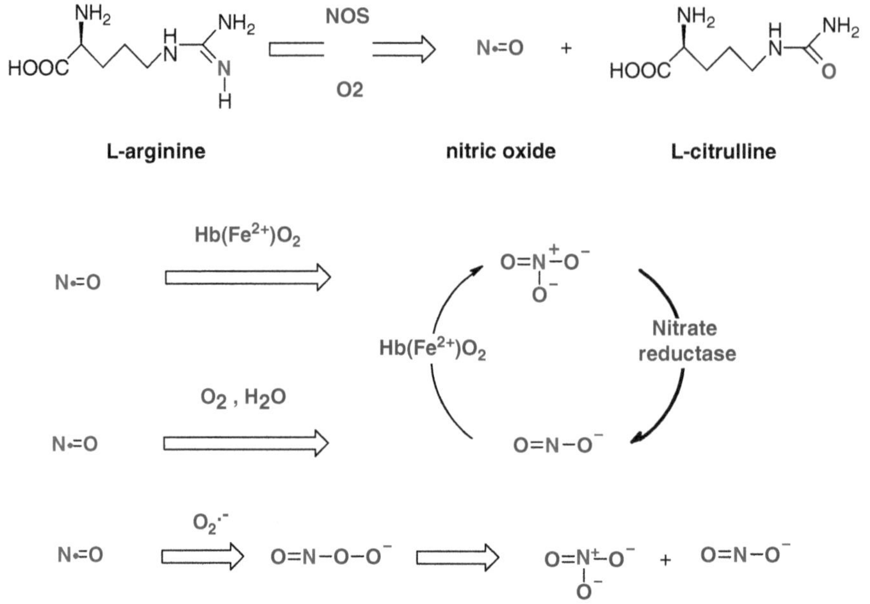

Fig. 1. Nitric oxide synthase (NOS) catalyzes the conversion of L-arginine to nitric oxide (NO) and L-citrulline. NO is oxidized to nitrite, nitrate, and peroxynitrite which decomposes to nitrite and nitrate. Nitrite is also oxidized to nitrate by oxyhemoglobin. Bacterial nitrate reductase activity reduces nitrate to nitrite. NO is biologically active; nitrite in the circulation can be reduced to NO and is therefore considered a storage form of NO bioactivity in the circulation. Nitrite and nitrate are excreted in the urine. For a comprehensive review of the L-arginine/NO pathway, see ref. 1.

authentic NO cannot be detected in human blood under basal conditions (3). The fraction of NO that survives oxyhemoglobin-catalyzed oxidation undergoes in part autoxidation to nitrite (ONO^-) (3) and in part oxidation to peroxynitrite ($ONOO^-$) by superoxide which is ubiquitous in human body (see Fig. 1). Peroxynitrite is highly unstable. It undergoes various reactions with numerous biomolecules, as well as decomposition and isomerization to nitrite and nitrate. Nitrite from autoxidized NO and from decomposed peroxynitrite, as well as nitrite produced from nitrate by bacterial nitrate reductase is oxidized to nitrate in red blood cells. It is worthy of mention that nitrite and nitrate are ubiquitous in nature, in food, in drinking water, and in the atmosphere as nitrogen oxides (NO_x) gases. Nitrite and nitrate from endogenous and exogenous sources are distributed throughout the body and are excreted in the urine. The transport of nitrite and nitrate in various cells and organs is scarcely investigated and poorly understood. Recent studies indicate that nitrite and nitrate are absorbed in the proximal tubulus of the nephron by a mechanism that likely involves carbonic anhydrase activity (4).

Under certain conditions, such as standardized low nitrite/nitrate diet, nitrite and nitrate in blood and urine are commonly used as indicators of NOS activity (5). Thus, plasma nitrite has been shown to originate mainly from L-arginine (6), and serum nitrite has been proposed as a measure of endothelial NOS activity in fasted humans (7). On the other hand, nitrate in urine is commonly used as a measure of whole body NO synthesis in humans (5).

In contrast to circulating nitrite, the origin and importance of urinary nitrite beyond bacteriuria (8) is step motherly investigated and poorly understood. A possible explanation may be analytical shortcomings due to the comparably low concentration of nitrite in the urine which is the same as in plasma, i.e., of the order of 1 μM, whereas nitrate concentration in the urine is about 20 times higher compared to plasma or serum (1). In bacteriuria, nitrate reductases reduce a considerable portion of nitrate to nitrite so that nitrite can be easily detected, for instance by spectrophotometric assays based on the Griess reaction (9).

Acute urinary tract infection (UTI) is one of the commonest acute bacterial infections among women. The problem with universal antibiotic use is the growing problem of antibiotic resistance. The key question is whether near patient tests can be used for better diagnosis and the targeting of antibiotic treatment. Dipsticks are the most widely used tests in primary care. A dipstick rule is based on having nitrite or both leucocytes and blood in mid-stream specimen urine. Mostly, urinary nitrite is found to be most predictive of UTI, followed by blood and based on leucocyte esterase (10). Taking into consideration, the potential contribution of diet, organic nitrates, diuretics and laboratory material to nitrite in urine, the present GC-MS method represents a reliable analytical tool for specific identification of UTI and its accurate quantification by measuring nitrite, and should be useful in the cost-effective use of antibiotics.

Recently, we found a considerable correlation between urinary nitrite and urinary 3-nitro-tyrosine, a biomarker of nitrosative stress (1), in patients with rheumatic diseases (11). This may indicate that nitrite in urine could be a useful parameter for nitrosative stress and myeloperoxidase activity. Administration of organic nitrates yields to elevation of serum concentrations of nitrite as well as to enhanced excretion of nitrite in the urine (12). In addition to nitrite, *S*-nitrosothiols (i.e., thionitrites) are considered storage and transport forms of NO bioactivity in the circulation (13), and these substances may also contribute to circulating and excretory nitrite. Dietary nitrite/nitrate and supplementation of other types of NO donors are likely to contribute to urinary nitrite and need to be considered.

Healthy adult humans with uncontrolled nitrate diet excrete into the urine approximately 0.5 μmol nitrite/mmol creatinine or 5 μmol nitrite/24 h (1, 13).

1.2. Measurement of Nitrite by Gas Chromatography-Mass Spectrometry

1.2.1. Brief Overview of Analytical Methods for Nitrite

There is a plethora of analytical methods for the quantitative determination of nitrite in urine, plasma and other biological methods. Nitrite can be analyzed in its native form, but the majority of the analytical methods use a chemical derivatization. For recent reviews of methods of analysis of nitrite, see refs. 1, 14–19. Among them gas chromatography-mass spectrometry (GC-MS) offers versatile facilities and highest analytical reliability in terms of specificity (19).

1.2.2. Derivatization and GC-MS of Nitrite

Chemical conversion of nitrite into thermally stable, volatile derivatives accessible to GC analysis is absolute required. Simultaneous derivatization and quantitation of nitrite and nitrate is possible by using the versatile derivatization reagent pentafluorobenzyl bromide (PFB-Br; 2,3,4,5,6-pentafluorobenzyl bromide or α-bromopentafluorotoluene) (20). PFB-Br undergoes substitution reactions with a variety of nucleophiles, including organic and inorganic anions (21). In these reactions, the leaving group Br of PFB-Br is substituted (most likely through an S_N2 mechanism) by nucleophiles to produce *N*-, *O*- and *S*-PFB derivatives. Nitrite ([^{14}N] nitrite) and its externally added internal standard [^{15}N]nitrite react with PFB-Br to form almost exclusively the nitro PFB derivatives, i.e., PFB-$^{14}NO_2$ and PFB-$^{15}NO_2$, (Fig. 2) (20–22). Derivatization of nitrite with PFB-Br can be performed in aqueous solutions of acetone, acetonitrile, or alcohols, such as methanol and ethanol. This derivatization can be performed by heating the sample at 50°C for 5–60 min (20). Because acetone can be more rapidly evaporated than other water-miscible organic solvents, acetone is superior over other solvents in this method. PFB-NO_2 is freely soluble in water-nonmiscible organic solvents, such as toluene. Quantitative extraction of PFB-NO_2, excess of PFB-Br and other PFB derivatives occurs within a few seconds by vortex-mixing.

PFB-$^{14}NO_2$ and PFB-$^{15}NO_2$ are volatile and thermally stable compounds. Using chemically bonded fused silica capillary columns of middle polarity such as Optima 17, PFB-$^{14}NO_2$ and PFB-$^{15}NO_2$ have almost identical retention times. The electrically

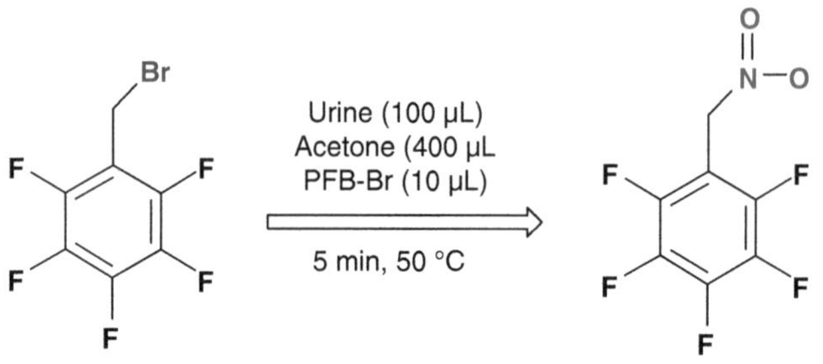

Fig. 2. Derivatization of nitrite in urine (100 µL) using acetone (400 µL) as the organic solvent and pentafluorobenzyl bromide (PFB-Br, 10 µL) as the derivatization reagent. Derivatization is complete after 5 min of heating at 50°C.

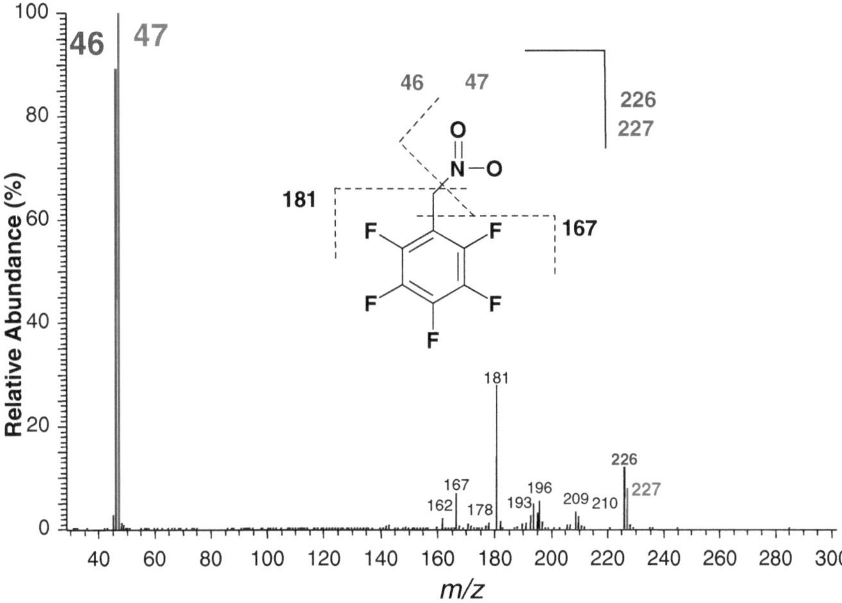

Fig. 3. ECNICI GC-MS spectrum of the PFB derivatives of approximatelly equal amounts of [^{14}N]nitrite and [^{15}N]nitrite using methane as the reagent/buffer gas. This spectrum was generated on the DSQ instrument (ThermoFisher, Dreieich, Germany).

uncharged gaseous molecules are released directly into the ion source of the GC-MS instrument, where they ionize in a manner depending upon the ionization mode and the GC-MS instrument. PFB derivatives are strongly electron-capturing substances. Under electron-capture negative-ion chemical ionization (ECNICI) conditions, e.g., by using methane as the reagent gas, PFB-NO_2 and PFB-$^{15}NO_2$ readily ionize to yield a few anions (Fig. 3). The most characteristic and abundant anions are $^{14}NO_2^-$, with a mass-to-charge ratio (*m/z*) of 46 and $^{15}NO_2^-$ with *m/z* 47. Thus, [^{14}N]nitrite and [^{15}N]nitrite can be discriminated mass spectrometrically.

Quantitative determination of nitrite as its PFB derivative with GC-MS instruments, e.g., with single-stage quadrupole instruments, is performed in the selected-ion monitoring (SIM) mode, because this scanning technique provides the most sensitive detection (Fig. 4). The mass spectrometer is set to pass alternately *m/z* 46 for NO_2^- and *m/z* 47 for $^{15}NO_2^-$. Representative GC-MS chromatograms from the quantitative determination of nitrite in unspiked and spiked human urine samples are shown in Fig. 5. PFB-$^{14}NO_2$ and PFB-$^{15}NO_2$ elute at 2.64 min. Calculation of concentrations of nitrite ([NO_2^-] in urine can be simply performed by using the known concentration of the internal standard [$^{15}NO_2^-$] added to the sample and the following equation:

$$\left[NO_2^-\right] = (PAR - PAR_0) \times [^{15}NO_2] \qquad (1)$$

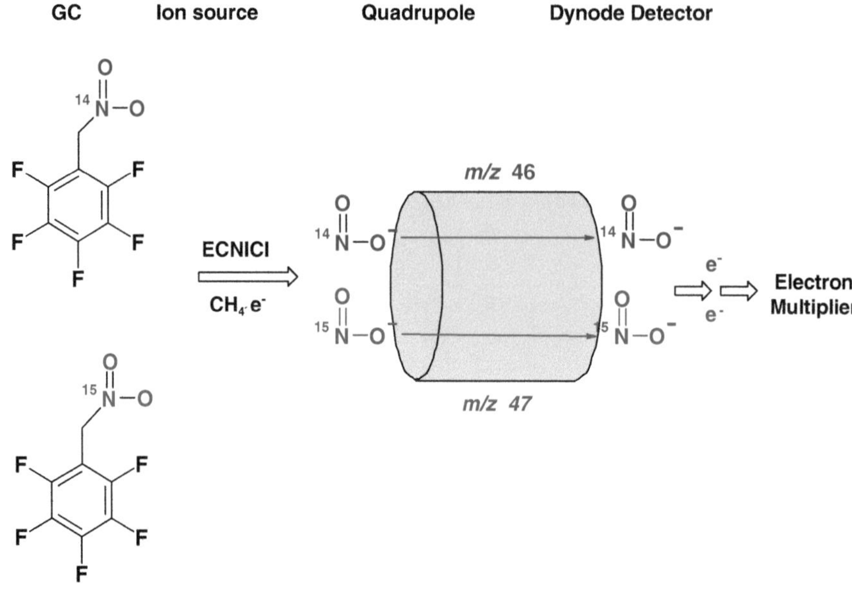

Fig. 4. Schematic of the quantitative determination of nitrite as PFB derivative on a quadrupole GC-MS instrument in the ECNICI mode by selected ion monitoring. The PFB derivatives of urinary nitrite, i.e., [^{14}N]nitrite, and of the internal standard [^{15}N]nitrite, i.e., PFB-$^{14}NO_2$ and PFB-$^{15}NO_2$, respectively, emerge from the GC column at the same time. In the ion-source of the GC-MS instrument, PFB-$^{14}NO_2$ and PFB-$^{15}NO_2$, ionize under ECNICI conditions by using methane as the reagent/buffer gas. The quadrupole is set alternately to two different voltages allowing the ions with *m/z* 46 and *m/z* 47 to pass the electric field, respectively, with a dwell time of 50 ms each. The conversion dynode behind the quadrupoles generates electrons which are multiplied by the electron multiplier detector. The current generated corresponds to the concentration of endogenous nitrite in the urine samples and of the externally added internal standard.

In this formula, PAR is the ratio of the peak area of *m/z* 46 to the peak area of *m/z* 47 for nitrite measured in the sample, and PAR_0 is the ratio of the peak area of *m/z* 46 to the peak area of *m/z* 47 for [^{15}N]nitrite measured in their stock solutions, for instance at a concentration of 8 mM. Typical PAR_0 values are of the order of 0.015 for nitrite.

2. Materials

1. Sodium [^{15}N]nitrite (98 atom% at ^{15}N) (Cambridge Isotope Laboratories; Andover; MA, USA) (see Note 1).
2. 2,3,4,5,6-Pentafluorobenzyl bromide, PFB-Br (Aldrich; Steinheim, Germany) (see Note 2).
3. GC-MS instrument. GC-MS analyses were performed on a single-stage quadrupole ThermoElectron DSQ mass spectrometer

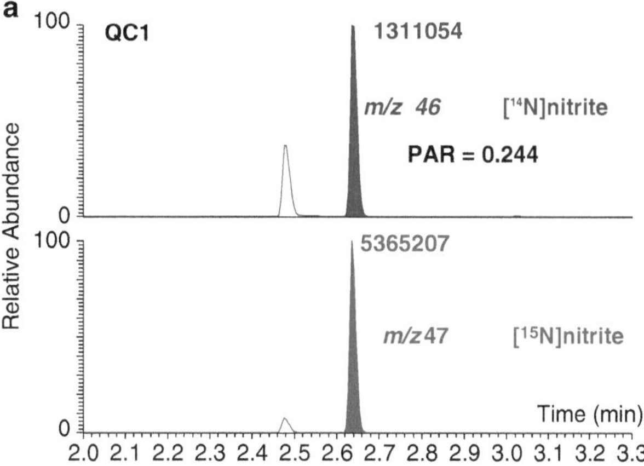

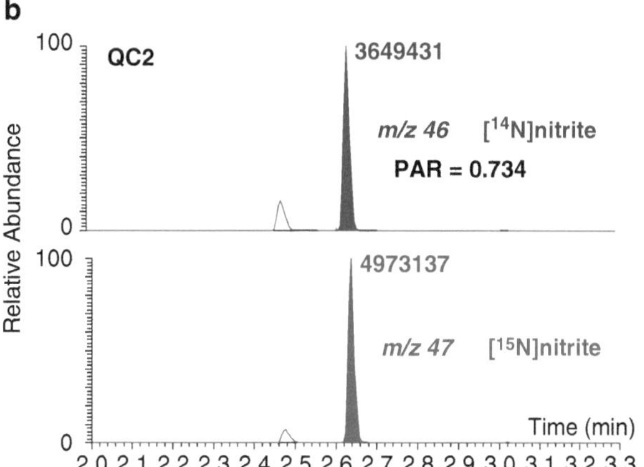

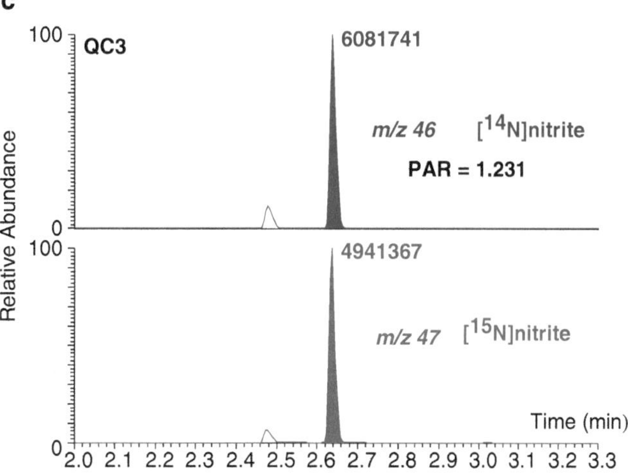

Fig. 5. Representative partial GC-MS chromatograms from the quantitative analysis of nitrite in quality control (QC) samples. The concentration of the internal standard [^{15}N]nitrite was 8 μM in all samples. The QC urine samples were analyzed (**a**) without addition of [^{14}N]nitrite (QC1) and after addition of (**b**) 4 μM [^{14}N]nitrite (QC2) and (**c**) 8 μM [^{14}N]nitrite (QC2). The peaks eluting in front of the nitrite peaks are due to [^{14}N]nitrate and [^{15}N]nitrate (20). PAR indicates the peak area ratio of *m/z 46* to *m/z* 47. See also Fig. 6.

(ThermoElectron, Austin, Texas, USA) directly interfaced with a Focus gas chromatograph (ThermoElectron, Milano, Italy) equipped with an autosampler AS 2000 (ThermoElectron, Milano, Italy).

4. GC high-resolution capillary columns. An Optima 17 (15 m × 0.25 mm i.d., 0.25-μm film thickness) from Macherey-Nagel (Düren, Germany) was used. Optima 17 is a mid-polar fused silica capillary column with immobilized phenylmethylpolysiloxane (50% phenyl) groups. Similar chemically bonded GC phases are: OV-17, DB-17, HP-50+, HP-17, SPB-50, SP-2250, Rtx-50, CP-SIL 24 CB, ZB-50 (see Note 3).
5. Thermostat (model Bioblock Scientifics; sample tray capacity for 60 samples; Thermolyne Corp., Iowa, USA).
6. Nitrogen evaporator (TurboVap LV Evaporator; sample tray capacity for 50 vials; Zymark, Idstein/Taunus, Germany).
7. Glass ware (Macherey-Nagel; Düren, Germany). Crimp vials N11 flat (1.5 mL, 11.6 × 32 mm); Crimp vials N11 conical (1.1 mL, 11.6 × 32 mm).
8. Autosampler/injector syringe (5-μL SGE syringe, SK-5F-HP-0.63; SGE International Pty. Ltd, Australia).
9. Vortex mixer (Model Reax 2000; Heidolph, Germany).

3. Methods

3.1. Derivatization and Extraction Procedures

Carry out nitrite analyses in a separate laboratory, where all steps of the method can be done. Arrange a set of pipettes and glass syringes for exclusive use in nitrite analyses. Avoid any contamination of working places and materials with nitrite and nitrate. Derivatization should be performed preferably in safely and tightly closed glass vials by using metal-block thermostats in a well-functioning fume hood.

1. Introduce a 100 μL aliquot of the urine sample into a glass snap ring vial.
2. Add a 10 μL aliquot of the internal standard, e.g., of a 80-μM [^{15}N]nitrite solution in distilled water (see Notes 4 and 5).
3. Add a 400 μL aliquot of acetone (see Note 6).
4. Add a 10 μL aliquot of PFB-Br (see Notes 2, 7 and 8).
5. Close the vial with a crimp cap.
6. Put the vial onto a thermostat at 50°C.
7. Incubate for 5 min (see Note 9).
8. After cooling to room temperature, evaporate acetone with a nitrogen stream (see Note 10).

9. Extract by vortexing for 60 s with an approximate aliquot of 1,000 μL of toluene.
10. Let separate phases or separate phases by short centrifugation.
11. Decant an approximate aliquot of 800 μL of the toluene phase into a crimp vial and close the vial with a crimp cap.

PFB-Br is insoluble in water, but freely soluble in most organic solvents, including acetone, acetonitrile, methanol, and ethanol. PFB-Br is also soluble in other organic solvents that are not miscible with water, such as toluene. Since derivatization of nitrite by PFB-Br must take place in urine, i.e., in an aqueous medium, a water-miscible organic solvent has to be used. The most suitable organic solvent is acetone. In order to obtain a homogenous liquid phase, which is indispensable for derivatization, acetone has to be used in the fourfold volume of the urine sample. For practical reasons, 400 μL of acetone are used for 100 μL of urine. Derivatization can be performed using various volumes of pure PFB-Br, ranging between 1 and 10 μL. The yield of the derivatization reaction products depends upon various experimental conditions, such temperature, time, and amount of PFB-Br. Optimum conditions for rapid analysis are 50°C, 5 min and 10 μL, respectively. Derivatization should be performed in autosampler air-tight glass vials in a well-ventilated fume hood.

Upon derivatization, the samples are put into a nitrogen evaporator and acetone is evaporated by means of gentle nitrogen stream (see Note 10). During this short-lasting process, the sample becomes cool and excess of PFB-Br remains as a small drop on the bottom. PFB-NO_2, other PFB derivatives and excess PFB-Br are subsequently extracted with toluene by vortex-mixing. Phase separation occurs spontaneously and quickly.

3.2. Quality Control for Urinary Nitrite

Analyze study urine samples alongside three quality control (QC1, QC2, QC3) urine samples. For this purpose, choose a pooled urine collected for 24 h of medium nitrite concentration, e.g., 2 μM (1, 13, 20). The concentration of the internal standard [^{15}N]nitrite to be added to all study and QC urine samples should be 8 μM. This permits accurate quantitation of nitrite in a wide concentration range, e.g., up to 100 μM.

1. Analyze QC1 sample without external addition of [^{14}N]nitrite.
2. Spike QC2 with 4 μM of [^{14}N]nitrite.
3. Spike QC3 sample with 8 μM of [^{14}N]nitrite.
4. Analyze QC samples in duplicate, study samples simply.
5. Determine accuracy (recovery, in %) and imprecision (relative standard deviation, RSD, in %) from the QC samples (see Notes 11–13).

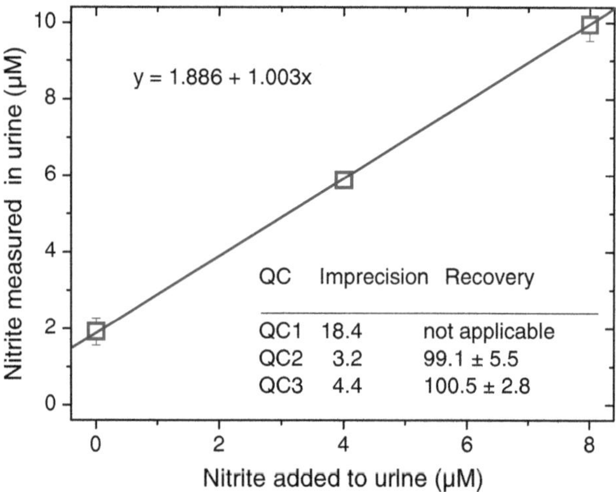

Fig. 6. Plot of the nitrite concentration measured in QC urine versus the nitrite concentration added to the QC urine sample at final concentrations of 0 μM (QC1), 4 μM (QC2), and 8 μM (QC3). See also Fig. 5. Inserts show the regression equation between measured (*y*) and added (*x*) nitrite concentration, the imprecision (%) values, and the recovery (%) values for QC2 and QC3.

6. Recovery and imprecision values should be close to 100% (for QC2 and QC3) and below 20% (for all QC samples), respectively.
7. Define own values for recovery and imprecision for the QC samples (e.g., recovery = 100 ± 20%; RSD ≤ 20%).
8. Consider only those data from study samples, the corresponding QC samples of which fullfil the criteria set, otherwise repeat analysis of study and QC samples.

An example from a quality control analysis performed during routine analysis of nitrite in 120 urine samples from healthy humans analyzed with four runs at four different days is shown in Fig. 6. GC-MS chromatograms from these analyses are shown in Fig. 5. The urine used for QC contained nitrite at a basal concentration of 1.886 μM (*y* axis intercept). Nitrite added to the QC urine sample at 4 μM (QC2) and 8 μM (QC3) was measured with a mean recovery of 99.1 and 100.5%, respectively, and with an imprecision (RSD, %) of 18.4, 3.2, and 4.4%, respectively.

3.3. GC-MS and Mode of Quantification

1. Inject 0.5 μL or 1 μL aliquots of the toluene phase in the splitless mode (see Notes 14 and 15).
2. Use an oven temperature program, starting at 70°C (see Note 16).
3. Perform SIM of *m/z* 46 and *m/z* 47 with a dwell-time of 50 ms for each ion in the ECNICI mode.

4. Separate chromatograms of the single ions and integrate peaks.
5. Calculate the peak area ratio (PAR) of *m/z* 46 and *m/z* 47.
6. Calculate concentrations by using Formula 1 (see Notes 17 and 18).

$$\left[NO_2^-\right] = (PAR - PAR_0) \times [^{15}NO_2] \qquad (1)$$

7. Divide the concentration of nitrite measured in the urine sample (in units of µM) by the concentration of creatinine measured in the same urine sample (in units of mM) and express the result in units of µmol nitrite per mmol creatinine (see Note 19).

Aliquots of the toluene phase are injected in the splitless mode and analyzed using an oven temperature program, e.g., the column is held at 70°C for 1 min and then increased to 180°C at a rate of 30°C/min and to 320°C at a rate of 70°C/min. Helium (1 mL/min) and methane (2.4 mL/min) are used as the carrier and the reagent gases, respectively, for ECNICI. Electron energy and electron current are set to 70 eV and 100 µA, respectively. Constant temperatures of 180, 260, 200°C (see Note 15) are kept at the ion source, interface, and injector of the instrument, respectively. SIM of *m/z* 46 for nitrite and *m/z* 47 for [^{15}N]nitrite is performed with a dwell-time of 50 ms for each ion. The electron multiplier voltage is set to about 1,400 V. Other GC-MS instruments may require different GC-MS conditions.

3.4. Special Considerations in Urinary Nitrite Analysis

3.4.1. Contribution of Laboratory Materials and Laboratory Air to Urinary Nitrite

Because of the ubiquity of nitrite all materials, chemicals, and solvents used in this method may contribute to endogenous nitrite in urine samples. Given the relatively low concentration of urinary nitrite, the extent of contribution of the materials used may falsely increase the concentration of endogenous nitrite in urine likely in plasma (23). Importantly, this contribution may vary greatly and may thus lead to both inaccurate and imprecise nitrite concentrations in urine. Therefore, all used materials should be tested for their contribution, and the least contributing materials should be finally used for nitrite analysis by the present GC-MS method.

Laboratory air contains nitrogen oxides which can be absorbed from the urine and contribute to endogenous nitrite (Fig. 7) (23). It is therefore necessary to keep closed both urine samples and all the other flasks used, including those containing the derivatization reagent PFB-Br and acetone, during sample preparation where applicable, i.e., to minimize exposition time to laboratory air and absorption of air NO_x.

Recently, we reported on an approach for the identification and quantification of contaminating analytes in biological fluids by stable-isotope dilution GC-MS methods (24). This approach can

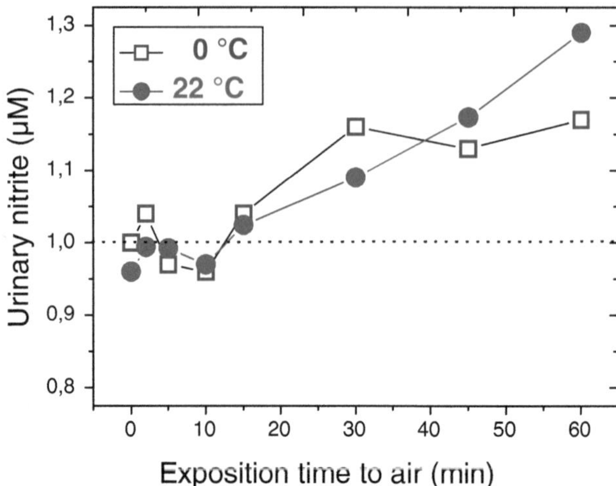

Fig. 7. Effect of exposition time of human urine samples (100 µL) placed in 1.8-mL glass vials incubated in an ice bath or at room temperature (approximately 22°C) to the laboratory air. This figure was constructed by using data reported elsewhere (23).

also be used for urinary nitrite and is based on the analysis of different urine volumes and determination of the PAR of [^{14}N]nitrite to [^{15}N]nitrite. PAR values that correlate inversely with the urine volume subjected to analysis would reveal contaminating nitrite. The extent of contamination is obtained by plotting the PAR of [^{14}N]nitrite to [^{15}N]nitrite versus the reciprocal of the urine volume analyzed.

3.4.2. Effects of Drugs on Urinary Nitrite

Effects of Organic Nitrates

Organic nitrates (R-ONO_2) were introduced into therapy of angina pectoris over 100 years ago and still are important for the symptomatic treatment of this disease. The metabolism of organic nitrates upon administration at therapeutically relevant doses leads to the formation of nitrate and nitrite which can considerably contribute to circulating and excretory nitrite (Fig. 8) (12).

Effects of Diuretics and Interference by Carbonate/Bicarbonate

Diuretics may affect the excretion of nitrite and nitrate into the urine, and acute changes in urinary levels of nitrite and nitrate may occur (4, 25) and may not reflect whole body NO synthesis (25, 26). Upon administration the diuretics acetazolamide, an inhibitor of carbonanhydrase activity, results in enhanced excretion of bicarbonate in the urine alkalinization of the urine. We found that exogenous and endogenous carbonate and bicarbonate may interfere with the analysis of nitrite by the present method (Fig. 9). The underlying mechanism of this interference is not fully understood. This type of interference can be overcome by slight urine acidification, preferably by diluted acetic acid, to pH values around 5 (Fig. 9). Most likely, elimination of the carbonate/bicarbonate interference by acids is due to their conversion to CO_2 and its

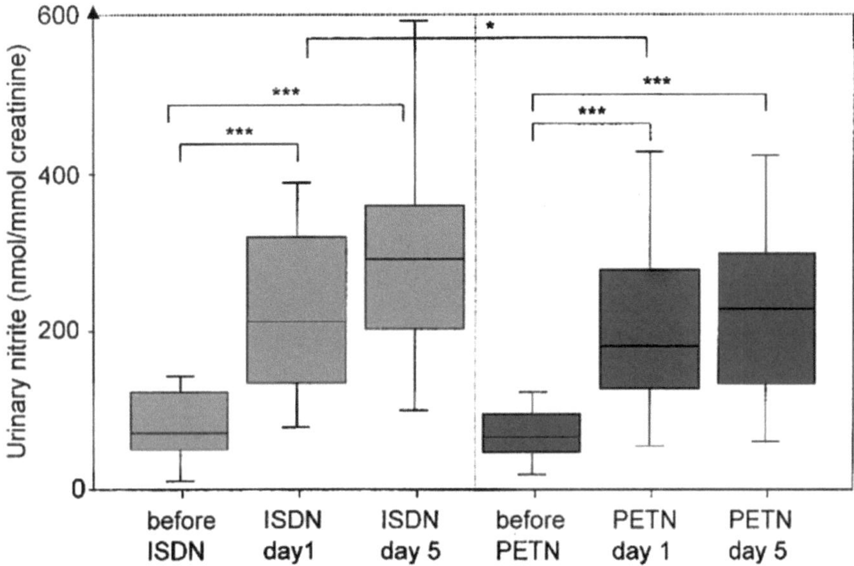

Fig. 8. Change in urinary nitrite excretion (nmol/mmol creatinine) in 18 healthy subjects receiving 30 mg isosorbide dinitrate (ISDN) TID or 80 mg pentaerythrityl tetranitrate (PETN) TID. Urinary nitrite excretions are given as box plots. *Asterisks* indicate significance using signed rank test. *$P<0.02$, ***$P<0.001$. Reproduced from ref. 12.

spontaneous evaporation, rather than due to the pH shift. The carbonate/bicarbonate system decreases the derivatization yield of [^{14}N]nitrite and [^{15}N]nitrite considerably so that blank nitrite increases with respect to urinary nitrite and its contribution to the measured nitrite becomes relevant (24).

4. Notes

1. The element nitrogen (N) consists of two stable, naturally occurring isotopes, i.e., ^{14}N and ^{15}N, with a natural abundance of 99.635 and 0.365%, respectively. The element oxygen, O, consists of three stable, naturally occurring isotopes, i.e., ^{16}O, ^{17}O, and ^{18}O, with a natural abundance of 99.759, 0.037, and 0.204%, respectively. Thus, naturally occurring nitrite contain all these isotopes. When monitoring the ions *m/z* 46 and *m/z* 47, the ^{18}O isotope does not contribute to them. Theoretically, the ratio of the abundances of [^{15}N]nitrite to [^{14}N]nitrite, can be calculated (27) and amounts to 0.00442. This ratio is constant, i.e., it does not depend upon the concentration of nitrite and nitrate in a certain matrix. However, for maximum sensitivity, chemicals, solutions, and materials of low nitrite should be used, and the isotopic purity of ^{15}N should be as high as available.

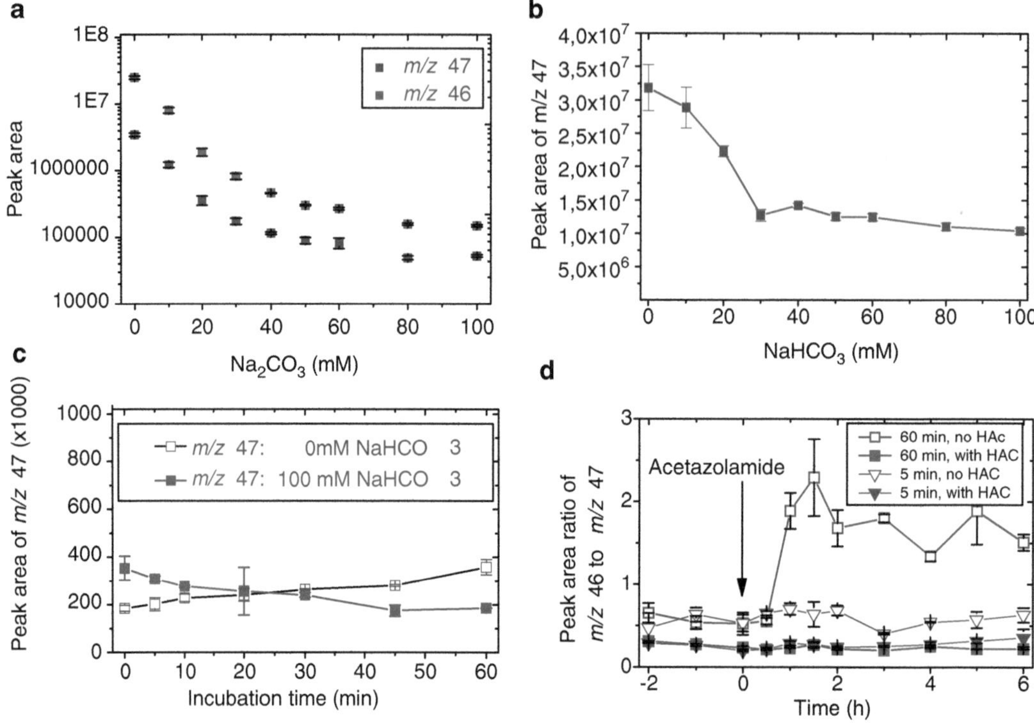

Fig. 9. Interference of carbonate (Na_2CO_3) and bicarbonate ($NaHCO_3$) in the GC-MS analysis of nitrite as PFB derivative. (**a**) Aqueous Na_2CO_3 decreases concomitantly the peak area of [^{14}N]nitrite (*m/z* 46) and [^{15}N]nitrite (*m/z* 47) in a concentration-dependent manner. Note the logarithmic scale of the y axis. (**b**) Aqueous $NaHCO_3$ decreases the peak area of [^{15}N]nitrite (*m/z* 47) in a concentration-dependent manner. (**c**) Effect of derivatization time on the peak area of [^{15}N]nitrite (*m/z* 47) in the absence and in the presence of aqueous $NaHCO_3$ at the fixed concentration of 100 mM. In the absence of $NaHCO_3$ the peak area of *m/z* 47 increases with increasing incubation time. In the presence of $NaHCO_3$ the peak area of *m/z* 47 decreases with increasing incubation time. (**d**) Effect of acidification by acetic acid (HAc) and incubation time (5 or 60 min) on the peak area ratio of *m/z* 46 (endogenous [^{14}N]nitrite in urine) to *m/z* 47 ([^{15}N]nitrite added to spot urine samples) before (−2 h, −1 h, 0 h) and up to 6 h after oral intake of a 500-mg tablet of the drug acetazolamide (see *arrow* at time zero) by a healthy volunteer. Urine acidification to pH 4–5 yielded almost identical PAR values across the whole observation period, whereas no acidification yielded diverging values depending upon the incubation time. The increase in the PAR in non-acidified urine samples was mainly due to the decrease of the peak area *m/z* 47 and the almost unchanged peak area of *m/z* 46 due to blank, i.e., nonurinary nitrite.

2. PFB-Br is corrosive and an eye irritant. Inhalation and contact with skin and eyes should be avoided. All work should be performed in a well-ventilated fume hood. After GC-MS analysis of the samples, remove the toluene phase from the glass vials and collect them in an appropriate container. Let all glass ware open in a fume hood until no liquid is visible. Do not dispose remaining PFB-Br into washbasin.
3. The PFB derivatives of nitrite (PFB-NO_2) and of nitrate (PFB-ONO_2) coelute on the DB-5 MS capillary column (DB-5 MS, 15 m×0.25 mm i.d., 0.25 μm film thickness, from J & W Scientific, Rancho Cordova, CA, USA). Because PFB-ONO_2 also ionizes to form *m/z* 46 (see Fig. 5) and nitrate is present in

urine at high molar excess over nitrite, accurate analysis of urinary nitrite requires gas chromatographic separation of PFB-NO_2 and PFB-ONO_2.

4. Use commercially available salts of [^{15}N]nitrite of the highest isotopic purity available. For quantitative measurements, prepare stock solutions (each 8 mM) of unlabeled nitrite as well as of [^{15}N]nitrite in distilled water and dilute with distilled water appropriately as needed.
5. Do not make solutions and dilutions of [^{14}N]nitrite and [^{15}N] nitrite in sodium phosphate buffer. Do not freeze and do not store frozen (e.g., at −20°C or −80°C) aqueous solutions and dilutions of [^{14}N]nitrite and [^{15}N]nitrite because nitrite oxidation to nitrate occurs under freezing conditions (28). Nitrite oxidation occurs more rapidly and to a higher extent in sodium phosphate buffer compared to potassium phosphate buffer of distilled water.
6. Accurate pipetting of 400 μL aliquots of acetone is difficult. There is no pressing need to accurately pipette 400 μL acetone aliquots because acetone is evaporated after derivatization. It is however important that urine and acetone are present in the mixture at an approximate proportion of 1:4 by volume.
7. PFB-Br is commercially available at different sizes. PFB-Br is liquid at room temperature (density, 1.728 g/mL) and solid in the refrigerator (e.g., at 8°C). Purchase the smallest available size, e.g., flasks containing 1 g or about 580 μL PFB-Br for the analysis of about 50 urine samples. Do not store and do not use remaining PFB-Br for forthcoming nitrite analyses. Do not let open the PFB-Br flask for long time.
8. Like acetone, accurate pipetting of 10 μL aliquots of pure PFB-Br is difficult. However, there is no pressing need to accurately pipette 10 μL PFB-Br aliquots because PFB-Br is used at very high excess over nitrite, nitrate, and other inorganic anions, including chloride in urine.
9. The derivatization time of urine samples can range between 5 and 60 min at 50°C. An incubation time of 60 min is required for simultaneous quantification of nitrite and nitrate in urine.
10. Do only evaporate acetone. Do not evaporate the aqueous phase because PFB-NO_2 is volatile and would be evaporated, too.
11. By definition, recovery applies to QC2 and QC3 and imprecision applies to all samples.
12. Recovery is calculated by using the formula:

    [100 × (mean concentration measured in QC2 or QC3) *minus* (mean concentration measured in QC1)/added concentration in QC2 or QC3].

13. Imprecision is calculated by the formula:

    [100 × (standard deviation divided by the corresponding mean concentration of QC1 or QC2 or QC3)].
14. Use fresh toluene to clean autosampler glass syringes with each run of samples.
15. The injector temperature should not exceed 200°C, because PFB-NO_2 degrades with increasing injector temperature, thus leading to loss of sensitivity.
16. By this method, urinary nitrite can also be determined isothermally, for instance at an oven temperature of 105°C. However, most accurate analysis of urinary nitrite is accomplished using an oven temperature program.
17. For the determination of the peak area ratio PAR_0, derivatize the 8 mM stock solutions of [^{14}N]nitrite and [^{15}N]nitrite. Prior to GC-MS analyses dilute the toluene extract with toluene (1:10, v/v). Set the electron multiplier voltage to a lower value, for instance at 1200 V.
18. Nitrite concentration in human urine is of the order of 1 μM.
19. Creatinine-corrected urinary excretion of nitrite is of the order of 0.5 μmol/mmol creatinine for healthy adults (1, 13).

References

1. Tsikas, D. (2008) A critical review and discussion of analytical methods in the L-arginine/nitric oxide area of basic and clinical research, *Anal Biochem* **379**, 139–163.
2. Kelm, M., Schrader, J. (1990) Control of coronary vascular tone by nitric oxide, *Circ. Res.* **66**, 1561–1570.
3. Tsikas, D. (2005) Methods of quantitative analysis of the nitric oxide metabolites nitrite and nitrate in human biological fluids, *Free Radic Res* **39**, 797–815.
4. Tsikas, D., Schwarz, A., Stichtenoth, D.O. (2010) Simultaneous measurement of [^{15}N]nitrate and [^{15}N]nitrite enrichment and concentration in urine by gas chromatography mass spectrometry as pentafluorobenzyl derivatives, *Anal Chem* **82**, 2585–2587.
5. Tsikas, D., Gutzki, F.M., Stichtenoth, D.O. (2006) Circulating and excretory nitrite and nitrate as indicators of nitric oxide synthesis in humans: methods of analysis, *Eur J Clin Pharmacol* **62**, 51–59.
6. Rhodes, P., Leone, A.M., Francis, P.L. et al. (1995) The L-arginine:nitric oxide pathway is the major source of plasma nitrite in fasted humans, *Biochem Biophys Res Commun* **209**, 590–596.
7. Kelm, M., Preik-Steinhoff, H., Preik, M. et al. (1999) Serum nitrite sensitively reflects endothelial NO formation in human forarm vasculature: evidence for biochemical assessment of the endothelial L-arginine-NO pathway, *Cardiovasc Res* **41**, 765–772.
8. Cruickshank, J., Moyes, J.M. (1914) The presence and significance of nitrites in urine, *Br Med J* **2**, 712–713.
9. Tsikas, D. (2007) Analysis of nitrite and nitrate in biological fluids by assays based on the Griess reaction: Appraisal of the Griess reaction in the L-arginine/nitric oxide area of research, *J. Chromatogr B* **851**, 51–70.
10. Little, P., Turner, S., Rumsby, K. et al. (2009) Dipsticks and diagnostic algorithms in urinary tract infection: development and validation, randomised trial, economic analysis, observational cohort and qualitative study, *Health Technol Assess* **13**, 1–96.
11. Pham, V.V., Stichtenoth, D.O., Tsikas, D. (2009) Nitrite correlates with 3-nitrotyrosine but not with the F_2-isoprostane 15(*S*)-8-*iso*-$PGF_{2\alpha}$ in urine of rheumatic patients, *Nitric Oxide* **21**, 210–215.
12. Keimer, R., Stutzer, F.K., Tsikas, D. et al. (2003) Lack of oxidative stress during sustained

therapy with isosorbide dinitrate and pentaerythrityl tetranitrate in healthy humans: A randomized, double-blind crossover study, *J Cardiovasc Pharmacol* **41**, 284–292.

13. Tsikas, D., Böger, R.H., Bode-Böger, S.M. et al. (1994) Quantification of nitrite and nitrate in human urine and plasma as pentafluorobenzyl derivatives by gas chromatography-mass spectrometry using their ^{15}N-labelled analogs *J Chromatogr B* **661**, 185–191.
14. Giustarini, D. Milzani, A. Dalle-Donne, I. et al. (2007) Detection of *S*-nitrosothiols in biological fluids: A comparison among the most widely applied methodologies, *J Chromatogr B* **851**, 124–139.
15. Boudko, D. (2007) Bioanalytical profile of the L-arginine/nitric oxide pathway and its evaluation by capillary electrophoresis, *J Chromatogr B* **851**, 186–210.
16. MacArthur, P.H., Shiva, S., Gladwin, M.T. (2007) Measurement of circulating nitrite and *S*-nitrosothiols by reductive chemiluminescence, *J Chromatogr B* **851**, 93–105.
17. Grau, M., Hendgen-Cotta, U.B., Brouzos, P. et al. (2007) Recent methodological advances in the analysis of nitrite in the human circulation: Nitrite as a biochemical parameter of the L-arginine/NO pathway, *J Chromatogr B* **851**, 106–123.
18. Jobgen, W.S., Jobgen, S.C., Li, H. et al. (2007) Analysis of nitrite and nitrate in biological samples using high-performance liquid chromatography, *J Chromatogr B* **851**, 71–82.
19. Helmke, S.M., Duncan, M.W. (2007) Measurement of the NO metabolites, nitrite and nitrate, in human biological fluids by GC-MS, *J Chromatogr B* **851**, 83–92.
20. Tsikas, D. (2000) Simultaneous derivatization and quantification of the nitric oxide metabolites nitrite and nitrate in biological fluids by gas chromatography/mass spectrometry, *Anal Chem* **72**, 4064–4072.
21. Wu, H.-L., Chen, S.-H., Lin, S.-J. et al. (1983) Gas chromatographic determination of inorganic anions as pentafluorobenzyl derivatives, *J Chromatogr* **269**, 183–190.
22. Tsikas, D. (2004) Measurement of nitric oxide synthase activity *in vivo* and *in vitro* by gas chromatography-mass Spectrometry, *Methods Mol Biol* **279**, 81–103.
23. Tsikas, D., Mitschke, A., Gutzki, F.M. et al. (2010) Evidence by gas chromatography-mass spectrometry of ex vivo nitrite and nitrate formation from air nitrogen oxides in human plasma, serum, and urine samples, *Anal Biochem* **397**,126–8.
24. Tsikas, D. (2010) Identifying and quantifying contaminants contributing to endogenous analytes in gas chromatography-mass spectrometry, *Anal Chem* **82**, 7835–7841.
25. Sütő, T., Losonczy, G., Qiu, C. et al. (1995) Acute changes in urinary excretion of nitrite + nitrate do not necessarily predict renal vascular NO production *Kidney Int* **48**, 1272–1277.
26. Baylis, C., Vallance, P. (1998) Measurement of nitrite and nitrate levels in plasma and urine – what does it measure tell us about the activity of the endogenous nitric oxide system? *Curr Opin Nephrol Hypertens* 7, 59–62.
27. Beynon, J.H. (1960) *Mass Spectrometry and Its Application to Organic Chemistry*. Elsevier Publishing Company, Amsterdam, 1960.
28. Tsikas, D., Frölich, J.C. (2004) Trouble with the analysis of nitrite, nitrate, *S*-nitrosothiols and 3-nitrotyrosine: freezing-induced artifacts? *Nitric Oxide* 11, 209–13; author reply 214–5.

Index

Robert B. Ashman (ed.), *Leucocytes: Methods and Protocols*, Methods in Molecular Biology, vol. 844, DOI 10.1007/978-1-61779-527-5, © Springer Science+Business Media, LLC 2012

MIX
Papier aus verantwortungsvollen Quellen
Paper from responsible sources
FSC® C105338

If you have any concerns about our products,
you can contact us on
ProductSafety@springernature.com

In case Publisher is established outside the EU,
the EU authorized representative is:
Springer Nature Customer Service Center GmbH
Europaplatz 3, 69115 Heidelberg, Germany

Printed by Libri Plureos GmbH
in Hamburg, Germany